Transdisciplinary Management of Ecological and Social Crises

Series Editors

Mathias Schüz, School of Management and Law, Zurich University of Applied Sciences (ZHAW), Winterthur, Switzerland

René Schmidpeter, Bern University of Applied Sciences, Bern, Switzerland

AF800993

Complex problems can only be solved with the aid of complex approaches. Transdisciplinarity is a promising candidate.

This book series explores the possibilities for overcoming global social and ecological crises by pursuing transdisciplinary approaches. In particular, it addresses the question of how responsible leadership can and should be practiced in politics, business and society, and how classical top-down leadership can be replaced in response to the complexity of today's ecological, social and economic problems.

The volumes in the series present successful alternative leadership models based on transdisciplinary experiential knowledge, as well as intra-, inter- and multidisciplinary insights. In this context, spiritual experiential knowledge, e.g. from ancient teachings or nature-based peoples, is also included, and is recognized as holding great potential for overcoming our world's problems. The series is intended as a platform for all professionals, policymakers and researchers who are interested in responsible leadership.

Mathias Schüz
Editor

Transdisciplinary Thinking and Acting

Fundamentals and Perspectives

Springer

Editor
Mathias Schüz
Zurich University of Applied Sciences
Winterthur, Switzerland

ISSN 2731-8842 ISSN 2731-8850 (electronic)
Transdisciplinary Management of Ecological and Social Crises
ISBN 978-3-031-70679-0 ISBN 978-3-031-70680-6 (eBook)
https://doi.org/10.1007/978-3-031-70680-6

Chapter "Methodology of Transdisciplinarity–Levels of Reality, Logic of the Included Middle and Complexity" is licensed under the terms of the Creative Commons Attribution 4.0 International License (http://creativecommons.org/licenses/by/4.0/). For further details see license information in the chapter.

© The Editor(s) (if applicable) and The Author(s), under exclusive license to Springer Nature Switzerland AG 2025
This work is subject to copyright. All rights are solely and exclusively licensed by the Publisher, whether the whole or part of the material is concerned, specifically the rights of translation, reprinting, reuse of illustrations, recitation, broadcasting, reproduction on microfilms or in any other physical way, and transmission or information storage and retrieval, electronic adaptation, computer software, or by similar or dissimilar methodology now known or hereafter developed.
The use of general descriptive names, registered names, trademarks, service marks, etc. in this publication does not imply, even in the absence of a specific statement, that such names are exempt from the relevant protective laws and regulations and therefore free for general use.
The publisher, the authors and the editors are safe to assume that the advice and information in this book are believed to be true and accurate at the date of publication. Neither the publisher nor the authors or the editors give a warranty, expressed or implied, with respect to the material contained herein or for any errors or omissions that may have been made. The publisher remains neutral with regard to jurisdictional claims in published maps and institutional affiliations.

This Springer imprint is published by the registered company Springer Nature Switzerland AG
The registered company address is: Gewerbestrasse 11, 6330 Cham, Switzerland

If disposing of this product, please recycle the paper.

Foreword to the New Book Series

Why a new book series? We believe that it addresses the prevalent unsolved global issues, such as species extinction, climate change, pandemics, and financial crises, which pose a threat to humanity's survival. As Albert Einstein suggested, *complex problems require complex solutions*. He stated that complex problems cannot be solved with the same thinking that created them. The resolution of these issues cannot be achieved solely through conventional scientific disciplines, as their methods of treating the world as an object, analyzing it in its components, quantifying it, and reconstructing it according to their own ideas have contributed to the problem. Most academic disciplines have a tendency to overlook important aspects of reality due to their narrow focus on their own methodologies. This can lead to the destruction of reality under certain circumstances. As Carl Friedrich von Weizsäcker once noted for physics, *scientific disciplines do not tell the untruth with what they reveal, but with what they ignore*.

The book series, *Transdisciplinary Management of Environmental and Social Crises,* addresses the urgent need for coping strategies in the face of humanity's current poly-crises in society and nature. It explores new approaches to successfully dealing with these crises through transdisciplinary methods that integrate knowledge from different disciplines, philosophies, and experiential knowledge. The series also draws on wisdom doctrines from different provenances. The latter includes the knowledge or intuitions of non-experts, such as amateurs, as well as the experiences of practitioners, such as farmers, craftsmen, artists, socially engaged individuals, clerics, or those with genuine spiritual experiences. Each perspective may contribute to the solution, and their differences can complement each other. It is important not to dismiss these perspectives simply because they do not align with mainstream academia.

Furthermore, the book series explores the conditions for responsible leadership in politics, business, and society and proposes alternative models that serve, integrate, and inspire all stakeholders. The classical form of top-down leadership is probably obsolete in view of the complexity of the problems we face. As alternative leadership models successfully demonstrate, leadership can also be seen as serving, integrating, and inspiring. As such, it involves all those affected in its decisions on an equal and respectful basis.

Although the majority of the peer-reviewed contributions should meet academic standards, such as the use and identification of reliable sources, the book series can and may also express other forms of knowledge and knowledge transfer, such as the form of essays and well-argued reflections, experience reports, etc. The transdisciplinary contributions do not necessarily have to reflect the research horizon of the reviewers, but they may and should go beyond it. They are expected to be comprehensible to the readers, to explain technical languages and to stimulate further discussion. In doing so, they broaden the prevalent tunnel vision of each discipline in order to provide perspectives on the whole.

Equally, they should expose and reject simplistic conspiracy theories. Throughout human history, these have often been the cause of pogroms against minorities or the declared enemy, who have been portrayed as scapegoats. Blame games such as those that took place during the Corona pandemic between pro- and anti-vaccination groups ought to be avoided.

The book series takes up what was initiated by *Basarab Nicolescu* more than two decades ago and disseminated over many years in numerous conferences and book publications. His work has been rediscovered in recent years and is now to be continued by the next generation of transdisciplinary thinkers and researchers. We therefore dedicate this series to this great pioneer, thinker, and bio-philanthropist. We hope to realize his vision, as expressed in his *Charter of Transdisciplinarity* (reprinted in this, the first volume).

The trigger for this book series was a *Global Transdisciplinarity Conference*, organized by the Austrian *University for Continuing Education* in Krems from 27 to 29 September 2021 under the direction of *Prof. Dr. habil. Gerald Steiner*. Furthermore, the work of the *International Center for Transdisciplinary Research* (CIRET) and its co-founder Basarab Nicolescu inspired the editors to launch this book series. Another conference took place at the Turkish *Capadocia University* in Mustafapasa/Ürgüp June 26 to 28, 2024.

The editors of the book series:
Mathias Schüz and René Schmidpeter

Gachnang, Switzerland
Ingolstadt, Germany
29 January 2025

Mathias Schüz
René Schmidpeter

Preface

As the editor of the first volume, I have been explicitly dealing with the topic of transdisciplinarity in science, philosophy, and economy for decades. Explicitly I have outlined the concept of "paradigmatology" in my interdisciplinary doctoral thesis about philosophical consequences of quantum physics. Different worldviews, whether philosophical, scientific, or consciousness-based, offer various solutions to problems related to a good life. However, reductionist solutions tend to reinforce these problems. My experience as manager in business enterprises for almost twenty years has shown that in the pursuit of shareholder value, world problems are often ignored and left to impact society and nature. Therefore, simple solutions are not sufficient to overcome these issues. *Transdisciplinary thinking and action* are necessary.

The first volume of our book series presents approaches for discussion. It aims to inspire readers to pursue transdisciplinarity and contribute to solving global issues.

I received support from many colleagues in compiling this work. I would like to express my gratitude to *René Schmidpeter*, who initially suggested the idea of creating an own book series at Springer Nature, *Gerald Steiner*, who brought to my attention Basarab Nicolescu's fundamental work on transdisciplinarity, *Marcus Schmieke*, who kindly provided his premises in the German Schloss Kränzlin near Neuruppin for our first symposium with the majority of the authors of this volume, and *Christian Halper*, whose generous donation enabled us to establish a *Swiss Association for the Promotion of Transdisciplinary Knowledge*.

Gachnang, Switzerland
29 January 2025

Mathias Schüz

Overview of This Volume

The first volume of our new book series gives an overview of *Transdisciplinary Thinking and Acting*. Accordingly, it is divided into two parts: the first being *foundations* and the second being *applications*. What makes transdisciplinary thinking and acting? Which philosophies, methods, scientific insights found transdisciplinarity? What nurtures beyond of them solutions to cope with our world problems respectively poly-crises?

In the spirit of the *Charter of Transdisciplinarity*, adopted at the First World Congress of Transdisciplinarity in Portugal, 1994, and reprinted here, these questions shall be answered. The Charter delivers the *constitution* and *methodological framework* for this. At first, it addresses the growing complexity and fragmentation of knowledge. To overcome the associated risks it fosters a transdisciplinary approach, which synthesizes and transcends individual disciplines to achieve a more comprehensive understanding of reality. It highlights the value of openness, rigor, and tolerance in transdisciplinary work. Transdisciplinarity should lead to an open attitude toward myths, religions, and diverse cultures and emphasizes the importance of dialogue, discussion, and shared understanding, rejecting any attitude that refuses these principles. It promotes a transcultural approach that values contextual, concrete, and global forms of knowledge. Not last, it revalues the role of intuition, imagination, sensibility, and the body in education and emphasizes the need for an economy that serves all living beings.

The *first part* examines the foundations of transdisciplinarity. The need for this *Mathias Schüz* recognizes in the DNA of the modern Enlightenment: the rational claim to dominion over nature. While it improved our living conditions unimaginably, it at the same time almost irrevocably destroyed important foundations of life. Therefore, with his holistic model, he demands a *transdisciplinary responsibility* for the consequences. *Basarab Nicolescu*, the pioneer of postmodern transdisciplinarity, describes the *history* and *methods* of transdisciplinarity, while *Marco Bischof* examines the *diversity of scientific approaches* to the world. *Johannes Heinrich* presents his model of a *transdisciplinary philosophy*. *Albrecht von Müller* sees in his idea of the *Constellatory Self-unfolding* the origin of all phenomena, while *Marcus Schmiecke* identifies the phenomenon of *Coherence* as a common thread for transdisciplinary thinking. *Salvatore Lavecchia* sees in the *Wondering I* the origin for whole worldviews that should be dedicated to the good. *Elfriede Neuhold* compares the epistemologies of occidental thought with the Indian-Vedic tradition. *Gerald Steiner* examines with his *science-practice-approach* the possibilities of pragmatically moving the most diverse social groupings (stakeholders) toward a transdisciplinary dialogue through mutual learning. Finally, *Yusuf Örnek* thinks the *idea of the university* further in the spirit of transdisciplinarity.

The *second part* presents practical applications of transdisciplinary thinking. *Petra Barthelmess* sees in the method of *cultural comparison* a possibility to promote peace with nature. *René Schmidpeter* calls for new foundations of economic thinking and acting and, for this, also questions artificial intelligence. In his analysis of democratic decision-making in Germany, *Karl-Rudolf Korte* shows the role played by *informal processes*. *Claudia Pölderl* elaborates the basics of an *art of listening*, while *Birgit Feldhusen* examines *transdisciplinary* forms of *organizational development*. *Nina Boogen* and *Fernando Obieta* highlight the importance of transdisciplinarity for *artistic forms of expression* by means of various projects. *Nandarani Meister* emphasizes the importance of transdisciplinarity for a *new health medicine*. *Christian Halper* illuminates the transdisciplinary aspects of *vegetarian and vegan nutrition*. *Patrick Nickisch* sees possible forms of *peace-building* in transdisciplinary bridge-building between *religions*. Finally, together with *Maika Steinborn*, *Swami Sacinandana* describes the *benefits of meditative practices* for the well-being of body, soul and spirit and for the transdisciplinary overcoming of the subject–object split.

Gachnang, Switzerland
29 January 2025

Mathias Schüz

The Charter of Transdisciplinarity

Basarab Nicolescu

Manifesto of Transdisciplinarity, 1994, Signed by B. Nicolescu, E. Morin and L. de Freitas[1]

Origin

The "Charter of Transdisciplinarity" offered here appears as the first appendix to Nicolescu's Manifesto of Transdisciplinarity (2002). Actually, however, the Charter has been adopted at the *First World Congress of Transdisciplinarity*, held in 1994. In the author's view, "transdisciplinarity" defines a space for synthesis "across, between and beyond" disciplines. It is worth noting the deep anthropological importance of transdisciplinarity (evident throughout the Charter), as well as the values of rigor, opening, and tolerance mentioned in Article 14. Nicolescu's promotion of transdisciplinarity is still in progress at present.

Preamble

Whereas, the present proliferation of academic and nonacademic disciplines is leading to an exponential increase of knowledge which makes a global view of the human being impossible;

[1] B. Nicolescu, Manifesto of Transdisciplinarity, translated by K. Claire Voss (Albany: State University of New York Press, 2002), pp. 147–152. Available in: https://inters.org/Freitas-Morin-Nicolescu-Transdisciplinarity, retrieved 26 Sep. 2023.

Whereas, only a form of intelligence capable of grasping the cosmic dimension of the present conflicts is able to confront the complexity of our world and the present challenge of the spiritual and material self-destruction of the human species;

Whereas, life on earth is seriously threatened by the triumph of a techno-science that obeys only the terrible logic of efficacy of efficacy's sake;

Whereas, the present rupture between increasingly quantitative knowledge and increasingly impoverished inner identity is leading to the rise of a new brand of obscurantism with incalculable social and personal consequences;

Whereas, a historically unprecedented growth of knowledge is increasing the inequality between those who have and those who do not, thus engendering increasing inequality within and between the different nations of our planet;

Whereas, at the same time, hope is the counterpart of all the afore-mentioned challenges, a hope that this extraordinary development of knowledge could eventually lead to an evolution not unlike the development of primates into human beings.

Therefore, in consideration of all the above, the participants of the *First World Congress of Transdisciplinarity* (Convento da Arrábida, Portugal, November 2–7, 1994) have adopted the present Charter, which comprises the fundamental principles of the community of transdisciplinary researchers and constitutes a personal moral commitment, without any legal or institutional constraint, on the part of everyone who signs this Charter.

Article 1

Any attempt to reduce the human being by formally defining what a human being is and subjecting the human being to reductive analyses within a framework of formal structures, no matter what they are, is incompatible with the transdisciplinary vision.

Article 2

The recognition of the existence of different levels of reality governed by different types of logic is inherent in the transdisciplinary attitude. Any attempt to reduce reality to a single level governed by a single form of logic does not lie within the scope of Transdisciplinarity.

Article 3

Transdisciplinarity complements disciplinary approaches. It occasions the emergence of new data and new interactions from out of the encounter between disciplines. It offers us a new vision of nature and reality. Transdisciplinarity does not strive for mastery of several disciplines but aims to open all disciplines to that which they share and to that which lies beyond them.

Article 4

The keystone of Transdisciplinarity is the semantic and practical unification of the meanings that traverse and lie beyond different disciplines. It presupposes an open-minded rationality by re-examining the concepts of "definition" and "objectivity." An excess of formalism, rigidity of definitions and a claim to total objectivity, entailing the exclusion of the subject, can only have a life-negating effect.

Article 5

The transdisciplinary vision is resolutely open insofar as it goes beyond the field of the exact sciences and demands their dialogue and their reconciliation with the humanities and the social sciences as well as with art, literature, poetry and spiritual experience.

Article 6

In comparison with interdisciplinarity and multidisciplinarity, transdisciplinarity is multireferential and multidimensional. While taking account of the various approaches to time and history, transdisciplinarity does not exclude a transhistorical horizon.

Article 7

Transdisciplinarity constitutes neither a new religion, nor a new philosophy, nor a new metaphysics, nor a science of sciences.

Article 8

The dignity of the human being is of both planetary and cosmic dimensions. The appearance of human beings on Earth is one of the stages in the history of the

Universe. The recognition of the Earth as our home is one of the imperatives of transdisciplinarity. Every human being is entitled to a nationality, but as an inhabitant of the Earth is also a transnational being. The acknowledgment by international law of this twofold belonging, to a nation and to the Earth, is one of the goals of transdisciplinary research.

Article 9

Transdisciplinarity leads to an open attitude toward myths and religions and also toward those who respect them in a transdisciplinary spirit.

Article 10

No single culture is privileged over any other culture. The transdisciplinary approach is inherently transcultural.

Article 11

Authentic education cannot value abstraction over other forms of knowledge. It must teach contextual, concrete, and global approaches. Transdisciplinary education revalues the role of intuition, imagination, sensibility, and the body in the transmission of knowledge.

Article 12

The development of a transdisciplinary economy is based on the postulate that the economy must serve the human being and not the reverse.

Article 13

The transdisciplinary ethic rejects any attitude that refuses dialogue and discussion, regardless of whether the origin of this attitude is ideological, scientistic, religious, economic, political, or philosophical. Shared knowledge should lead to a shared understanding based on an absolute respect for the collective and individual Otherness united by our common life on one and the same Earth.

Article 14

Rigor, opening and tolerance are the fundamental characteristics of the transdisciplinary attitude and vision. Rigor in argument, taking into account of all existing data, is the best defense against possible distortions. Opening involves an acceptance of the unknown, the unexpected and the unpredictable. Tolerance implies acknowledging the right to ideas and truths opposed to our own.

Final Article

The present *Charter of Transdisciplinarity* was adopted by the participants of the *First World Congress of Transdisciplinarity*, with no claim to any authority other than that of their own work and activity.

In accordance with procedures to be agreed upon by transdisciplinary-minded persons of all countries, this Charter is open to the signature of anyone who is interested in promoting progressive national, international, and transnational measures to ensure the application of these Articles in everyday life.

Convento de Arrábida, November 6, 1994.
Editorial Committee: Lima de Freitas, Edgar Morin, Basarab Nicolescu.

Contents

Part I
Foundations

Transdisciplinary Management of Polycrises: A Philosophical Approach

Mathias Schüz

1 Introduction

A "*polycrisis*" in nature, society, and economy threatens the globalized living world of today's mankind. Climate change and energy crisis, extinction of species, pandemics, overpopulation, growing gap between rich and poor, over-indebtedness of entire national economies, and armed conflicts—all these are factors of the polycrisis. As a whole, it is more dangerous than "the sum of its parts". This is because the individual crises reinforce each other.[1] No aspect is isolated from the other.

For example, the war in Ukraine reveals conflicts between systems of politics that were thought to have been overcome long ago. Does the *right of the strongest* apply, as practiced in the system of an *autocracy,* or the *strength of the right,* as intended in the system of a *democracy?* In the sense of Friedrich Hegel (1770–1831), right should realize the freedom of all, the "general will", against the "caprice" of a few, the "arbitrary will".[2] In both systems, however, a sharp distinction is hardly possible, since each party to the conflict carries within itself parts of the opponent and also uses them to enforce its interests.[3] In a similar sense, the legal scholar Katharina

[1] Thus economic historian Adam Tooze, (2022: 2), who takes the concept of polycrisis from complexity theorist Edgar Morin and applies it to current world problems. Gerald Steiner delves into the topic of polycrisis and transdisciplinary coping strategies in this volume from a systems theory perspective (see in this volume G. Steiner, Chapter 9: "Transdisciplinarity for Mastering Poly-Crises by Mutual Learning Between Science and Practice").

[2] Hegel, (1952 [1820]), cf. §§ 3, 15, 75, 104, 219.

[3] With Hegel dialectical "sublation, *Aufhebung*" of the opposites (2013, 80, § 2).

M. Schüz (✉)
Zurich University of Applied Sciences, Winterthur, Switzerland
e-mail: mathias.schuez@zhaw.ch

© The Author(s), under exclusive license to Springer Nature Switzerland AG 2025
M. Schüz (ed.), *Transdisciplinary Thinking and Acting*, Transdisciplinary Management of Ecological and Social Crises,
https://doi.org/10.1007/978-3-031-70680-6_1

Pistor has pointed out that the free, libertarian market economy of the democratic West uses the right to enrich the few at the expense of the many. It contributes to a growing gap between rich and poor.[4] Conversely, China, which tends to be viewed as autocratic, has lifted hundreds of millions of people out of poverty within a few decades.[5] Therefore, it is important to understand that the world cannot be simply divided into two opponent spheres, because both sides bear the kernel of the opponent.

In the long run, the system that maintains or even promotes the diversity of its members, as in nature, instead of fighting against it, is likely to be more viable. Monocultures, on the other hand, have lower chances of survival. Exponentially growing species extinction is a major driver of the polycrisis. It leads to a loss of buffer zones in nature and thus also reduces its resistance. This exacerbates climate change through excess CO_2, but also promotes pandemics through increased opportunities for contact with pathogens from the animal world (so-called zoonoses)[6]. The "impact structure"[7] as well as the "resilience" of nature" is massively disturbed by this.[8]

In order for a therapy to be successful, an anamnesis and diagnosis is necessary. How could it come so far that mankind gained *power over nature as* well as a small *minority* gained *power over the majority* with the help of science, technology, and economy? Which consciousness in mankind has led to the anthropogenic, i.e., man-made, catastrophic potential? How big is the chance of a change of consciousness that enables a way out of the crisis?

There are many indications that the scientific-technical and economic goals and methods of the modern Enlightenment, despite all its achievements, have contributed decisively to present-day polycrisis. Steven Pinker could still prove interdisciplinarily that in the last 10,000 years the interhuman violence potential may have decreased: exponentially thanks to the modern Enlightenment.[9] The dark side, however, that the extinction of species in nature has increased exponentially is thereby concealed.[10]

The claim to dominion over nature was already raised thousands of years ago in the ancient myths of mankind and realized, for example, in the enslavement of the weaker fellow human beings by the strong.[11] Nonetheless, its systematic disciplining became possible only in modern times—thanks to newly discovered methods. Thus, Karl Marx (1818–1883) could claim: "*Philosophers have only* interpreted *the world*

---

[4] Pistor, (2019).

[5] The World Bank, (2022).

[6] Cf. Kaiser & Tominski, (2021).

[7] Cf. Lorenz, (1984: 313–323); Wilson, (1992).

[8] Hanna et al. (2019).

[9] Pinker, (2011: 42 ff).

[10] Gorke, (1999: 11); cf. Tollefson, (2019).

[11] Cf. Blom, (2022).

differently, the point is to transform *it.*"[12] Marx thus summed up the modern paradigm: instead of philosophical contemplation (contemplatio) of how man best fits into the world, the world was now to be made, conversely, to fit man. Modernity went even further by declaring man himself the object of making, even degrading him to a spare parts store[13]. In any case, it was only with the help of science, technology, and economics that the "wholeness of life"[14] consisting of man and nature was finally destroyed.

This chapter explores ways in which man's self-destructive claim to dominion over nature and his weaker fellow human beings might be more responsibly managed with the help of transdisciplinary consciousness and action; that is, how the destructive potential of scientific disciplining of the world might be neutralized, perhaps even compensated, with the help of transdisciplinary approaches.

To this end, the emergence and powerful unfolding of the modern paradigm is illuminated with the help of philosophical reflections. For millennia, thinkers have warned against an understanding of truth that does not leave the world in its so-being. Some of them will be presented here as pioneers of transdisciplinary thought and action and located in a holistic model of transdisciplinarity.

In the first step, the survival crisis of humanity in nature is thematized and its roots in the Enlightenment's claim to dominion over nature are uncovered. Secondly, philosophical thinkers who have developed approaches to transdisciplinarity are presented as examples. Thirdly, a holistic model of transdisciplinarity serves as orientation for the necessary thinking and acting, which respects the diversity of possible constructions of reality, but at the same time ties it back in the cognition and ethical demand of a universal (transdisciplinary) responsibility.

2 The Survival Crisis of Mankind in Nature

Karl Marx's demand to no longer interpret the world, but to reshape it, had its antecedents. Already in the sixteenth century it was raised by Francis Bacon (1561–1626) in the sense of *knowledge is power* ("*true ends of knowledge . . . to the sovereignty and power [of man]*")[15] and methodologically substantiated by René Descartes (1596–1650) only a few decades later. The foundations for man's claim to domination over nature, however, go back much further into human history. In the Old Testament, with the so-called "Dominium Terrae" commandment in Genesis, it

[12] Marx, (2022 [1845]).

[13] Cf. Kimbrell, (1998); la Mettrie & Julien, (1912).

[14] On the concept of "wholeness of life," see G. Schüz, (2001).

[15] Bacon, (1984: 42).

was even regarded as a mandate of divine omnipotence to mankind: "*Subdue the earth!*" (Gen. 1, 28).[16]

In fact, according to *philosophical anthropology*, man—whether as a supposed creature of God or as an "arriviste ape"[17]—has "accepted" this mission in the course of the millennia and realized it in different ways. The reason for this lies in his "specific position" (Adolf Portmann[18]) as a "non-fixed animal" (Friedrich Nietzsche[19]) or as a "flawed being" (Arnold Gehlen,[20] derived from Johann Gottfried Herder[21]) whose instincts are reduced (Konrad Lorenz[22]). Biologically seen, a human being actually is a life form of nature, thus a part of it. But they are not bound in a given ecological niche; instead, they have to create their own (super)-niches of life first. In this sense, a human is "world-open" (Max Scheler[23]). He or she can survive in the Arctic as well as in the desert. This succeeds however only by providing cultural achievements with which it compensates for its "deficiencies." By considering the consequences of his actions and using them purposefully to satisfy his needs, she may also be able to responsibly justify them.[24]

2.1 Problem Solving through Cultural Achievements

Such specific achievements of man, however, presuppose a spiritual understanding of the nature surrounding him. Without considering the interrelationships between food and hunger, clothing and cold, shelter and severe weather, purposeful interventions in nature to defend against its threats are not possible. Without the deployment of natural resources, the dangers of nature cannot be overcome. Without cooperation of the individual human being with other members of a community, survival would hardly be possible. Man's nature thus enforces the satisfaction of his biological needs through *understanding* and *modifying* nature as well as *collaborating* with others—in other words through *culture*.

---

[16] Whether this command actually means "careful treatment," as many theologians today seek to justify the commandment, cannot be clearly proven in the overall context of the Old Testament. In any case, the original Hebrew text lifts man "out of his environment" with the commandment to rule. Other passages seem to relativize this (Jüngling, 1994: 29 ff; cf. Blom, 2022: 83).

[17] Arnold Gehlen, (1978: 9).

[18] Cf. Landmann, (1982: 157).

[19] Nietzsche, (1997, II: 62)

[20] Gehlen, (1978: 20, 24, 34 ff).

[21] Herder, (1993, p. 24, cf. 20).

[22] Gehlen, (1978: 26).

[23] Scheler, (2009).

[24] In his philosophical anthropology, Richard Wisser characterizes the basic human condition as "critical-crisical". Crises force man again and again to critically deal with his survival strategies and, if necessary, to look for new ones and to take responsibility for them. (1997: 334 ff.)

This term comes from the Latin *colere* = to cultivate, to work, to dwell, to feed, but also to cultivate, to preserve, to revere, to cherish, to worship, to hold sacred.[25] The Romans knew therefore besides *agricultura* = agriculture also *cultura animi* = education and *cultura ingenii* = care of the spirit. As early as 1939, the ethnologist Bronislaw Malinowski (1884–1942) interpreted *culture* as a function of *coping with problems* of human survival:

> Culture is essentially an instrumental apparatus by which man is put in a position the better to cope with the concrete specific problems that face him in his environment in the course of the satisfaction of his needs.[26]

It consists of "implements and consumers' goods, of constitutional charters for the various social groupings, of human ideas and crafts, beliefs and customs,"[27] with "the construction of a new, secondary, or artificial environment."[28] It satisfies basic human needs such as nutrition, reproduction, housing, security, growth, and health. Different forms of nutrition, family bonds, housing, protection, education, and hygienic measures are developed as cultural achievements.[29]

Culture therefore means an active change of the threatening nature by human activities. Depending on the relationship of man to nature and how they perceive this, they develop different ways of dealing with it and solution patterns (paradigms) for the problems in dealing with it. As in the case of biodiversity, we can assume that a *diversity of cultures* is more conducive to the long-term survival of humanity as a whole than a reduction to a *monoculture*. Because different solutions are required depending on the changing circumstances. The latter creates a chariot mentality through which complex problems are approached in a highly one-sided manner and are thus reinforced.

2.2 Knowledge and Ethical Responsibility

To what extent can, may, and should humans intervene with their cultural achievements in the ecological structure of nature in order to ward off dangers and satisfy their needs for life? An answer given to this question is always also an ethical one. For ethics, as an *art of "getting along well with each other,"*[30] examines how human actions affect other forms of life and seeks, through rational reflection, to optimize how we deal with them. Seen in this light, any *theoretical* understanding of nature

[25] Stowasser et al., (1979: 86)

[26] Malinowski, (1960: 150).

[27] Ibid: 36.

[28] Ibid: 37

[29] Cf. Ibid: 36 f, 91 ff.

[30] Schüz, (2021: 117), cf. (2019: 99).

should also imply *ethical* behavior toward nature.[31] Those who recognize nature as an "organic-lifeworld" context in which they see themselves as embedded will deal with it differently than those who see it as a mathematically calculable and technically exploitable total object. The former will approach it only with reverence[32], the latter with an attitude of entitlement toward nature as a resource to be exploited.

Exemplarily, the bison hunting of (North) American natives represents the attitude of the first. Before they went hunting, they connected in appropriate rituals with the spirits of nature, such as the sun, which is supposed to give them the necessary strength and insight, or with the spirit of the buffalo herd, to ask for a sacrifice of, for example, a weak or sickly member, which they were allowed to hunt for their own survival.[33] In return, they offered care to the herd, but also to the surrounding nature. *Reciprocity* was and is the basis of survival for Native Americans: No taking without giving and thanking[34], because they knew that they themselves were part of the nurturing Mother Nature ("Pachamama" among the Andean peoples[35]), which would have been suicidal to destroy.

On the other hand, the professional hunters of the white immigrants succeeded in killing millions of bison within a very short time in order to get hold of their hides and to market them in a one-time but extremely profitable way, especially in Europe. The hunters were under the protection of the army, as one of their commanders, Lieutenant General Philip Henry Sheridan (1831–1888), explained:

> These men have done . . . to settle the vexed Indian question than the entire regular army . . . They are destroying the Indians' commissary [. . .] let them kill, skin and sell until the buffaloes are exterminated. Then your prairies can be covered with speckled cattle and the festive cowboy, who follows the hunter as a second forerunner of an advanced civilization.[36]

In attitudes like these Konrad Lorenz saw an "inability to feel any reverence, . . . a dangerous disease of our culture."[37] Today, this disease has had such a global impact that by August 1st of 2022 (*Earth Overshoot Day*[38]), humanity has already used up the Earth's annual available resources to provide a comfortable life through technological advancements. The dark side of this progress has brought mankind to the edge of the abyss.

---

[31] Cf. Karen Gloy, (1995: 9).

[32] Cf. Albert Schweitzer's ethics of reverence for life (2002) or Konrad Lorenz (1984, p. 321) with reference to William H. Thorpe's *Science, Man and Morals,* who sees there three different ways of man "to approach the truths of the universe: scientific inquiry, artistic intuition, and religious experience" (Ibid: 321).

[33] Cf. Mails, (2002); Brink, (2008): 105 ff.

[34] Cf. Mauss, (2002): 18 f; Blom, (2022): 331 f. Similarly, Konrad Lorenz refers to the farmer "who not only 'sticks to his clod' but also *loves* it" (translated by me, 1984: 319); Simmel (1989: 86) describes exchange ("Tausch") as the middle between robbery ("Raub") und charity ("Geschenk").

[35] Cf. Bossinger, (2020).

[36] Cited in: Hampf, (2019: 211)

[37] Lorenz, (1984: 321).

[38] Global Footprint Network, (2023).

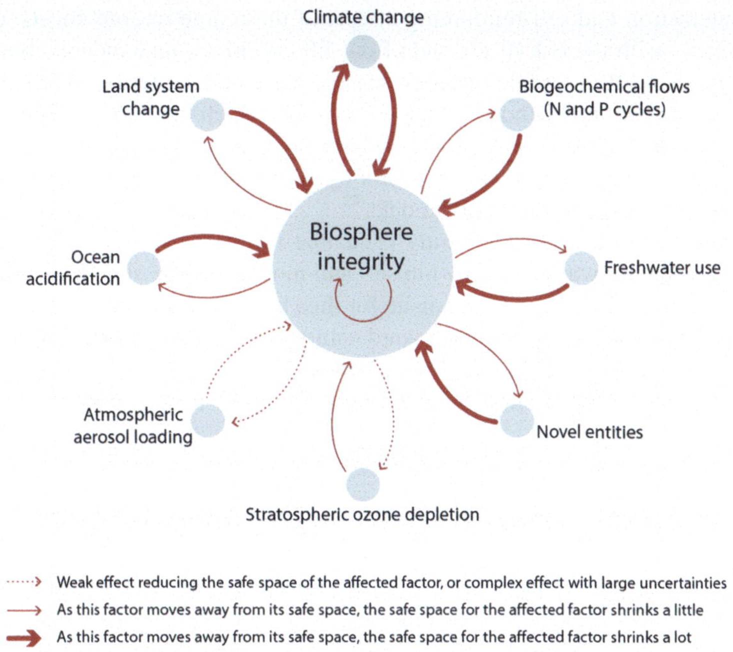

Fig. 1 Factors of the ecological polycrisis. Source: Steffen et al., 2015

For example, the Doomsday Watch, co-initiated by Albert Einstein, is no longer set at 17 min to twelve, as it was in 1991, but only at 90 seconds (as of February 2023),[39] after statement of the scientists, who readjust the clock for decades in accordance with newest scientific data again and again. In doing so, they include not only man-made interventions in the ecosphere, but also global political crises and armed conflicts. They see the COVID-19 pandemic as a wake-up call for global crisis management.[40]

Other scientists see concretely the greatest threat to our livelihoods besides (1) climate change, (2) loss of biodiversity, which accounted for 80–90% in the last 300 years, (3) massive land use by human intervention, which just last year for the first time permanently destroyed more than half of the existing nature reserves, (4) the over-fertilization, which is accompanied by a devaluation of soils for arable farming, (5) acidification of the oceans, (6) ozone hole, (7) fresh water consumption, (8) aerosol pollution of the atmosphere, and last but not least (9) the introduction of novel substances and life forms (such as plastics, chemicals, and genetically modified organisms) into our ecosystem (see Fig. 1).[41]

[39] https://weltuntergangsuhr.com/

[40] Cf. Mathias Schüz, (2021: 306).

[41] Rockström, (2021: 106).

The interaction and self-reinforcing effects of these nine reasons for the ecological polycrisis with the risk of the end of our life-world are unavoidable, but due to their complexity they are hardly foreseeable in concrete terms. The scientists involved see only a time window of 8 to 10 years in which the train speeding toward the abyss can perhaps still be deflected. Otherwise, at least from the point of view of climate change, there is a threat of a "hot age" that will call into question, if not destroy, our cultural achievements to date.[42]

It should have become sufficiently clear that the survival crisis of man in the ecosystem of nature was caused by himself and must therefore also be answered for and overcome by himself. A prerequisite for this, however, is a fundamental change in his understanding of and attitude toward nature. As I said, understanding of nature generates corresponding behavior toward nature. If we want to change the latter, the latter has to be developed anew or another one has to be (re)discovered.

3 In the Tunnel Vision of Occidental Scientific Disciplines

Human consciousness is naturally characterized by tunnel vision. This means that the horizon of perception is limited. One sees only what one wants to, should, or can see. According to a postulate of Immanuel Kant, one goal of the modern Enlightenment was to free oneself "from self-imposed immaturity"[43], i.e., to become capable of acting on one's own responsibility. A person who is immature cannot assume responsibility because he is in a relationship of dependence on other powers. He only carries out what others tell him to do. This dependence is self-inflicted, insofar as everyone has the potential to free himself from it, namely by using his intellect.

In fact, mankind has gained new degrees of freedom with the help of modern science and technology. It has largely emancipated itself from its dependence on nature, and at the same time from the dictates of the medieval church as the representative of transcendent powers. The power thus gained, however, has its price, because the activities that have been set free must now also be answered for. One can no longer attribute the consequences to the uncontrollable powers of nature or transcendence.

In this sense, the tunnel vision of the immature, unenlightened citizen has been replaced by the tunnel vision of disciplinary science. For every scientific discipline is based on the focused gaze of a certain method, in whose spotlight new insights shine, but at the same time others are faded out. In fact, therein lies its strength. With the focus on a research object, the researcher divides the world into objects, who can analyze them according to certain methods, count them (quantify them), and,

[42] Ibid: 102–106; cf. Schüz, (2021: 304).

[43] Translated by me: Kant, (1978: 53).

according to the methodology of René Descartes, also put them back together (synthesize them) at will.[44] By systematically dissecting and reconstructing nature in this way, modern man makes it subservient to him. Thus, as Descartes prophesied, he becomes the "master and owner of nature" who can reshape the world at will according to his own purposes, thus subjecting it to a new rationality of purpose.[45]

The objectification of nature[46] in the focus of scientific methodological dissection is accompanied by a "disenchantment of the world" (Max Weber), in which something essential, precisely the power over nature, is gained, but also a great deal is lost.[47] Epistemologically, the old ways of accessing the world, such as magic and myth, are discredited as pipe dreams, although today their value is also recognized again. Thus, Hartmut Böhme has unmasked the myth of modernity as "self-enchantment."[48] It acknowledges only bare facts, but at the same time suppresses the still existing "energies and needs" of pre-modern magic, myth, and cultus,[49] if it only acknowledges naked facts. In this process, it neutralizes and isolates the cognizant subject from world occurrence, as if it were its neutral observer—*the* "illusion" of modernity.[50]

From the ethical point of view, the objectifying approach of modernity has not only brought the systematic use of the forces of nature, but in return has generated a tremendous force of destruction of our living world, as indicated above. In the seventeenth century, Descartes motivated his followers for vivisections, i.e., cutting up animals alive[51]—with the purpose of exploring the mechanism of life. What his successors paradoxically want to investigate until today, however, is destroyed by this method.

This paradox of modern scientific practice, that the living whole can only be investigated by fatally breaking it down into its parts, has always been the object of criticism—combined with the call to question the truth claims of the scientific disciplines.

---

[44] Cf. Descartes, (1978: 15 f).

[45] Ibid: 50.

[46] A profound and extensive *history of objectivity* is provided by Daston and Galison (2021).

[47] Weber, (1988: 594).

[48] Böhme, (2014: 51).

[49] Ibid: 8.

[50] Ibid: 51 f.

[51] Descartes, (1978: 38); cf. Shepherd, (2015).

4 Criticism of Disciplinary Thinking

Martin Heidegger (1889–1976) summed up his criticism succinctly: "Science does not think."[52] Its "claim to power" consists in the "objectification of all areas and domains" of nature.[53] As an independent subject exploring the world as an object, the scientist loses contact with his own origin, which lies in nature. Accordingly, the whole "withdraws" from the grasp of the researcher. This creates a "gap . . . between thought and the sciences."[54] The effects on the objectivity and thus also truth content of scientific disciplines will be worked out in the following on the basis of some masterminds.

4.1 Ludwig Fleck

The physician and science theorist Ludwik Fleck (1896–1961) may have been the first to elaborate the typical "thought styles" *and associated* limitations of scientific communities, initially on the basis of medical theories on syphilis, for example, as early as 1935 in his work *Genesis and Development of a Scientific Fact.* In general, science is not a formally objective construct for him, as the logical empiricists or positivists of the Vienna Circle claimed.[55] Scientific knowledge is a collaborative research product of "*thought collectives*"[56] each with culturally specific languages, "tenets of faith and superstition"[57] with their own worlds of "feelings,"[58] logics, and social hierarchies. They are also expressions of "propaganda, imitation, authority, rivalry, solidarity, enmity, and friendship."[59] Last but not least, they are also dependent on "public opinion"[60]

> Both thinking and facts are changeable, if only because changes in thinking manifest themselves in changed facts. Conversely, fundamentally new facts can be discovered only through new thinking.[61]

Thus, according to Fleck, there is no presuppositionless viewing and observing in any scientific discipline. They are historically conditioned and bound to thinking styles, decisions, and habits. The respective state of knowledge generates a closed,

---

[52] Heidegger, (1971: 4, 154).

[53] Ibid: 155.

[54] Ibid: 4 f.

[55] Fleck, (1981: 50).

[56] Ibid: 39.

[57] Ibid: 44.

[58] Ibid: 49, 142.

[59] Ibid: 43.

[60] Ibid: 105.

[61] Ibid: 50 f.

"style-permeated system of opinions" as a "dual relationship" between "knowing subject" and "object to be known."[62] In this way, sciences would "discipline" our patterns of experience[63]. Thus also the established occidental logic, foundation of the modern sciences, does not apply to the whole mankind. There is no reason to reject "different logic," as it characterizes the worldviews of so-called "primitive races," "mystics," or "gnostics":

> The Concept of a thought collective comprising the whole of Homo Sapiens is of little use, because the intellectual interactions between different types of human society are too weak.[64]

4.2 *Thomas S. Kuhn*

Decades later, Fleck inspired Thomas S. Kuhn (1922–1996) to his influential *The Structure of Scientific Revolutions*, which elaborated the typical elements of scientific paradigms (from Greek paradeigma = archetype, pattern) and their sequence. Accordingly, every scientific theory—even in physics, for example, the theory of Newtonian mechanics and Einstein's theory of relativity—presents a typical pattern of solutions to problems that research follows.[65]

"Anomalies" are ignored or bent until a new theory prevails, but usually only when the "crisis" of the old established theory can no longer be ignored or its adherents have died out.[66] The paradigms are usually "incommensurable to each other"[67]; i.e. they each constitute incomparable worldviews. Unlike the science theorist Karl Popper (1902–1994) with his approximation thesis[68], Kuhn concludes from his analyses that no absolute truth can be obtained with scientific methods.[69] In this respect, Kuhn has further developed the foundations of transdisciplinary thinking with his paradigm theory.

Connections, hierarchies, differences, and commonalities of paradigmatic approaches to the world would have to be defined in more detail by a "paradigmatology," which I have proposed[70] as a transdisciplinary attempt in 1986. Paradigmatological awareness transcends existing paradigms without claiming to be free of paradigmatic influences.[71]

---

[62] Ibid: 38.
[63] Cf. Ibid: 145.
[64] Ibid: 174 (FN 41).
[65] Cf. Kuhn, (2012: 24).
[66] Ibid: 150.
[67] Ibid:103.
[68] Popper, (1979: 58).
[69] Kuhn, (2012: 169); cf. Planck, (1979: 13).
[70] Schüz, (1986: 14–25, 88 ff, 246).
[71] Ibid: 87–95, cf. 98 ff.

4.3 Paul Feyerabend

The science theorist Paul Feyerabend (1924–1994) goes even further. In his *Against Method: Outline of an Anarchistic Theory of Knowledge* in 1975, he laid the foundations for a transdisciplinary understanding of the world without using the term transdisciplinarity. He elaborates how normal science works. Theories determine what researchers grasp as experience in the world, and these seem to confirm those. Their "appearance of success *cannot in the least be regarded as a sign of truth and correspondence with* nature."[72] In this circle of empirical verification, alternative forms of perception or theories are in a sense bitten off by the scientific community. Similar to Fleck before him he states:

> That interests, forces, propaganda and brainwashing techniques play a much greater role than is commonly believed in the growth of our knowledge and in the growth of science, can also be seen from an analysis of the relation between idea and action.[73]

The search for "intellectual security in the form of clarity, precision, 'objectivity', 'truth'" with the help of scientific methodology or rationality is obsolete according to Feyerabend. He therefore pleads for an open relativism with the slogan "*Anything goes*"[74].

Feyerabend, Fleck, and Kuhn substantiate their criticism of the sociologically conditioned dogmatic blindness of the established sciences with numerous examples. Even today the criticism has not been silenced[75]. Thus, Rupert Sheldrake attacks the *Science Delusion* with many more or less convincing arguments. His own theory of the "morphogenetic fields" is an example for it. It is not recognized by the established science. This is because it is directed against a mechanistic biology which still adheres to the idea of a genetic program controlling all forms of life—despite empirical contradictions.[76] In general, Sheldrake criticizes the handling of empirical data, which, if they do not fit the given theories or the interests of the financier, are simply ignored. Thus, they convey an illusory objectivity.[77]

The fact that such illusions arise from distortions of individual and collective forms of perception is, after all, also the self-critical result of scientific research. Thus, numerous sources of socio-psychological, emotional,[78] political, economic[79]

---

[72] Feyerabend, (1986: 50) (translated by author from the German version revised by Feyerabend); cf. Hossenfelder, (2018).

[73] Feyerabend, (1988: 17).

[74] Ibid: 19.

[75] Examplarily cf. Longino, (1990); Horgan, (2015); Daston & Galison, (2021); Smolin, (2008); Lamont, (2009); Goldacre, (n.d.)

[76] Sheldrake, (2020: chap. 6), According to epigenetics, many genes only become effective in interaction with environmental factors (cf. Baedke, 2018).

[77] Sheldrake, (2020: chap. 11).

[78] Cf. e.g. Barret and Bar, (2009).

[79] Political and economic interests also play a role in the selection of scientific theories (cf. Socol et al., 2019). For example, the financial support of university research projects by industry (so-called third-party funding) can significantly distort the results (cf. Lamont, 2009; Krimsky, 2013; Reutlinger, 2020). Conversely, Duff McDonald's study The Golden Passport (2017) uses

but above all logical and cognitive biases have been acknowledged. Thereby it relativizes the objective truthfulness of its insights.[80]

4.4 *Logical and Cognitive Biases of Truth*

In his advocacy for *Rationality*, social psychologist Steven Pinker describes 88 such "biases and fallacies."[81] However, he is convinced that they can be avoided by freeing the rational core from the many shells of irrationality. In this way, rationality actually succeeds in getting at "objective truths."[82] He is thus still completely in the tradition of the Enlightenment philosopher René Descartes, who also calls on us to separate the confused and dark from our ideas.[83] Only those who understand things by "clear and distinct conclusions"[84] reach the truth.[85] Descartes justifies this with the fact that man, in contrast to all other natural beings, can think and thus has a spirit that is connected with the all-reason of God.[86] If he distances himself from the deceptive sense illusions of nature, then he can attain truth.[87]

Pinker does not need a God for this. He does not presuppose a dualistically split world of mind and matter. He argues evolutionary-biological: The spirit of rational thinking has developed in the course of human history as an epiphenomenon from the material world. Reasoning proves itself as a result of the "selective pressure" in the direction of a successful survival.[88] "Reason is valid that it *works*"[89] even in morality,[90] for example, by creating more tolerant and peaceful communities with less interpersonal violence.[91]

In such a rational thinking, *subjective* statements about *objective* facts or realities are judged according to the *two-valued* logic, namely whether they are either *true* or *false* according to the *principle of the excluded middle*—there is no third. Whether

---

Harvard Business School as an example to show that universities can also influence social institutions, such as the notion of a one-sided profit-oriented capitalism.

[80] The fact that economic interests also play a role in this process is illustrated by Duff McDonald (2017) using the example of the history of the Harvard Business School and its research.

[81] Pinker, (2021: 389 f).

[82] Ibid: 41 f.

[83] Descartes, (1978: 31).

[84] Ibid: 23.

[85] Ibid: 27.

[86] Cf. Ibid: 28.

[87] Cf. Ibid: 29.

[88] Cf. Pinker, (2021: 87). Within the "evolutionary epistemology" there are conflicting interpretations—from realist to radical constructivist—e.g., concerning the concept of "fitness" (cf. Roth, 1997: 344 ff).

[89] Pinker, (2021: 41).

[90] Ibid: 67 f.

[91] Cf. Pinker, (2011: 85 ff).

for example the statement "There are paranormal phenomena like out-of-body consciousness in near-death experiences, telepathy, precognition or telekinesis" is true or false depends now on the fact that such phenomena can be actually observed or reproduced.

This requires *first* of all an unbiased (= objective) examination. But this does not exist independently from accepted beliefs of a scientific community (e.g., "Everything that is, is physically explainable").[92] Who rejects from the beginning "paranormal" phenomena on the basis of his three-dimensional, materialistic or mechanistic worldview as "impossible" will lock them away in the poison cupboard of the "unthinkable and irrational" and deny their existence. Who on the other hand, like certain representatives of modern physics, assumes a multidimensional world, in which the spatio-temporal world is embedded, can absolutely accept paranormal phenomena as explainable and, thus, accept them in their observability.[93]

Secondly, the assessment of the statement depends on the applied criteria for "true or false." These are by far not unambiguous, since in the history of philosophy a multiplicity of truth theories was produced. For example, as suggested by the *evolutionary theory of knowledge*, the subjective statement (mind) and the objective fact (matter) can be understood as circular: They are mutually dependent: The mind (re-)constructs a material fact, which has (co-)created it. Then truth is self-referential and constructivist. What is to be determined as objectively true reflects the logic and thus the search pattern (paradigm) of the observer.[94] Furthermore, scientific truth can at best grasp *how* the world functions, but *never why* the world exists at all.

Pinker himself rejects the "miracles" mentioned in the above statement as unreasonable[95] and thus untrue. In contrast, a transdisciplinary research group at Virginia University's Division of *Perceptional Studies* (DOPS) is convinced of the opposite. They document their scientifically sound observations on paranormal phenomena and understanding of truth in extensive anthologies with titles such as *Irreducible Mind*, *Beyond Physicalism*, and *Consciousness Unbound*.[96]

Let us conclude: Different conceptions of rationality and logic produce different conceptions of truth. They are never unconditioned. John Manoogian III has collected over 180 such "cognitive biases," grouped them by types and main triggers (see Fig. 2), and presented them graphically.[97] They point out possible traps that

[92]Cf. Roth, (1997: 355 ff). According to him, there is no objective knowledge, only intersubjective knowledge (357).

[93]The astrophysicist Bernard Carr (2015), a disciple of Stephen Hawking, traces the various approaches to multidimensional physics historically and summarizes the state of the current research. (For example, the M-theory of the string theory calculates with 11 dimensions.) In doing so, he develops his own theory to explain paranormal phenomena.

[94]Cf. Joldersma, (2011); Stapp, (2015). Skirbekk (2019) discusses different theories of truth.

[95]Pinker, (2021: 158 ff, cf. 310 f).

[96]Kelly et al., (2007); Kelly et al., (2015); Kelly & Marshal, (2021). Cf. Radin, 2018; or the discussion between two representatives of the two presumably contradicting paradigms: Steven Pinker and the Indian mystique Sadhguru at Harvard University (Pinker & Sadhguru, 2023).

[97]Cf. the "Cognitive Bias Codex" by Manoogian (2023).

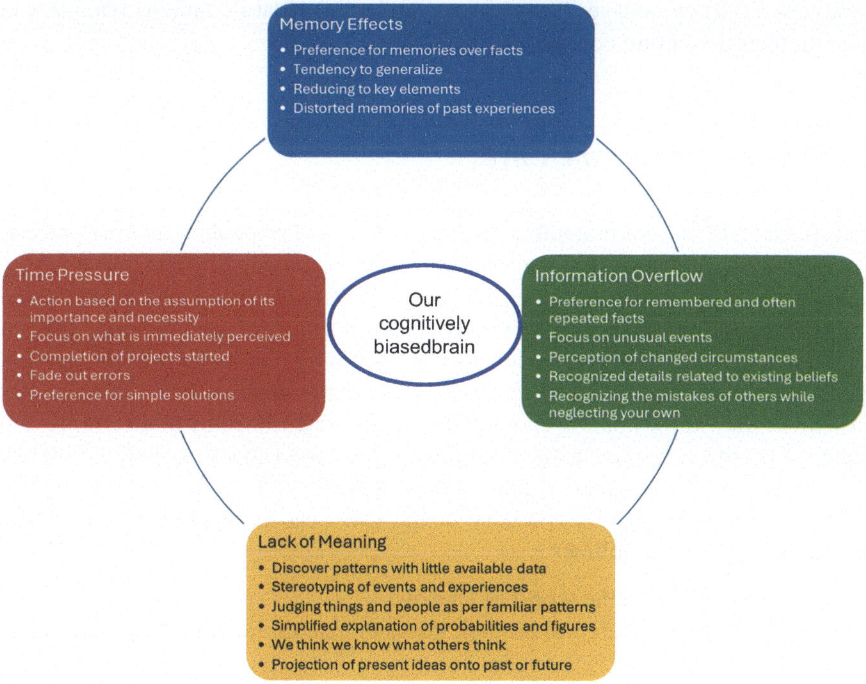

Fig. 2 Main triggers for cognitive biases (based on the systematics of Manoogian, 2023). Source: The author

researchers fall into when perceiving or constructing facts and accordingly defend faulty theories.

As an example for the "information overflow" (see Fig. 2), detailed knowledge of an established scientific guild in the so-called "Semmelweis reflex"[98] prevents new findings. In an evidence-based study, the Hungarian physician Ignaz Semmelweis (1818–1865) had identified a lack of hygiene on the part of the treating physicians as the reason for the fatal puerperal fever. Instead of adopting his new hygiene recommendations, his colleagues dismissed the attack on their existing belief system "as non-scientific." Due to an intrigue, he was sent to a mental hospital at the age of 47 and died there, possibly after being subjected to violence.[99]

Even if Semmelweis is an extreme example of scientific ignorance, present-day representatives of a discipline are not immune to defaming adherents of a different

[98] Ibid.

[99] Seidl, (1985); Bührke, (2012: 168–173).

worldview.[100] They behave like the cave residents in Plato's famous parable who discredit their dissenting contemporaries.[101]

4.5 *Plato's Myth of the Cave*

In his parable, Plato sees humankind locked in a cave. The people can only perceive the shadow plays in front of them on a wall. They are created by a fire behind their backs, in front of which objects move back and forth. This remains hidden to the humans because their bodies are bound and therefore they can only perceive the wall in front of them. Therefore, they consider the shadow plays as the actual reality, do scientific research about it, develop theories, make predictions, and admire the best ones. Since they know nothing else and their life is determined by it, they refuse any change of perspective or even the exit from the cave as an attack on their worldview. Thus, access to a "truer" reality remains closed to them.[102]

The lessons from Plato's Allegory of the Cave can be applied to our current scientific disciplines as follows:

1. We do science from the world of shadows.
2. Our knowledge is therefore limited; the wise man knows, like Socrates, that he knows nothing.
3. We resist new ways of knowing.
4. Only after unleashing the shackles, and usually only by force, can we be pulled out of our rut and open ourselves to the "true" reality.
5. Once we have recognized the "truth" (show of ideas), we want to convey it to those "left behind."
6. However, the sage experiences resistance from the cavemen and even runs the risk of being killed. The established scientists, as little as Plato's cave dwellers, want to give up their painstakingly gained knowledge about the world. The one who disturbs their circles is often rejected or even killed. The latter takes place in the modern scientific establishment, at least symbolically, when new ideas from young scientists are rejected in the "double peer review" process and therefore not placed in renowned journals.[103]

[100] For example, the renowned philosopher Thomas Nagel was attacked by fellow academics for his attempt to prove the neo-Darwinian theory wrong (Nicholson, 2012). In German-speaking countries, the number of professors who have been dismissed for unknown reasons or publicly demoted seems to be growing (Egner & Uhlenwinkel, 2021).

[101] In this context, Plato himself can be seen as a source of transdisciplinarity, or even as a source for the whole of European philosophy. Alfred North Whitehead actually called the latter "a series of footnotes to Plato" (1978: 39).

[102] Plato, (1978), Book VII, 514a-517a. About Plato's and later the influence of his pupil Aristotle on the modern world with its disciplines Herman, (2013).

[103] Cf. Carroll, (2018).

7. The tunnel vision of the scientist blanks out essential aspects of the real and considers his view of the world (shadow plays) to be the only true one.
8. He resists change based on habit.
9. Change is only possible through recognizing and discarding cherished habits—a painful process, as Plato confirms.
10. Good change leads to the realization that the world is multidimensional.
11. Only modern physics (quantum theory) confirms Plato's view.[104]

The last three points indicate by what and how our modern worldview can be overcome. They indicate an essential basic trait of transdisciplinarity that in order to overcome the omnipresent tunnel vision, we first recognize the distortions with which our mind perceives the world.

4.6 *The Concealed Lie of the Disciplines*

Basically, we can state that most scientific disciplines objectify the world according to their respective methods, i.e., make it measurable, and thereby reconstruct it. The world itself, as Immanuel Kant already stated, is not captured in the process. Therefore, the quantum physicist and philosopher Carl Friedrich v. Weizsäcker (1912–2007) stated in 1943: Even the "physical world view … is not wrong with what it claims, but with what it conceals."[105] According to what has been said so far, this statement applies to all scientific disciplines. It brings up again what already the pre-Socratic thinker Heraclitus (520–460 B.C.) warned against: "Nature loves to conceal itself" ("Physis kryptesthai philei").[106] The more you want to grasp it, the more it eludes you.[107]Gregory Bateson (1904–1980), another transdisciplinary thinker, warns against the one-sided worldview of modern times because its "mere purposive rationality unaided by such phenomena as art, religion, dream, and the like, is necessarily pathogenic and destructive of life."[108]

[104] See especially Weizsäcker, (1980: 89, 379 ff); Schüz, (1986: 52 f).

[105] Weizsäcker, (1976: 25) (translated by the author into English).

[106] Heraklit No 27 in Mansfeld & Primavesi, (2012: 259).

[107] This may remind us of Heisenberg's uncertainty principle in quantum physics: the more precisely one wants to determine the position of an atomic particle, the less precise the measurement of its momentum will be, and vice versa. For more details, see Weizsäcker, 1980: 120 ff; Schüz, 1986: 214 ff.

[108] Bateson, (1972: 155); cf. Berman, (1981: 235).

4.7 Summary

1. The scientific disciplines capture the world horizontally in the sense that they objectify it like an object, measure it, and derive general formulas from it.
2. Each discipline reconstructs its own worldview according to its culturally determined, methodological and utilitarian approach, destroying original aspects of the world in the process.
3. Therefore, essentials of the world elude the grasp of the disciplines.
4. We can assume that this objectifying understanding of nature often promotes a wrong, i.e., life-destroying, behavior in and toward nature in the long term, and thus creates a fundamental ethical problem.

How can transdisciplinarity avoid the destructive potential of one-sided use of the scientific disciplines? How can it, at the same time, give voice to, or at least raise awareness of, what is concealed in nature? Is it possible to re-enchant the world disenchanted by science and technology?[109] The next section will outline a holistic approach to transdisciplinarity.

5 Transdisciplinarity as a Paradigmatological Way out of the Modern Enlightenment Trap

Since every established scientific discipline takes a look at something, but at the same time fades out others, consistently thought through, also inter-disciplinary thinking would only mean a (horizontal) change of shape ("Gestaltwechsel") from tunnel vision to tunnel vision. Therefore, it is obvious to complement the disciplinary gained knowledge on the one hand *horizontally* by the views of the manifold ways of experience, for example of so-called "lay people" or traditional wisdom teachings, and to remain thereby in the immanent world experiences.

Furthermore, it is still necessary to grasp their *origin* "transcendentally" (not only in the sense of Immanuel Kant), i.e., to reflect their "conditions of their possibility" in a *vertical* thinking movement. Perhaps then one will grasp at least a breath of the *origin of being* itself. Finally, still further, other forms of consciousness should be admitted, which go beyond scientific thinking and "common understanding" (Kant) and integrate, for example, existential, artistic, magical, shamanic, mythical, mystical, and other spiritual modes of experience in an encompassing experience of the world.

In this way, transdisciplinary knowledge represents a bridge between immanent and transcendent worlds. Without a reference to transcendence we will hardly be able to solve our immanent world problems—precisely because the hidden is ignored in the disciplinary grasp of science and technology. Thereby, the unity of

[109] Cf. Berman, (1981: 235 ff).

nature is irrevocably destroyed in a dualistic worldview, which threatens the continued existence of mankind itself.

5.1 Transcendence Reference

Martin Heidegger, in the famous interview published after his death in the German weekly news magazine *Der Spiegel*, emphasized why, without reference to transcendence, no philosophy offers a way out of the life-threatening crisis of humanity:

> Philosophy will not be able to bring about an immediate change in the present state of the world. This is true not only of philosophy, but of all merely human senses and aspirations. Only a God can still save us.[110]

According to Heidegger, such a claim to be able to causally effect something with thinking would again correspond to a claim to power. Therefore, he demands more humility toward and integration into the larger whole of being.

Heidegger's counterpart and co-player Karl Jaspers (1883–1969), who as a physician, psychologist, philosopher, and political thinker fits seamlessly into our transdisciplinary approach with his extensive body of work, argues similarly. He starts from three ways of experiencing the world: (a) philosophical and scientific world orientation,[111] (b) communicative illumination of human existence[112], and (c) reading the ciphers of transcendence.[113] Truth shows itself for him in different "ways of the encompassing,"[114] in which the world as "totality of the immanent [. . .] hovers like a sphere between transcendence and existence."[115] In this field of tension the search for truth becomes a challenge. Because truth eludes finite consciousness in view of the numerous "sources of untruth"[116].

Without naming the term itself, transdisciplinarity shows up in Jaspers' understanding of truth. It can never be attained by science or by philosophy, but it can be experienced in the "philosophical faith,"[117] which is founded in the origin and aims at it. The experience is then unique for each individual and never completed:

> Only from the philosophical faith, at any time originally, [. . .] it is possible to find the way to the truth, which has been absorbed into the jumble of aberrations in the history of philosophy.[118]

---

[110] Heidegger, (1988: 99 f) (translated by the author).

[111] Jaspers, (1973a).

[112] Jaspers, (1973b).

[113] Jaspers, (1973c); id. (1977).

[114] Jaspers (1983: 47).

[115] Ibid: 143.

[116] Ibid: 492.

[117] Jaspers (1974: 24).

[118] Jaspers (1974: 24) (translated by the author).

5.2 Aim of Transdisciplinarity

Even without the onto-theological confessions of great thinkers and philosophers, our transdisciplinary approach makes clear the necessity of a reintegration of the human mind into the original wholeness of the universe as the *aim of transdisciplinarity*.

Some findings of modern physics show that the universe does not consist of "dead matter," but since its origin, the big bang, is an expression of an interplay of matter, energy, and information.[119] Without information, i.e., that which brings into form, the moving force of energy would not produce structures conducive to life. One could even say that all life forms are *children of the Big Bang*, as they have the same origin and are therefore *irrevocably related to each other*. In the information field of the universe consequently also the potential solution ways for their survival are "callable." This, however, only if the human mind opens itself to them—*methodically* with the help of expanded forms of consciousness, holistic styles of thinking, artistic forms of expression, ethical as well as spiritual experiences, is made possible by practicing mindfulness, such as the Buddhists have cultivated for more than two millennia. The mystics from all epochs and cultures bear witness to this in any case.

5.3 Concept of Transdisciplinarity

First, we resort to the notion of transdisciplinarity as outlined primarily by quantum physicist Basarab Nicolescu in the "Charter of Transdisciplinarity" (in this volume: p. XI). According to him, the term transdisciplinarity was first coined by the Swiss developmental psychologist Jean Piaget (1896–1980), the Austrian astrophysicist Erich Jantsch (1929–1980), and the French mathematician André Lichnerowitz (1915–1998). Nicolescu himself understood this to be "beyond disciplines."[120]

Nicolescu repeatedly refers to the findings of quantum physics, which was generally admired as probably the most exact of all *scientific disciplines,* because it revealed the necessity of a transdisciplinarity "beyond all disciplines."[121] For it was precisely this that abruptly made us aware of the limits of all disciplinary ways of approaching the world. The idea of a neutral observer of the world, which is objectively spread out in front of him as a subject in order to be measured in its truth content, has turned out to be theoretically and experimentally wrong with *quantum physics*. There is no neutral observer who could describe the world in itself with

---

[119]Weizsäcker (1980: 274–294).

[120]Nicolescu (1985); id. (2010), id. in this volume Chapter 2: "Methodology of Transdisciplinarity–Levels of Reality, Logic of the Included Middle and Complexity".

[121]For Nicolescu "multidisciplinarity and "interdisciplinarity" remain disciplinary in contrast to "transdisciplinarity", id. (2010); id. in this volume Chapter 2: "Methodology of Transdisciplinarity–Levels of Reality, Logic of the Included Middle and Complexity"; cf. id. (2006: 147).

reason and scientific methodology. Whenever an observer measures something on the atomic level, he produces, depending on the experimental question, two different phenomena contradicting each other according to classical logic: once as a particle concentrated in space, once as a force field spread far over space ("Wave-Particle Duality"). The modern vision of a subject completely separated from the object, which describes the world by reasonable means, has proved to be illusionary.[122] Subject and object are not separate entities, but in principle intertwined with each other.[123]

This also means that each scientific discipline with its methodology constitutes its own world. Some appear completely separate and actually have nothing to do with each other. One can take the example of a flowering tree. For the biologist it belongs to a plant species with special characteristics, for the chemist it is an oxygen factory with photosynthesis, for the painter it is an interesting figure, for the fruit grower it is a supplier of fruit, or for the poet it is the mediator of a special mood. The hiker sees it as a provider of shade, the sociologist as a gathering place, and the physicist as a collection of atomic processes. Each discipline sees the blossoming tree quite as a whole, but under a methodically conditioned interest. That is why Martin Heidegger once said: thinking [we add: of the disciplines] has "never yet left the blossoming tree where it stands."[124]

Transdisciplinarily, we should consider methods of perceiving the tree holistically beyond disciplinary narrowing of the view. For example, Canadian forest ecologists have explored the importance of fungi for the survival of trees. They use mycelial networks to share resources, build defenses, and communicate with each other.[125] In addition to such rather multidisciplinary approaches, Zen Buddhism developed methods for experiencing nature originally again beyond thinking with an "empty mind" in a spiritual experience of unity.[126]

Nicolescu speaks in this context of the "interplanetary and intergalactic vacuum" that we should reach with our minds in order to exceed (transcend) the limits of the visible world. Thus, he sees "Plurality which is not reduced to unity is confusion;

---

[122] Immanuel Kant already worked out in his *Critique of Pure Reason that* the world in itself is unknowable for human beings, because they constitute the perception of the world in the first place with the help of the methods of understanding and the forms of perception of space and time. However, Kant—standing completely in the tradition of the great Enlightenment philosopher René Descartes—saw the recognizing subject as res cogitans facing the objective, in principle measurable world of matter as res extensa. The world of reason is separated from the world of nature. With the findings of quantum physics the artificial wall between these two worlds falls down. The philosopher and quantum physicist and long-time friend of Werner Heisenberg, Carl Friedrich v. Weizsäcker, in a certain sense, has critically turned Kant's critical approach upside down in view of the findings of quantum physics itself, by making plausible a unity of the real beyond the separation of mind and nature in a transdisciplinary way (cf. Schüz, 1986: 105–177).

[123] Cf. Schüz (1986: 125, 137).

[124] Heidegger (1971: 18).

[125] Simard et al. (2011). Cf. Blom, (2022: 317 f).

[126] Hisamatsu, (1993); cf. Kapleau, (1989).

unity which does not depend on plurality is tyranny" as constitutive of transdisciplinarity.[127] Thus, Blaise Pascal (1623–1662) once stated: "Plurality without unity means anarchy, and unity without diversity means tyranny."[128]

5.4 *Holistic Model of Transdisciplinarity*

In this sense, we can also certainly preserve the knowledge from the disciplines of the natural sciences and the humanities, but at the same time we can also overcome it by raising it transdisciplinarily to a new level of knowledge. We can, in the sense of Hegel's dialectic, firstly grasp the visible and measurable world *intra-, inter-, and multidisciplinarily* (Thesis). Secondly, we take into account everything that must inevitably elude scientific methodology (antithesis). Third, only in the complementary completion of both perspectives (synthesis) does the actual *transdisciplinary movement of thought* become apparent, and this in two ways:

(a) By *horizontally* accessing *the immanent world* through scientific research as well as extra-scientific experiential knowledge—generated for instance through tradition, aesthetic or religious experience—in a complementary *way* (*Science-Practice Approach of Transdisciplinarity*, cf. contribution by Gerald Steiner in this volume, cf. Chap. 8);
(b) By *vertically* plumbing *the invisible depth dimensions of the transcendent world*[129] for example through graded philosophical reflection on its reasons, meaning or even unitary origin *(Transcendental-Metaphysical Approach of Transdisciplinarity).*

Both dimensions of transdisciplinarity complement each other, especially since the first one cannot do without the explanation of the second one. Because the visible universe manifests itself in the intersection of the horizontal and vertical thinking movement. If the second is ignored, the world threatens to disintegrate into random manifestations without a reason for meaning. These can then be unhesitatingly objectified without deeper ethical justification and degraded to the object of making.

The following graphic traces the path of transdisciplinary thinking as a possible approach to solving our world problems. It ignores that the base in the horizontal itself should actually be thought of in two directions; i.e. it represents a surface

[127] Nicolescu (2010).

[128] Pascal (1660, No. 871).

[129] The Latin concept of the transcendent (from the Latin "transcendere = to surmount, to exceed, to transcend" (Stowasser, 1979: 468)) goes back to Plato's myth of the cave. He demands the necessary ascent of cognition from the illusory world of visible being (world of shadows) via (intermediate) stages to the world of truly being ideas up to the absolute, unconditional, original ground of being, the transcendent (cf. Halfwassen, 1998, p. 1443). Without the mostly forced ascent, the realizations get lost in the world of appearances or shadow plays.

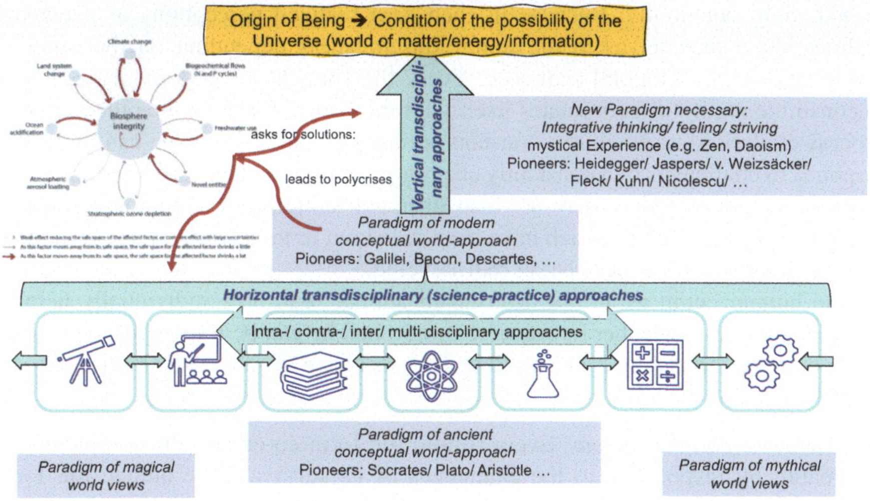

Fig. 3 Holistic model of transdisciplinarity Source: The author with use of Fig. 1

spanned by scientific disciplinary and non-scientific empirical knowledge. The complete picture should look more like a pyramid than the one shown (see Fig. 3).

5.5 *Axioms of Transdisciplinarity*

Basarab Nicolescu has now formulated three axioms for transdisciplinarity[130]. They are commented on in more detail in his article, but for the sake of completeness they will be cited again here and extended with a fourth axiom:

1. The *ontological axiom*: There are different levels of reality in nature and in our perception of nature, and accordingly different levels of perception.
2. The *logical axiom*: The change from one level to another level of the reality is reasoned by the logic with its *theorem* of the "*included third*."
3. The *complexity axiom*: The structure of the totality of all levels of reality or perceptions has a complex structure: Each level is what it is because all levels exist at the same time.

With the *ontological axiom* (from Greek on = being) the being of our universe is grasped as a *multiplicity of realities*. According to this, the being shows itself in manifold ways, depending on which form of consciousness perceives it. With the *logical axiom* the *realities are* grasped *with our finite consciousness*. The *complexity axiom* does assume that *all levels of reality exist simultaneously*, but, as I would like

[130]For more details, see Nicolescu, (2006, 2010, Chapter 2: "Methodology of Transdisciplinarity–Levels of Reality, Logic of the Included Middle and Complexity" in this volume).

to add, only potentially. Because in each case realized perceptions or perceived realities show themselves to the finite consciousness depending on the stage of development as a temporal sequence. Every life form has only a *limited capacity* to constitute reality. It coordinates itself with the reality forms of the others. In this process evolution arises, thus a constant argument, which life niche can outlast. It responds to challenges of a constantly changing environment and co-environment. It is therefore obvious that there are also different, culturally conditioned forms of scientific access to being, which have been faded out in the Eurocentrism of modern science, but are increasingly being rediscovered.[131]

The human being as a "not determined animal" can now individually help to shape this process, whether he wants to or not: i.e., he can be responsible, constructively or destructively. So for him still applies, as I have to add to Nicolescu's axioms above:

4. The *responsibility axiom*: Every (finite) life form co-creates the complexity of reality by responding to the challenges of its survival. The human life form experiences degrees of freedom, which compels it to responsibility.

5.6 *Transdisciplinary Responsibility*

If every formation of reality is co-constituted from an interplay of being and consciousness, then man is co-responsible for the consequences: He himself has co-caused them and has also been co-affected by them. Thus, the recognition of complex diversity by transdisciplinary logic demands at the same time the recognition of a transdisciplinary ethics, in which "universal responsibility"[132] functions as a guiding idea. This concept assumes that humans and their institutions are part of the universe and interact with it. The physicist and philosopher Heinz von Foerster (1911–2002) once expressed this in the following way:

> When I ask myself 'Am I part of the universe?' and answer 'Yes, I am!', I decide here that whenever I act, not only I change, but the universe changes as well. Something, it's not just me that changes, but the universe, too ... This position [...] ties my actions inseparably to my responsibility.[133]

In other words, there are limits to the freedom of the human mind to create new realities with its cultural achievements. Feyerabend's *anything goes* finds its limit in the destruction of the foundations of life. Not everything that is feasible is also meaningful. After all, the consequences fall back on the causers—for instance in the form of the described polycrises. In the meantime, the whole of humanity suffers

[131] See in particular the contribution by Marco Bischof in this volume, Chapter "History of Science in a Transdisciplinary Perspective: From Transdisciplinarity to Transcultural Science".

[132] The concept of "universal responsibility," which goes back to Georg Picht, was further developed in a holistic model by Mathias Schüz (2019: 64 ff).

[133] Förster (1991: 66, 68); Schüz (2019: 81 f).

from them—with an uncertain outcome for its continued existence. As perpetrator and victim at the same time, each participant bears the *responsibility* for the consequences of their actions. This can be done in a very *limited* way, by perceiving them only in relation to his own benefit, or in a *universal* way, by justifying them before all affected life forms. The latter requires as a measure an *ethic* that teaches the art of *getting along well with all forms of life*[134]—in the long run a precondition for well-being and happiness.[135]

The related *limitless responsibility* (Albert Schweitzer)[136] must be transdisciplinary. Because the awareness of the consequences and of all concerned depends on the specific search patterns of the disciplines. For example, virologists, epidemiologists, sociologists, social psychologists, political scientists, ethicists, etc., perceive the consequences of the COVID-19 pandemic for those affected differently and propose correspondingly divergent coping strategies. A multidisciplinary approach would adopt the equal view of all. In a transdisciplinary way, the concerns of the "normal" population or the insights of complementary physicians among others could be taken into account, and changes of perspective through mindfulness methods could also be applied.[137] It could even be accepted the unity of the actual,[138] according to which perpetrator and victim are only seemingly separate. In this unity, they carry cause and effect in each other: "*What you do to others, you do to yourself. So put yourself in the shoes of all concerned.*" (I call it: the golden rule of transdisciplinary ethics.)

No human being, due to his or her finite consciousness, is able to have a complete overview of the causal networks they trigger. However, with sensitization, they can initially more easily (a) *identify* and, if not (b) *avoid*, then (c) *mitigate* and (d) *compensate* the associated *risks*.[139] Despite being embedded in a larger whole, each agent retains a *certain freedom* to identify, evaluate, and select his available options for action, i.e., to assume responsibility in this context. For example, they can make a concrete decision to consume meat which is harmful to the climate or to switch to a vegan diet.

In this sense, transdisciplinarity serves the individual to become aware of his or her embeddedness as part of the whole and to perceive the responsibility associated with it. The individual thus succeeds only in making a small contribution to overcoming the polycrisis. In the sum of the contributions of all, however, the moving train could still avoid the impending abyss in time and ensure a good survival of mankind.[140]

---

[134] Cf. Schüz (2019: 65 f, 292).

[135] Cf. Schüz (2024).

[136] Schweitzer (1949: 239, 274).

[137] Cf. Schüz (2020: 83 f).

[138] Cf. Schüz (1986: 262–266).

[139] For the stages of risk management cf. Haller (1975: 43).

[140] In *Applied Business Ethics* (2019), I provide concrete instructions and justifications on how such responsibility can be exercised holistically in combination with *utilitarian*, *deontological*, and *aretological ethics* during private and professional life. In addition, there are already numerous positive examples of joint efforts to overcome today's polycrises (e.g., Fichtner 2023; Freeman et al., 2020).

References

Bacon, F. (1984). *Valerius Terminus–or the interpretation of nature [1603], English-German*. Königshausen + Neumann.
Baedke, J. (2018). *Above the gene, beyond biology: Toward a philosophy of epigenetics*. University of Pittsburgh.
Barret, L. F., & Bar, M. (2009). See it with feeling: Affective predictions during object perception. *Philosophical Transactions of the Royal Society, 364*, 1325–1334. Available: https://doi.org/10.1111/10.1098/rstb.2008.0312
Bateson, G. (1972). *Steps to an ecology of mind—collected essays in anthropology, psychiatry, evolution and epistemology*. Jason Aronson.
Berman, M. (1981). *The Reenchantment of the world*. Cornell University Press.
Blom, P. (2022). *Die Unterwerfung—Anfang und Ende der menschlichen Herrschaft über die Natur*. Hanser.
Böhme, H. (2014). Fetishism and culture. In *A different theory of modernity*. (translated by Anna Galt). De Gruyter.
Bossinger, W. (2020). *Pachamama–Über die Liebe zwischen Natur und Mensch: Eine musikalisch-ökologische Liebeserklärung an die Erde*. BoD.
Brink, J. W. (2008). *Imagining head-smashed-in–aboriginal Buffalo hunting on the Northern Plains*. AU Press.
Bührke, T. (2012). *Genial gescheitert: Schicksale großer Entdecker und Erfinder*. dtv.
Carr, B. (2015). Hyperspatial models of matter and mind. In E. F. Kelly, A. Crabtree, & P. Marshall (Eds.), *Beyond physicalism—Toward reconciliation of science and spirituality* (pp. 227–273). Rowman and Littlefield.
Carroll, A. (2018, Nov 5). Peer review: The worst way to judge research, except for all the others. *New York Times*. Available: https://www.nytimes.com/2018/11/05/upshot/peer-review-the-worst-way-to-judge-research-except-for-all-the-others.html
Daston, Lorraine/ Galison, Peter (2021): Objectivity. : Princeton University Press. (5th paperback printing).
la Mettrie, D., & Julien, O. (1912). *Man a machine*. The Open Court Publishing.
Descartes, R. (1978). *Von der Methode des richtigen Vernunftgebrauchs und der wissenschaftlichen Forschung*. Meiner, 1978: pp. 15 f.
Egner, H., & Uhlenwinkel, A. (2021). Zur Rechtsstaatlichkeit universitätsinterner Verfahren bei Entlassung oder öffentlicher Degradierung von Professor*innen. *Ordnung der Wissenschaft, 3*, 173–184. Available: https://ordnungderwissenschaft.de/wp-content/uploads/2022/01/03_03_Egner_Uhlenwinkel.pdf
Feyerabend, P. (1986). *Wider den Methodenzwang*. Suhrkamp.
Feyerabend, P. (1988). *Against method–outline of an anarchistic theory of knowledge* (revised ed.). Verso. Available: https://archive.org/details/againstmethod0000feye_l3f7/page/n5/mode/2up
Fichtner, U. (2023). *Geboren für die großen Chancen—Über die Welt, die unsere Kinder und uns in Zukunft erwartet*. DVA.
Fleck, L. (1981). In T. J. Trenn & R. K. Merton (Eds.), *Genesis and development of a scientific fact*. The University of Chicago Press. Available: https://worldpece.org/sites/default/files/artifacts/media/pdf/fleck_et_al._-_2008_-_genesis_and_development_of_a_scientific_fact.pdf
Foerster, H. (1991). Through the eyes of the other. In F. Steier (Ed.), *Research and reflexivity* (pp. 63–75). Sage Publications. Available: http://cepa.info/1729
Gehlen, A. (1978). *Der Mensch–Seine Natur und seine Stellung in der Welt* (12th ed.). Athenaion.
Global Footprint Network. (2023). *Earth overshoot day*. Available: https://www.footprintnetwork.org/our-work/earth-overshoot-day/
Gloy, K. (1995). *The history of scientific thought–the understanding of nature*. Beck.
Goldacre, B. *Bad Science*. Fourth Estate.
Gorke, M. (1999). *Artensterben—Von der ökologischen Theorie zum Eigenwert der Natur*. Klett Cotta.

Halfwassen, J. (1998). Transzendenz, Transzendieren. In *Historisches Wörterbuch der Philosophie* (Vol. 10, pp. 1442–1447). Schwabe.

Hampf, M. M. (2019). *Empire of liberty–the United States from reconstruction to the Spanish-American war*. Walter de Gruyter.

Hanna, L., et al. (2019). Resilience to disturbance is a cross-scale phenomenon offering a solution to the disturbance paradox. *Ecosphere, 10*(4), 02682. https://doi.org/10.1002/ecs2.2682. Available: https://esajournals.onlinelibrary.wiley.com/doi/epdf/10.1002/ecs2.2682

Haller, M. (1975). *Sicherheit durch Versicherung?* Herbert Lang.

Hegel, G. W. F. (1952). *Hegel's philosophy of right [1820] (translated by T. M. Knox)*. Oxford University Press. Available: https://www.marxists.org/reference/archive/hegel/works/pr/philosophy-of-right.pdf

Hegel, G. W. F. (2013). *Hegel's logic, part 1 of encyclopedia of the philosophical sciences [1830] (translated by William Wallace)*. Marxists Internet Archive. Available: https://www.marxists.org/admin/books/hegels-logic/Hegels-Logic.pdf

Heidegger, M. (1971). *Was heißt Denken?* Tübingen: Niemeyer (3. Aufl.).

Heidegger, M. (1988). Das Spiegel-Interview. In G. Neske & E. Kettering (Eds.), *Antwort–Martin Heidegger im Gespräch* (pp. 79–114). Pfullingen.

Herder, J. G. (1993). *Abhandlung über den Ursprung der Sprache*. Reclam.

Herman, A. (2013). *The cave and the light—Plato versus Aristotle, and the struggle for the soul of Western civilization*. Random House.

Hisamatsu, H. S.'i. (1993). *The fullness of nothing*. Il Melangolo.

Horgan, J. (2015). *The end of science–facing the limits of knowledge in the twilight of the scientific age* (2nd ed.). Perseus Books.

Hossenfelder, S. (2018). *Lost in math: How beauty leads physics astray*. Basic Books.

Jaspers, K. (1973a). *Philosophie I—Philosophische Weltorientierung* (4th ed.). Springer.

Jaspers, K. (1973b). *Philosophie II–Existenzerhellung* (4th ed.). Springer.

Jaspers, K. (1973c). *Philosophie III–Metaphysik* (4th ed.). Springer.

Jaspers, K. (1974). *Der philosophische Glaube* (6th ed.). Piper.

Jaspers, K. (1977). *Chiffren der Transzendenz* (3rd ed.). Piper.

Jaspers, K. (1983). *Von der Wahrheit*. Wissenschaftliche Buchgesellschaft.

Joldersma, C. W. (2011). Ernst von Glasersfeld's radical constructivism and truth as disclosure. *Educational Theory, 61*(3), 275–293. https://doi.org/10.1111/j.1741-5446.2011.00404.x

Jüngling, H.-W. (1994). 'Macht Euch die Erde untertan' (Gen 1:28). Der geschaffene Mensch und die Schöpfung. In P. Schmitz (Ed.), *Macht Euch die Erde untertan* (pp. 9–38). Echter.

Kaiser, P., & Tominski, K. (2021, March 21). Entfesselt: Die Corona-Pandemie als Preis für die Ausbeutung der Natur. *mdr-Wissen*. Available: https://www.mdr.de/wissen/corona-pandemie-folge-ausbeutung-natur-100.html

Kant, I. (1978). *Beantwortung der Frage: Was ist Aufklarung?* [1784]. In *Theorie-Werkausgabe* (Vol. XI, 2nd ed., pp. 51–61). Insel.

Kapleau, P. (1989). *The three pillars of Zen: Teaching, practice, and enlightenment*. Anchor Books.

Kelly, E. F., Kelly, E. W., & Crabtree, A. (Eds.). (2007). *Irreducible mind: Toward a psychology for the 21st century*. Littlefield Publishers.

Kelly, E. F., Adam, C., & Marshal, P. (Eds.). (2015). *Beyond physicalism: Toward reconciliation of science and spirituality*. Littlefield Publishers.

Kelly, E. F., & Marshal, P. (2021). *Consciousness unbound—Liberating mind from the tyranny of materialism*. Littlefield Publishers.

Kimbrell, A. (1998). *The human body shop: The cloning, engineering, and Marketing of Life* (2nd ed.). Regnery Publishing.

Krimsky, S. (2013). Do Financial conflicts of interest bias research?: An inquiry into the 'funding effect' hypothesis. *Science, Technology, & Human Values, 38*(4), 566–587. https://doi.org/10.1177/0162243912456271

Kuhn, T. S. (2012). *The structure of scientific revolutions–with an introductory essay by Ian hacking [1962]* (4th ed.). The University of Chicago Press.

Lamont, M. (2009). *How professors think–inside the curious world of academic judgement*. Harvard University Press.
Longino, H. E. (1990). *Science as social knowledge–values and objectivity in scientific Theory*. Princeton University Press.
Lorenz, K. (1984). In I. Eibl-Eibesfeldt (Ed.), *Das Wirkungsgefüge der Natur und das Schicksal des Menschen–Gesammelte Arbeiten* (4th ed.). Piper.
Mails, T. E. (2002). *The Mystic warriors of the plains: The culture, arts, crafts and religion of the plains Indians*. Da Capo Press.
Malinowski, B. (1960). *A scientific theory of culture and other essays*. Oxford University Press. Available: https://monoskop.org/images/f/f5/Malinowski_Bronislaw_A_Scientific_Theory_of_Culture_and_Other_Essays_1961.pdf
Manoogian, J. III (2023). *Cognitive Bias Codex*. Available: https://upload.wikimedia.org/wikipedia/commons/a/a4/The_Cognitive_Bias_Codex_-_180%2B_biases%2C_designed_by_John_Manoogian_III_%28jm3%29.png
Mansfeld, J./ Primavesi, O. (2012): *Die Vorsokratiker*, griechisch-deutsch. : Reclam.
Marx, K. (2022). *Theses on Feuerbach* (3rd ed.). Carlos Bendana-Pedroza.
Mauss, M. (2002). *The gift–the form and reason for exchange in archaic societies*. Routledge.
McDonald, D. (2017). *The Golden passport–Harvard Business School, the limits of capitalism, and the moral failure of the MBA elite*. Harper Collins.
Nicholson, M. T. (2012, October 23). Thomas Nagel is not crazy. In *Prospect magazine*. Available: https://www.prospectmagazine.co.uk/culture/50518/thomas-nagel-is-not-crazy
Nicolescu, B. (1985). *Nous, la particule et le monde*. Le Mail. (2nd ed. Monaco: Le Rocher 2002).
Nicolescu, B. (2006). Transdisciplinarity—Past, present and future. In B. Haverkort & C. Reijntjes (Eds.), *Moving worldviews—Reshaping sciences and practices for endogenous sustainable development* (pp. 142–166). Compas Editions.
Nicolescu, B. (2010). Methodology of transdisciplinarity–levels of reality, logic of the included middle and complexity. *Transdisciplinary Journal of Engeneering & Science, 1*, 17–32.
Nietzsche, Friedrich (1997): Jenseits von Gut und Böse. Gesammelte Werke (K. Schlechta), Vol. II, : pp. 563–759.
Pascal, Blaise (1660). *Pensées* (translated by W. F. Trotter). Available: https://ccel.org/ccel/p/pascal/pensees/cache/pensees.pdf
Pinker, S. (2011). *The better angels of our nature. Why violence has declined*. Penguin.
Pinker, S. (2021). *Rationality—What it is, why it seems scarce, why it matters*. Penguin.
Pinker, S., & Sadhguru. (2023). *Is consciousness a miracle? A discussion betweenl Harvard's cognitive scientist prof.* Steven Pinker & Sadhguru at Harvard University Theater. Available: https://www.youtube.com/watch?v=pCGDQPZYmQM
Pistor, K. (2019). *The code of capital. How the law creates wealth and inequality*. Princeton University Press.
Planck, M. (1979). Wissenschaftliche Selbstbiographie bzw. Persönliche Erinnerungen aus alten Zeiten. In *Vorträge und Erinnerungen* (7th ed.). Wissenschaftliche Buchgesellschaft.
Plato. (1978). *Politeia*. Sämtliche Werke, vol. 3 (translated by Friedrich Schleiermacher). *Hamburg, 1978*, 67–310.
Popper, K. (1979). *Objective knowledge–an evolutionary approach* (rev ed.). Clarendon Press.
Radin, D. (2018). *Real magic—Ancient wisdom, modern science and a guide to the secret power of the universe*. Harmony.
Reutlinger, A. (2020). What is epistemically wrong with research affected by sponsorship bias? The evidential account. *European Journal for Philosophy and Science, 10*, 15. https://doi.org/10.1007/s13194-020-00280-2
Rockström, J. (2021). Dieses Jahrzehnt bietet die letzte Gelegenheit, das Ruder herumzureißen. *Der Spiegel, 20*, 102–106.
Roth, G. (1997). *Das Gehirn und seine Wirklichkeit—Kognitive Neurobiologie und ihre philosophischen Konsequenzen*. Suhrkamp.

Scheler, M. (2009). *The human place in the cosmos (translated by Manfred Frings)*. Northwestern University Press.

Schüz, G. (2001). *Lebensganzheit und Wesensoffenheit des Menschen–Otto Friedrich Bollnows hermeneutische Anthropologie*. Königshausen & Neumann.

Schüz, M. (2024). Teaching and implementing ethics for Well-being in business–a field report. In M. Mühlböck (Ed.), *Sustainable business transformation and Well-being*. Springer. in print.

Schüz, M. (2021). *Angewandte Unternehmensethik—Grundlagen für Studium und Praxis* (2nd ed.). Pearson.

Schüz, M. (2020). Ethics approaches in dealing with the Corona pandemic. In M. Woesler & H.-M. Sass (Eds.), *Medicine and ethics in times of Corona* (pp. 75–86). Lit Verlag.

Schüz, M. (2019). *Applied business ethics—Foundations of study and practice* (1st ed.). World Scientific Publishing.

Schüz, M. (1986). *Die Einheit des Wirklichen–Carl Friedrich v. Weizsäckers Denkweg*. Neske.

Schweitzer, A. (1949). *Civilization and ethics* (3rd ed.). Adam & Charles Black. Available: https://archive.org/stream/in.ernet.dli.2015.191195/2015.191195.The-Philosophy-Of-Civilization-And-Ethics#page/n0/mode/2up

Schweitzer, A. (2002). In M. Meyer & K. Bergel (Eds.), *Reverence for life—The ethics of Albert Schweitzer for the twenty-first century*. Syracuse University Press.

Seidl, G. S. (1985). *The Semmelweis affair*. Herold Verlag.

Sheldrake, R. (2020). *The science delusion–freeing the Spirit of enquiry*. Hodder & Stoughton. (new ed.).

Shepherd, K. R. (2015). *Descartes and Vivisection*. Available: https://kevinrdshepherdcommentaries.info/2015/02/descartes-and-vivisection_15.html

Simard, S. W., et al. (2011). Understanding relationships among multiple ectomycorrhizal symbionts of Douglas-fir: Implications for community assembly and ecosystem functioning. *Science, 333*(6045), 880–882.

Simmel, G. (1989). *Philosophie des Geldes [1900]*. Suhrkamp Taschenbuch Wissenschaft.

Skirbekk, G. (Ed.). (2019). *Wahrheitstheorien—Eine Auswahl aus den Diskussionen über Wahrheit im 20. Jahrhundert*. Frankfurt a. M., Suhrkamp.

Smolin, L. (2008). *The trouble with physics–the rise of string theory, the fall of a science and what comes next*. Penguin.

Socol, Y., Shaki, Y. Y., & Yanovskiy, M. (2019). Interests, bias, and consensus in science and regulation. *Dose Response, 17*(2), 1559325819853669. https://doi.org/10.1177/1559325819853669

Stapp, H. P. (2015). A quantum mechanical theory of the mind/brain connection. In E. F. Kelly, A. Crabtree, & P. Marshall (Eds.), *Beyond physicalism—Toward reconciliation of science and spirituality* (pp. 157–192). Rowman and Littlefield.

Steffen et al. (2015). Supplementary Materials—Download in: *Planetary boundaries: Guiding human development on a changing planet*. Available: www.sciencemag.org/cgi/content/full/science.1259855/DC1

Stowasser, J. M., Petschenig, M., & Skutsch, F. (1979). *Der Kleine Stowasser–Latin-German dictionary*. Freytag Verlag.

Tollefson, J. (2019). Humans are driving one million species to extinction. *Nature, 569*, 171. Available: https://www.nature.com/articles/d41586-019-01448-4

The World Bank. (2022). *Four decades of poverty reduction in China–drivers, insights for the world, and the way Ahead*. Development Research Center. Available: https://openknowledge.worldbank.org/server/api/core/bitstreams/e9a5bc3c-718d-57d8-9558-ce325407f737/content

Tooze, A. (2022). Kawumm! In *Die Zeit* (Vol. 29, p. 2).

Weber, M. (1988). Science as a profession. In J. C. B. Mohr (Ed.), *Gesammelte Aufsätze zur Wissenschaftslehre* (7th ed., pp. 582–613).

Weizsäcker, C. F. (1976). *Zum Weltbild der Physik*. Hirzel.

Weizsäcker, C. F. (1980). *The Unity of nature*. (translated by Francis J. Zucker). Farrar–Straus–Giroux.

Whitehead, A. N. (1978). In D. R. Griffin & D. W. Sherburne (Eds.), *Process and reality—An essay in cosmology*. Gifford lectures 1927–28. The Free Press.
Wilson, E. O. (1992). *The diversity of life*. Harvard University Press.
Wisser, R. (1997). *Kein Mensch ist einerlei – Spektrum und Aspekte kritisch-krisischer Anthropologie*. Königshausen & Neumann.

Schüz, Mathias, Prof. Dr. phil., studied physics, philosophy and pedagogics at the Johannes Gutenberg University Mainz (Germany). His doctoral thesis on the philosophical consequences of quantum physics already dealt with aspects of transdisciplinarity in the form of a new approach to paradigmatology. After his studies, he was a trainee and key account manager at IBM. Subsequently, together with the insurance entrepreneur Dr. Rolf Gerling, he co-initiated the *Gerling Academy for Risk Research* in Zurich, where he taught and researched for 16 years as a member of the management board, as well as serving on the management board of Gerling Academy Publishing. From 2006-2022 he was senior-lecturer and professor for "Responsible Leadership" at the "School of Management and Law" of the *Zurich University of Applied Sciences* (ZHAW) in Winterthur, Switzerland. He teaches and conducts research on business ethics, strategic management and sustainable corporate responsibility. As Emeritus and Visiting Professor, he has been teaching on Responsible Leadership and business ethics at numerous universities in Switzerland and abroad since 2022. He is the author of numerous books (e.g. a transdisciplinary textbook about Applied Business Ethics published at Pearson, Munich, and World Scientific, Singapore) and articles on his research areas. Together with Prof. Dr. René Schmidpeter, he edits the book series Transdisciplinary Management of Social and Ecological Crises at Springer Nature. He is co-founder of the *Association for the Promotion of Transdisciplinary Knowledge* in Switzerland. For decades he was a trainer of the Korean martial art Taekwondo (5th Dan) and is still a practitioner of Yoga and Zen.

Methodology of Transdisciplinarity–Levels of Reality, Logic of the Included Middle and Complexity

Basarab Nicolescu

Abstract The concept of levels of Reality, formulated in 1982, is the key concept of transdisciplinarity1. The introduction of the levels of Reality induces a multidimensional and multi-referential structure of Reality, signifying the coexistence between complex plurality and open unity. Every level is characterized by its incompleteness; the laws governing this level are just a part of the totality of laws governing all levels. And even the totality of laws does not exhaust the entire Reality; we have also to consider the interaction between Subject and Object. The zone between two different levels and beyond all levels is a zone of non-resistance to our experiences, representations, descriptions, images, and mathematical formulations. The Gödelian structure of levels of Reality implies the impossibility of a selfenclosed complete theory. Knowledge is forever open. The unity of levels of Reality of the Object and its complementary zone of non-resistance defines the transdisciplinary Object. The unity of levels of Reality of the Subject and this complementary zone of non-resistance defines the transdisciplinary Subject. The zone of non-resistance plays the role of a third between the Subject and the Object, an interaction term which allows the unification of the transdisciplinary Subject and the transdisciplinary Object. This interaction term is called the Hidden Third. The ternary partition (Subject, Object, Hidden Third) is, of course, radically different from the binary partition (Subject vs. Object) of classical realism.

Keywords Reality · Levels · Transdisciplinarity · Classical realism · Quantum realism · Logic of the included middle · Complexity · Disciplinary boundaries · Spirituality · Homo economicus · Homo religious

---

This article was first published open-source by Nicolescu at the following location in December 2010: *Transdisciplinary Journal of Engeneering & Science*, Vol. 1, pp. 17-32, ISSN: 1949-0569, online: doi: 10.22545/2010/0009. Copyright © 2010 by the author.

B. Nicolescu (✉)
International Centre for Transdisciplinary Studies and Research (Centre international de recherches et études transdisciplinaires (CIRET)), Paris, France

© The Author(s), under exclusive license to Springer Nature Switzerland AG 2025

M. Schüz (ed.), *Transdisciplinary Thinking and Acting*, Transdisciplinary Management of Ecological and Social Crises,
https://doi.org/10.1007/978-3-031-70680-6_2

1 The War of Defination

1.1 How Transdisciplinarity was Born

Transdisciplinarity is a relatively young approach; Swiss philosopher and psychologist Jean Piaget (1896–1980) developed the concept seven centuries after disciplinarity had evolved. The word itself first appeared in France, in 1970, in the talks of Jean Piaget, Erich Jantsch, and André Lichnerowicz at the international workshop *Interdisciplinarity—Teaching and Research Problems in Universities*, organized by the *Organization for Economic Cooperation and Development* (OECD) in collaboration with the French Ministry of National Education and University of Nice.[1]

In his contribution, Piaget gives the following description of *transdisciplinarity*: "Finally, we hope to see succeeding to the stage of interdisciplinary relations a superior stage, which should be 'transdisciplinary,' i.e. which will not be limited to recognize the interactions and/or reciprocities between the specialized researches, but which will locate these links inside a total system without stable boundaries between the disciplines."[2] While this description is vague, it has the merit of pointing to a new space of knowledge "without stable boundaries between the disciplines." However, the idea of a "total system" opens the trap of transforming transdisciplinarity into a super- or hyperdiscipline, a kind of "science of sciences." In other words, the description of Piaget leads to a closed system, in contradiction with his own requirement of the instability of boundaries between disciplines. The key point here is the fact that Piaget retained only the meanings "across" and "between" from the Latin prefix "trans", eliminating the meaning "beyond." Understood in such a way, transdisciplinarity is just a new, "superior stage of interdisciplinarity. I think Piaget was fully conscious of this alteration of transdisciplinarity, but the intellectual climate was not yet prepared for receiving the shock of contemplating the possibility of a space of knowledge beyond the disciplines. The proof is that in his introduction to the Proceedings of the workshop, Pierre Duguet honestly recognized that some experts wanted to see the word "transdisciplinarity" in the title of the workshop, but authorities of the OECD refused to do so because they were afraid to confuse some representatives of the member countries.[3] In his contributions, Erich Jantsch, an Austrian thinker living in California, falls in the trap of defining transdisciplinarity as a hyperdiscipline. He writes that transdisciplinarity is "the coordination of all disciplines and interdisciplines of the teaching system and the innovation on the basis of a general axiomatic approach."[4] He clearly situates transdisciplinarity in the disciplinary framework. However, the historical merit of

[1]L. Apostel et al. (1972).

[2]Piaget (1972, p. 144).

[3]Duguet (1972, p. 13).

[4]Jantsch (1972a, p. 108). The same ideas are expressed in Jantsch (1972b).

Jantsch was to underscore the necessity of inventing an axiomatic approach for transdisciplinarity and also of introducing values in this field of knowledge.

Finally, the approach of Andr'e Lichnerowicz, a known French mathematician, is radically mathematical. He sees transdisciplinarity as a transversal play to describe "the homogeneity of the theoretical activity in different sciences and techniques, independently of the field where this activity is effectuated."[5] And, of course, this theoretical activity can be formulated, he thinks, only in mathematical language. Lichnerowicz writes: "The Being is put between parentheses, and it is precisely this non-ontological character which confers to mathematics its power, its fidelity, and its polyvalence."[6] The interest of Lichnerowicz for transdisciplinarity was accidental, but his remark about the non-ontological character of mathematics has to be remembered.

I described in some detail the three different positions of Piaget, Jantsch, and Lichnerowicz concerning transdisciplinarity because they can be found again, a quarter of a century later, in what I call "the war of definitions." The word "war" does not belong in the transdisciplinary vocabulary. However, I use it purposely because it appeared in the issue "Guerre et paix entre les sciences: disciplinarité et transdisciplinarité (War and Peace Between Sciences: Disciplinarity and Transdisciplinarity)"[7] of a French magazine. In this issue, one of the authors asked for the interdiction of the word "transdisciplinarity." His desire was obviously not satisfied.

I would like to add to this discussion about the incipient phase of transdisciplinarity the name of Edgar Morin. A short time after the Nice meeting, Morin begins to use the word "transdisciplinarity," and he even leads a transdisciplinary laboratory in human sciences within the framework of a prestigious French research institution. It is true that Morin did not give a definition of transdisciplinarity. For him, transdisciplinarity was, in that period, a kind of messenger of the freedom of thinking, a go-between discipline.

1.2 Beyond Disciplines

I proposed the inclusion of the meaning "beyond disciplines" in 1985 and have since developed this idea over the years in articles, books, and various official international documents. Many other researchers over the world contributed to this development of transdisciplinarity. A key date in this development is 1994, when the *Charter of Transdisciplinarity*[8] was adopted by the participants at the *First World*

[5] Lichnerowicz (1972, pp. 130–131).

[6] Ibid., pp. 127

[7] Alain Caill, in Guerre (1997).

[8] The Charter (1994); reprinted in this volume.

Congress of Transdisciplinarity (Convento da Arrbida, Portugal). This idea did not come from heaven or just from the pleasure of respecting the etymology of the word "trans" but from my long practice of quantum physics. For an outsider, it might seem paradoxical that it is from the very core of exact sciences that we arrive at the idea of limits of disciplinary knowledge. But from within, it provides evidence of the fact that, after a very long period, disciplinary knowledge has reached its own limitations with far-reaching consequences not only for science but also for culture and social life.

The crucial point here is the status of the Subject. Modern science was born through a violent break with the ancient vision of the world. It was founded on the idea—surprising and revolutionary for that era—of a total separation between the knowing subject and Reality, which was assumed to be completely independent from the subject who observed it. This break allowed science to develop independently of theology, philosophy, and culture. It was a positive act of freedom. But today, the extreme consequences of this break, incarnated by the ideology of scientism, pose the potential danger of self-destruction of our species.

On the spiritual level, the consequences of scientism have been considerable: the only knowledge worthy of its name must therefore be scientific, objective; the only reality worthy of this name must be, of course, objective reality, ruled by objective laws. All knowledge other than scientific knowledge is thus cast into the inferno of subjectivity, tolerated at most as a meaningless embellishment or rejected with contempt as a fantasy, an illusion, a regression, or a product of the imagination. Even the word "spirituality" has become suspect and its use has been practically abandoned.

Objectivity, set up as the supreme criterion of Truth, has one inevitable consequence: the transformation of the Subject into an Object. The death of the Subject is the price we pay for objective knowledge. The human being became an object—an object of the exploitation of man by man; an object of the experiments of ideologies that are proclaimed scientific; an object of scientific studies to be dissected, formalized, and manipulated. The Man—God has become a Man—Object, of which the only result can be self-destruction. The two world massacres of this century, not to mention the multiple local wars and terrorism, are only the prelude to self-destruction on a global scale.

In fact, with very few exceptions—Husserl, Heidegger, Gadamer, or Cassirer, for example—modern and post-modern thinkers gradually transformed the Subject in a grammatical subject. The Subject is today just a word in a phrase.[9]

The quantum revolution radically changed this situation. The new scientific and philosophical notions it introduced—the principle of superposition of quantum "yes" and "no" states, discontinuity, nonseparability, global causality, quantum indeterminism—necessarily led the founders of quantum mechanics to rethink the problem of the complete Object/Subject separation. For example, Werner Heisenberg, Nobel Prize winner of Physics, thought that one must suppress any rigid

[9]Descombes (2004).

distinction between the Subject and Object, between objective reality and subjective reality. "The concept of 'objective' and 'subjective,' " writes Heisenberg, "designate [s] two different aspects of one reality; however we would make a very crude simplification if we want to divide the world in[to] one objective reality and one subjective reality. Many rigidities of the philosophy of the last centuries are born by this black and white view of the world."[10] Heisenberg also asserts that we have to renounce the privileged reference to the exteriority of the material world. "The too strong insistence on the difference between scientific knowledge and artistic knowledge comes from the wrong idea that concepts describe perfectly the 'real things.' [...] All true philosophy is situated on the threshold between science and poetry."[11]

My line of thinking is in perfect agreement with that of Heisenberg. For me, "beyond disciplines" precisely signifies the Subject, and, more precisely, the Subject-Object interaction. The transcendence inherent in transdisciplinarity is the transcendence of the Subject. The Subject cannot be captured in a disciplinary camp.

The meaning "beyond disciplines" leads us to an immense space of new knowledge. The main outcome was the formulation of the methodology of transdisciplinarity, which I will analyze in the next section. It allows us also to clearly distinguish between multidisciplinarity, interdisciplinarity, and transdisciplinarity.

Multidisciplinarity concerns itself with studying a research topic in not just one discipline but in several simultanously. From this perspective, any topic will ultimately be enriched by incorporating the perspectives of several disciplines. Multidisciplinarity brings a plus to the discipline in question, but this plus is always in the exclusive service of the home discipline. In other words, the multidisciplinary approach overflows disciplinary boundaries while its goal remains limited to the framework of disciplinary research.

Interdisciplinarity has a different goal than multidisciplinarity. It concerns the transfer of methods from one discipline to another. Like multidisciplinarity, interdisciplinarity overflows the disciplines, but its goal still remains within the framework of disciplinary research. Interdisciplinarity even has the capacity of generating new disciplines, such as quantum cosmology and chaos theory.

Transdisciplinarity concerns that which is at once between the disciplines, across the different disciplines, and beyond all disciplines. Its goal is the understanding of the present world, of which one of the imperatives is the unity of knowledge.[12]

As one can see, there is no opposition between disciplinarity (including multidisciplinarity and interdisciplinarity) and transdisciplinarity, but there is instead a fertile complementarity. In fact, there is no transdisciplinarity without disciplinarity. In spite of this fact, the above considerations provoked, around 1990, a more or less violent war of definitions. This war is not yet finished.

[10] Heisenberg (1998: p. 269).

[11] Ibid., pp. 263–364.

[12] Nicolescu (1996a).

There is a specific different approach of transdisciplinarity that is characterized by the refusal of formulating any methodology and by its exclusive concentration on joint problem-solving of problems pertaining to the science-technology-society triad. This approach is represented by figures like Michael Gibbons[13] and Helga Nowotny.[14] The point of view of this transdisciplinary current was largely expressed at the Zürich Congress, held in the year 2000.[15]

This version of transdisciplinarity does not exclude the meaning "beyond disciplines" but reduces it to the interaction of disciplines with social constraints. The social field necessarily introduces a dimension "beyond disciplines," but the individual human being is conceived of as part of a social system only. The spiritual dimension is therefore absent in this approach.

It is difficult for us to understand why "joint problem solving" must be the unique aim of transdisciplinarity. It is certainly one of the important aims but not the only aim. The use of such a narrow characterization seems to us dangerous, as in religion, allowing unnecessary wars and unproductive dogmatism. Is transdisciplinarity concerning only society as a uniform whole, or, the human being who is (or has to be) in the center of any civilized society? Are we allowed to identify knowledge with production of knowledge? Why does the potential of transdisciplinarity have to be reduced to produce "better science"? Why does transdisciplinarity have to be reduced to "hard science"? In other words, the Subject-Object interaction seems to us to be at the very core of transdisciplinarity and not the Object alone.

I think the unconscious barrier to a true dialogue comes from the inability of certain transdisciplinary researchers to think about discontinuity. I will give an image in order to express what I have in mind. For them, the boundaries between disciplines are like boundaries between countries, continents, and oceans on the surface of the Earth. These boundaries are fluctuating in time, but a fact remains unchanged: the continuity between territories. We have a different approach of the boundaries between disciplines. For us, they are like the separation between galaxies, solar systems, stars, and planets. It is the movement itself that generates the fluctuation of boundaries. This does not mean that a galaxy intersects another galaxy. When we cross the boundaries, we meet the interplanetary and intergalactic vacuum. This vacuum is far from being empty; it is full of invisible matter and energy. It introduces a clear discontinuity between territories of galaxies, solar systems, stars, and planets. Without the interplanetary and intergalactic vacuum, there is no Universe.

It is my deep conviction that our formulation of transdisciplinarity is both unified (in the sense of unification of different transdisciplinary approaches) and diverse: unity in diversity and diversity through unity is inherent to transdisciplinarity. Much confusion arises by failing to recognize that there is a *theoretical transdisciplinarity*, a *phenomenological transdisciplinarity*, and an *experimental transdisciplinarity*.

---

[13] Gibbons et al. (1994).

[14] Nowotny (1994): "The Potential of Transdisciplinarity".

[15] Thomson Klein et al. (2001).

The word "theory" implies a general definition of transdisciplinarity and a well—defined methodology (which has to be distinguished from "methods"; a single methodology corresponds to a great number of different methods). The word "phenomenology" implies building models that connect the theoretical principles with the already observed experimental data in order to predict further results. The word "experimental" implies performing experiments following a well-defined procedure, allowing any researcher to get the same results when performing the same experiments.

I classify the work done by Michael Gibbons and Helga Nowotny as phenomenological transdisciplinarity. While I define my own work,[16] as well as that of Jean Piaget and Edgar Morin,[17] as theoretical transdisciplinarity. In its turn, experimental transdisciplinarity concerns a large number of experimental data already collected not only in the framework of knowledge production but also in fields such as education, psychoanalysis, the treatment of pain in terminal diseases, drug addiction, art, literature, history of religions, etc. The reduction of transdisciplinarity to only one of its aspects is very dangerous because it will transform transdisciplinarity into a temporary fashion, which I predict will disappear soon just as many other fashions in the field of culture and knowledge have indeed vanished. The huge potential of transdisciplinarity will never be accomplished if we do not accept the simultaneous and rigorous consideration of the three aspects of transdisciplinarity. This simultaneous consideration of theoretical, phenomenological, and experimental transdisciplinarity will allow both a unified and nondogmatic treatment of the transdisciplinary theory and practice, coexisting with a plurality of transdisciplinary models. ATLAS [The Academy of Transdisciplinary Learning & Advanced Studies[18]] seems to me an ideal place to practice all three aspects of transdisciplinarity in a fruitful manner.

2 Formulation of the Methodology of Transdisciplinarity

2.1 The Axiomatic Character of the Methodology of Transdisciplinarity

The most important achievement of transdisciplinarity in present times is, of course, the formulation of the methodology of transdisciplinarity, accepted and applied by an important number of researchers in many countries around the world. In the absence of a methodology, transdisciplinarity would be just talking, an empty discourse and therefore a short-term living fashion. The axiomatic character of the

[16]Nicolescu (1986, 1991, 1996a, 1996b, 1998a, 1998b, 2000, 2002, 2004, 2005, 2007, 2008a, 2008b, 2008c, 2009).

[17]Morin (1999).

[18]https://theatlas.org

methodology of transdisciplinarity is an important aspect. This means that we have to limit the number of axioms (or principles or pillars) to a minimum number. Any axiom that can be derived from the already postulated ones would have to be rejected. This fact is not new. It already happened when disciplinary knowledge acquired its scientific character due the three axioms formulated by Galileo Galilei in his *Dialogue on the Great World Systems*[19]:

1. There are *universal laws*, of a mathematical character.
2. These laws can be discovered by *scientific experiment*.
3. Such experiments can be perfectly *replicated*.

It should be obvious that if we try to build a mathematical bridge between science and ontology, we will necessarily fail. Galileo himself makes the distinction between human mathematics and divine mathematics.[20] Human mathematics constitutes, he says (via Salvati), the common language of human beings and God, while divine mathematics is connected with the direct perception of the totality of all existing laws and phenomena. Transdisciplinarity tries to seriously take this distinction into account. A bridge can be built between science and ontology only by taking into account the totality of human knowledge. This requires a symbolic language, different from mathematical language and enriched by specific new notions. Mathematics is able to describe repetition of facts due to scientific laws, but transdisciplinarity is about the singularity of the human being and human life. The key point here is, once again, the irreducible presence of the Subject, which explains why transdisciplinarity cannot be described by a mathematical formalism. The dream of the mathematical formalization of transdisciplinarity is just a phantasm, the phantasm induced by centuries of disciplinary knowledge.

After many years of research, we have arrived.
[21] at the following three axioms of the methodology of transdisciplinarity:

1. **The ontological axiom**: There are, in Nature and society and in our knowledge of Nature and society, different levels of Reality of the Object and, correspondingly, different levels of Reality of the Subject.
2. **The logical axiom**: The passage from one level of Reality to another is ensured by the logic of the included middle.
3. **The complexity axiom**: The structure of the totality of levels of Reality or perception is a complex structure: every level is what it is because all the levels exist at the same time.

The first two get their experimental evidence from quantum physics, but they go well beyond exact sciences. The last one has its source not only in quantum physics but also in a variety of other exact and human sciences. All three are in agreement with traditional thinking present on the earth since the beginning of historical times.

[19]Galileo (1956, 1992).

[20]Galileo (1992, p. 192).

[21]Nicolescu (1996b).

Axioms cannot be demonstrated; they are not theorems. They have their roots in experimental data and theoretical approaches, and their validity is judged by the results of their applications. If the results are in contradiction with experimental facts, they have to be modified or replaced.

Let me note that, in spite of an almost infinite diversity of methods, theories, and models that run throughout the history of different scientific disciplines, the three methodological postulates of modern science have remained unchanged from Galileo. Let us hope that the same will prove to be true for transdisciplinarity and that a large number of transdisciplinary methods, theories, and models will appear in the future.

Let me also note that only one science has entirely and integrally satisfied the three Galilean postulates: physics. The other scientific disciplines only partially satisfy the three methodological postulates of modern science. However, the absence of rigorous mathematical formulation in psychology, psychoanalysis, history of religions, law theory, and a multitude of other disciplines did not lead to the elimination of these disciplines from the field of science. At least for the moment, not even an exact science like molecular biology can claim a mathematical formulation as rigorous as that of physics. In other words, there are degrees of disciplinarity which can more or less completely take into account the three methodological postulates of modern science. Likewise, the process of more or less taking completely into account the three methodological pillars of transdisciplinary research will generate different degrees of transdisciplinarity. Large avenues are open for a rich and diverse transdisciplinary research.

The above three axioms give a precise and *rigorous definition of transdisciplinarity*. This definition is in agreement with the one sketched by Jean Piaget.

Let me now describe the essentials of these three transdisciplinary axioms.

2.2 The Ontological Axiom: Levels of Reality and Levels of Perception

The key concept of the transdisciplinary approach to Nature and knowledge is the concept of *levels of Reality*.

Here, the meaning we give to the word "Reality" is pragmatic and ontological at the same time. By "Reality," we intend first of all to designate that which resists our experiences, representations, descriptions, images, or even mathematical formulations.

Insofar as Nature participates in the being of the world, one has to assign also an ontological dimension to the concept of Reality. Reality is not merely a social construction, the consensus of a collectivity, or some inter-subjective agreement. It also has a trans-subjective dimension; for example, experimental data can ruin the most beautiful scientific theory.

Of course, one has to distinguish the words "Real" and "Reality." Real designates that which is, while Reality is connected to resistance in our human experience. The "Real" is, by definition, veiled forever, while "Reality" is accessible to our knowledge.

By "level of Reality," I designate a set of systems that are invariant under certain laws. For example, quantum entities are subordinate to quantum laws, which depart radically from the laws of the macro-physical world. That is to say that two levels of Reality are different if, while passing from one to the other, there is a break in the applicable laws and a break in fundamental concepts (like, for example, causality). Therefore there is a discontinuity in the structure of levels of Reality, similar to the discontinuity reigning over the quantum world.

Every level of Reality has its associated space-time, different from one level to the other. For example, the classical realism is associated with the 4-dimensional space-time (three dimensions of space and one dimension of time), while the quantum realism is associated with a space-time whose number of dimensions is greater than four. The introduction of the levels of Reality induces a multidimensional and multireferential structure of Reality.

A new *Principle of Relativity*[22] emerges from the coexistence between complex plurality and open unity in our approach: no level of Reality constitutes a privileged place from which one is able to understand all the other levels of Reality. A level of Reality is what it is because all the other levels exist at the same time. This Principle of Relativity is what originates a new perspective on religion, politics, art, education, and social life. And when our perspective on the world changes, the world changes. The great Brazilian educator Paulo Freire asserts in his *Pedagogy of the Oppressed*[23] that saying a true word is equivalent to the transformation of the world.

In other words, our approach is not hierarchical. There is no fundamental level. But its absence does not mean an anarchical dynamic but a coherent one of all levels of Reality, both those already discovered and those that will be discovered in the future.

Every level is characterized by its incompleteness: the laws governing this level are just a part of the totality of laws governing all levels. And even the totality of laws does not exhaust the entire Reality; we have also to consider the Subject and its interaction with the Object.

The zone between two different levels and beyond all levels is a zone of non-resistance to our experiences, representations, descriptions, images, and mathematical formulations. Quite simply, the transparency of this zone is due to the limitations of our bodies and of our sense organs—limitations that apply regardless of what measuring tools are used to extend these sense organs. We therefore have to conclude that the topological distance between levels is finite. However, this finite distance does not mean a finite knowledge. Take a segment of a straight line—it contains an infinite number of points. In a similar manner, a finite topological

[22] Ibid: pp. 54–55.

[23] Freire (1968).

distance could contain an infinite number of levels of Reality. We have work to do till the end of time.

This open structure of the unity of levels of Reality is in accord with one of the most important scientific results of the twentieth century concerning arithmetic, the theorem of Kurt Gödel, which states that a sufficiently rich system of axioms inevitably leads to results that are either undecidable or contradictory. The implications of Gödel's[24] theorem have considerable importance for all modern theories of knowledge, primarily because it concerns not just the field of arithmetic but all of mathematics that include arithmetic. The Gödelian structure of levels of Reality implies the impossibility of a self—enclosed, complete theory. Knowledge is forever open.

The zone of non-resistance corresponds to the sacred—to that which does not submit to any rationalization. Proclaiming that there is a single level of Reality eliminates the sacred, and self-destruction is generated.

The unity of levels of Reality and its complementary zone of non-resistance constitutes what we call the transdisciplinary Object.

Inspired by the phenomenology of Edmund Husserl,[25] I assert that the different levels of Reality of the Object are accessible to our knowledge thanks to the different levels of Reality of the Subject. They permit an increasingly general, unifying, encompassing vision of Reality without ever entirely exhausting it.

As in the case of levels of Reality of the Object, the coherence of levels of Reality of the Subject presupposes a zone of non-resistance to perception.

The unity of levels of Reality of the Subject and this complementary zone of non-resistance constitutes what we call the transdisciplinary Subject.

The two zones of non-resistance of transdisciplinary Object and Subject must be identical for the transdisciplinary Subject to communicate with the transdisciplinary Object. A flow of consciousness that coherently cuts across different levels of perception must correspond to the flow of information coherently cutting across different levels of Reality.

The two flows are interrelated because they share the same zone of non-resistance.

Knowledge is neither exterior nor interior; it is simultaneously exterior and interior. The studies of the universe and of the human being sustain one another. Without spirituality, the knowledge is a dead knowledge.

The zone of non-resistance plays the role of a third between the Subject and the Object, an Interaction term, which acts like a secretly included middle that allows for the unification of the transdisciplinary Subject and the transdisciplinary Object while preserving their difference. I will call this Interaction term the Hidden Third.

Our ternary partition (Subject, Object, Hidden Third) is, of course, different from the binary partition (Subject vs. Object) of classical realism.

---

[24]Nagel and Newman (1958).

[25]Husserl (1966).

The emergence of at least three different levels of Reality in the study of natural systems—the macrophysical level, the microphysical level, and the cyber-space-time (to which one might add a fourth level, that of superstrings, unifying all physical interactions)—is a major event in the history of knowledge. Based upon our definition of levels of Reality, we can identify other levels than just the ones in natural systems. For example, in social systems, we can speak about the individual level, the geographical and historical community level (family, nation), the cyber-space-time community level, and the planetary level.

Levels of Reality are radically different from levels of organization as these have been defined in systemic approaches.[26] Levels of organization do not presuppose a discontinuity in the fundamental concepts; several levels of organization can appear at the same level of Reality. The levels of organization correspond to different structures of the same fundamental laws.

The levels of Reality and the levels of organization offer the possibility of a new taxonomy of the more than 8000 academic disciplines existing today. Many disciplines coexist at the same level of Reality even if they correspond to different levels of organization. For example, Marxist economy and classical physics belong to one level of Reality, while quantum physics and psychoanalysis belong to another level of Reality.

The existence of different levels of Reality has been affirmed by different traditions and civilizations, but this affirmation was founded either on religious dogma or on the exploration of the interior universe only.

The transdisciplinary Object and its levels of Reality, the transdisciplinary Subject and its levels of perception, and the Hidden Third define the transdisciplinary model of Reality. Based on this ternary structure of Reality, we can deduce other ternaries of levels that are extremely useful in the analysis of concrete situations by contextualization:

- Levels of organization—Levels of structuring—Levels of integration
- Levels of confusion—Levels of language—Levels of interpretation
- Physical levels—Biological levels—Psychical levels
- Levels of ignorance—Levels of intelligence—Levels of contemplation
- Levels of objectivity—Levels of subjectivity—Levels of complexity
- Levels of knowledge—Levels of understanding—Levels of being
- Levels of materiality—Levels of spirituality—Levels of non-duality

I formulated the idea of levels of Reality in 1976 during a post-doctoral stay at the Lawrence Berkeley Laboratory following stimulating discussions with Geoffrey Chew, the founder of the bootstrap theory, and other colleagues. My main motivation was the fact that this idea offered a logical solution to the incompatibility between the theory of relativity and quantum mechanics. I interpreted this

[26] Camus et al. (1998).

[27] d'Espagnat (1981).

incompatibility as the necessity of enlarging the field of Reality by abandoning the classical idea of a single level of Reality.

In 1981, I was intrigued by the idea of a veiled reality of Bernard d'Espagnat,[27] but I realized that his solution was not satisfactory, and I therefore decided to publish my findings in an article published in 1982[28] and later, in an elaborated form, in 1985, in the first edition of my book *We, the particle and the world.*

In 1998, I was surprised to discover the idea of levels of Reality expressed in a different form, in a book by Werner Heisenberg, *Philosophy—The manuscript of 1942*. This book had a quite astonishing history: it was written in 1942[29] but it was published in German-only in 1984. I read the French translation of the book in 1998.

For Heisenberg, reality is "the continuous fluctuation of the experience as captured by consciousness. In that sense, it can never be identified to a closed system."[30] By "experience," he understands not only scientific experiments but also the perception of the movement of the soul or of the autonomous truth of symbols. For him, reality is a tissue of connections and of infinite abundance without any ultimate founding ground.

"One can never reach an exact and complete portrait of reality,"[31] writes Heisenberg. The incompleteness of physical laws is therefore present in his philosophy, even if he makes no explicit reference to Gödel.

Heisenberg asserts many times, in agreement with Husserl, Heidegger, and Cassirer (whom he knew personally), that one has to suppress any rigid distinction between the Subject and Object. He also writes that one has to renounce the privileged reference to the exteriority of the material world and that the only way to understand the nature of reality is to accept its division in regions and levels. The similarity to my own definition of reality is striking, but the differences are also important. By "region of reality," Heisenberg understands a region characterized by a specific group of relations. His regions of reality are, in fact, strictly equivalent to the levels of organization of contemporary systemic thinking.

Heisenberg classifies the numerous regions of reality in only three levels, in terms of the different proximity between the Object and the Subject.[32] He deduces that the rigid distinction between exact and human sciences has to be abandoned, a fact which sounds very, very transdisciplinary.

Heisenberg's first level of reality corresponds to fields that embody objectivity in an independent way from the knowledge process. Classical physics, electromagnetism, and the two theories of relativity of Einstein belong in this level.

---

[28] Nicolescu (1982, pp. 68-77).

[29] Heisenberg (1998).

[30] Ibid., p. 166.

[31] Ibid., p. 258.

[32] Ibid., p. 372.

The second level corresponds to fields inseparable from the knowledge process: quantum mechanics, biology, and the sciences of consciousness (like psychoanalysis), for example.

Finally, the third level corresponds to fields created in connection with the knowledge process. He situates there philosophy, art, politics, the metaphors concerning God, the religious experience, and the artistic creative experience.

If the first two levels of Heisenberg totally correspond to my own definition, the third one mixes levels and non-levels (in other words, the zones of non–resistance). The religious experience and the artistic creative experience cannot be assimilated to levels of Reality. They merely correspond to crossing levels in the zone of non-resistance. The absence of resistance and especially the absence of discontinuity in the philosophy of Heisenberg explain the difference between his approach and mine. A rigorous classification of regions in levels cannot be obtained in the absence of discontinuity.

Heisenberg insists on the crucial role of intuition: "Only an intuitive thinking," writes Heisenberg, "could bridge the abyss between old and new concepts; the formal deduction is impotent in realizing this bridge [...][33]". But Heisenberg did not draw the logical conclusion concerning this impotence of formal thinking; only the non-resistance to our experiences, representations, descriptions, images, or mathematical formalisms can bridge the abyss between two levels. This non-resistance restores the continuity broken by levels.

2.3 The Logical Axiom: The Included Middle

The incompleteness of the general laws governing a given level of Reality signifies that, at a given moment of time, one necessarily discovers contradictions in the theory describing the respective level: one has to assert A and non-A at the same time. This Gödelian feature of the transdisciplinary model of Reality is verified by all the history of science: a theory leads to contradictions and one has to invent a new theory solving these contradictions. It is precisely the way in which we went from classical physics to quantum physics.

However, our habits of mind, scientific or not, are still governed by the classical logic, which does not tolerate contradictions. The classical logic is founded on three axioms:

1. The *axiom of identity*: A is A.
2. The *axiom of non-contradiction*: A is not non-A.
3. The *axiom of the excluded middle*: There exists no third term T ("T" from "third") which is at the same time A and non-A.

---

[33] Ibid., p. 161.

Knowledge of the coexistence of the quantum world and the macrophysical world and the development of quantum physics have led, on the level of theory and scientific experiment, to pairs of mutually exclusive contradictories (A and non-A): wave and corpuscle, continuity and discontinuity, separability and non-separability, local causality and global causality, symmetry and breaking of symmetry, reversibility and irreversibility of time, and so forth.

The intellectual scandal provoked by quantum mechanics precisely consists in the fact that the pairs of contradictories that it generates are actually mutually exclusive when they are analyzed through the interpretive filter of classical logic.

However, the solution is relatively simple: one has to abandon the third axiom of the classical logic, imposing the exclusion of the third, the included middle T.

History will credit Stéphane Lupasco (1900–1988)[34] with having shown that the logic of the included middle is a true logic, mathematically formalized, multivalent (with three values: A, non-A, and T) and non-contradictory.[35]

In fact, the logic of the included middle is the very heart of quantum mechanics: it allows us to understand the basic principle of the superposition of "yes" and "no" quantum states.

Heisenberg was fully conscious of the necessity of adopting the logic of the included middle. "There is"—writes Heisenberg—"a fundamental principle of classical logic which seems to need to be modified: in classical logic, if one assertion has a meaning, one supposes that either this assertion or its negation has to be true. Only one of the sentences 'There is a table here' and 'There is no table here' is true: tertium non datur, i.e. there is not a third possibility and this is the principle of the excluded middle.[...] In quantum theory, one has to modify this law of the excluded middle. If one protests again any modification of this basic principle, one can immediately argue that this principle is implicated in the ordinary language [...]. Consequently, the description in ordinary language of a logical reasoning which does not apply to this language would mean simply a self-contradiction."[36]

Our understanding of the axiom of the included middle—there exists a third term T which is at the same time A and non-A—is completely clarified once the notion of "levels of Reality", not existing in the works of Lupasco, is introduced.

In order to obtain a clear image of the meaning of the included middle, let us represent the three terms of the new logic—A, non-A, and T—and the dynamics associated with them by a triangle in which one of the vertices is situated at one level of Reality and the two other vertices at another level of Reality. The included middle is in fact an included third. If one remains at a single level of Reality, all manifestation appears as a struggle between two contradictory elements. The third dynamic,

---

[34] Badescu and Nicolescu (ed.) (1999).

[35] Lupasco (1951).

[36] Heisenberg (1971, pp. 241-242). [Remark from the editor: Nicolescu references this quote with a publication by Heisenberg from 1971, without listing it in the bibliography. The only English-language publication by Heisenberg from 1971 that I know of is *Physics and Beyond*. However, the quote cannot be found there. The source of the quote therefore remains unknown here, as it could not be found even after intensive research.]

that of the T-state, is exercised at another level of Reality, where that which appears to be disunited is in fact united, and that which appears contradictory is perceived as non-contradictory.

It is the projection of the T-state onto the same single level of Reality which produces the appearance of mutually exclusive, antagonistic pairs (A and non-A). A single level of Reality can only create antagonistic oppositions. It is inherently self-destructive if it is completely separated from all the other levels of Reality. A third term which is situated at the same level of Reality as that of the opposites A and non-A, cannot accomplish their reconciliation. Of course, this conciliation is only temporary. We necessarily discover contradictions in the theory of the new level when this theory confronts new experimental facts.

In other words, the action of the logic of the included middle on the different levels of Reality induces an open structure of the unity of levels of Reality. This structure has considerable consequences for the theory of knowledge because it implies the impossibility of a self-enclosed complete theory. Knowledge is forever open.

The logic of the included middle does not abolish the logic of the excluded middle: it only constrains its sphere of validity. The logic of the excluded middle is certainly valid for relatively simple situations, for example, driving a car on a highway: no one would dream of introducing an included middle in regard to what is permitted and what is prohibited in such circumstances. On the contrary, the logic of the excluded middle is harmful in complex cases, for example, within the economical, social, cultural, religious or political spheres. In such cases it operates like a genuine logic of exclusion: good or evil, right or left, heaven or hell, alive or dead, women or men, rich or poor, whites or blacks. It would be revealing to undertake an analysis of xenophobia, racism, apartheid, anti-semitism, or nationalism in the light of the logic of the excluded middle. It would also be very instructive to examine the speeches of politicians through the filter of that logic.

There is certainly coherence among different levels of Reality, at least in the natural world. In fact, an immense self-consistency—a cosmic bootstrap—seems to govern the evolution of the universe, from the infinitely small to the infinitely large, from the infinitely brief to the infinitely long. A flow of information is transmitted in a coherent manner from one level of Reality to another in our physical universe. The included middle logic is a tool for an integrative process: it allows us to cross two different levels of Reality or of perception and to effectively integrate, not only in thinking but also in our own being, the coherence of the Universe. The use of the included third is a transformative process. But, at that moment, the included third ceases to be an abstract, logical tool: it becomes a living reality touching all the dimensions of our being. This fact is particularly important in education and learning.

2.4 The Complexity Axiom: The Universal Interdependence

There are several theories of complexity. Some of them, like the one practiced at the Santa Fe Institute, with the general guidance of Murray Gell-Mann, Nobel Prize of Physics, are mathematically formalized, while others, like the one of Edgar Morin, widely known in Latin America, are not. In the context of our discussion, what is important to be understood is that the existing theories of complexity do not include neither the notion of levels of Reality nor the notion of zones of non-resistance.[37] However, some of them, like the one of Edgar Morin,[38] are compatible with these notions. It is therefore useful to distinguish between the horizontal complexity, which refers to a single level of reality and vertical complexity, which refers to several levels of Reality. It is also important to note that transversal complexity is different from the vertical, transdisciplinary complexity. Transversal complexity refers to crossing different levels of organization at a single level of Reality.

From a transdisciplinary point of view, complexity is a modern form of the very ancient principle of universal interdependence. This recognition allows us to avoid the current confusion between complexity and complication. The principle of universal interdependence entails the maximum possible simplicity that the human mind could imagine, the simplicity of the interaction of all levels of reality. This simplicity cannot be captured by mathematical language, but only by symbolic language. The mathematical language addresses exclusively to the analytical mind, while symbolic language addresses to the totality of the human being, with its thoughts, feelings and body.

It is interesting to note that the combined action of the ontological, logical and complexity axiom engenders values. Therefore, there is no need to introduce values as a 4th axiom.[39] The transdisciplinary values are neither objective nor subjective. They result from the Hidden Third, which signifies the interaction of the subjective objectivity of the transdisciplinary Object and the objective subjectivity of the transdisciplinary Subject.

3 Building a New Spirituality

"Spirituality" is a completely devaluated word today, in spite of its etymological meaning as "respiration", in an act of communion between us and the cosmos. There is a big spiritual poverty present on our Earth. It manifests as fear, violence, hate and dogmatism. In a world with more than 10.000 religions and religious movements and more than 6.000 tongues, how can we dream about mutual understanding and peace? There is an obvious need for a new spirituality, conciliating technoscience

[37] Nicolescu (1996a, 1996b, 1998a, 1998b, 2000).

[38] Morin (1977, 1980, 1986, 1991, 2001, 2004).

[39] Cicovacki (2003).

and wisdom. Of course, there are already several spiritualities, present on our Earth from centuries and even millennia. One might ask: why is there a need for a new spirituality if we have them all, here and now?

Before answering to this question, we must face a preliminary question: is a Big Picture still possible in our post-modern times? Radical relativism answers in a negative way to this question. However its arguments are not solid and logical. They are in fact very poor and obviously linked to the totalitarian aspect of the political and philosophical correctness expressed by the slogan "anything goes" [cf. Paul Feyerabend, see chap. 1, Schüz, this volume, p. 14.]. For radical relativists, after the death of God, the death of Man, the end of ideologies, the end of History (and, perhaps, tomorrow, the end of science and the end of religion) a Big Picture is no more possible. For transdisciplinarity, a Big Picture is not only possible but also vitally necessary, even if it will never be formulated as a closed theory. We are happy that the well-known art critic Suzi Gablik, in her book *Has Modernism Failed?,*[40] joined recently our point of view. The last chapter of her book is entitled *Transdisciplinarity–Integralism and the New Ethics*. For her, the essential intellectual change of the last two decades is precisely transdisciplinarity. This change was anticipated by the big quantum physicist Wolfgang Pauli (1900–1958), Nobel Prize of Physics, who wrote fifty years ago: "Facing the rigorous division, from the 17th century, of human spirit in isolated disciplines, I consider the aim of transgressing their opposition [. . .] as the explicit or implicit myth of our present times."[41]

The first motivation for a new spirituality is technoscience, with its associated fabulous economic power, which is simply incompatible with present spiritualities. It drives a hugely irrational force of efficiency for efficiency sake: everything which can be done will be done, for the worst or the best. The second motivation for a new spirituality is the difficulty of the dialogue between different spiritualities, which often appear as antagonistic, as we can testify in our everyday life. The new phenomenon of a planetary terrorism is not foreign to these two problems.

In simple words, we need to find a spiritual dimension of democracy. Transdisciplinarity can help with this important advancement of democracy, through its basic notions of "transcultural" and "transreligious".[42]

The transcultural designates the opening of all cultures to that which cuts across them and transcends them, while the transreligious designates the opening of all religions to that which cuts across them and transcends them.[43] This does not mean the emergence of a unique planetary culture and of a unique planetary religion, but of a new transcultural and transreligious attitude. The old principle "unity in diversity and diversity from unity" is embodied in transdisciplinarity.

[40] Gablik (2004). The first edition was published in 1984.

[41] Pauli (1999), chapter "Science and Western Thinking", p. 178. This chapter was first published in 1955, in *Europa Erbe und Aufgabe, Internationaler Gelehrtehkongress*, Mainz.

[42] Nicolescu (1996a, 1996b).

[43] Nicolescu (1996a, 1996b).

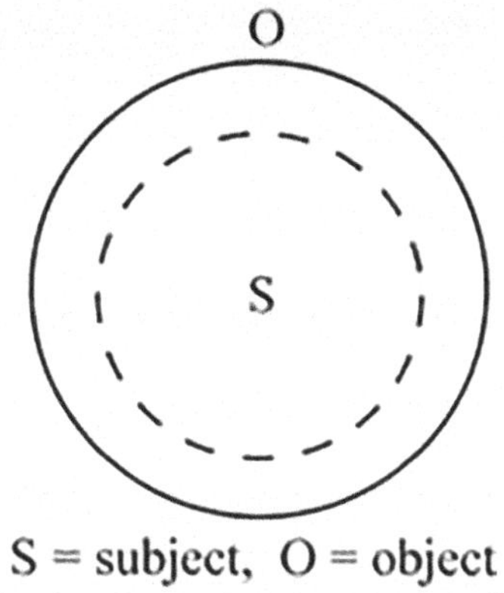

Fig. 1 Pr-Modernity

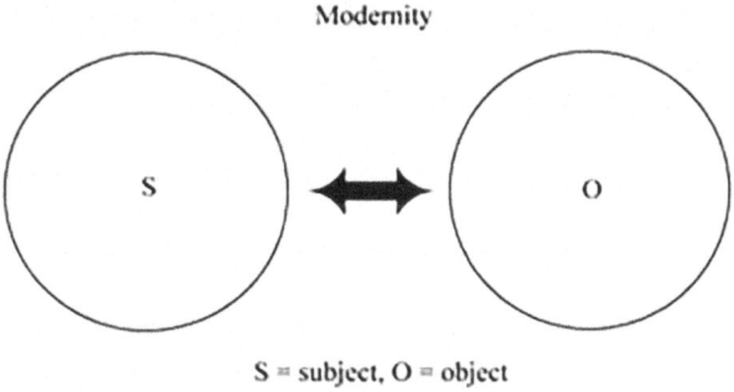

Fig. 2 Modernity

Through the transcultural, which leads to the transreligious, the spiritual poverty could be eradicated and therefore render the war of civilizations obsolete. The transcultural and transreligious attitude is not simply a utopian project—it is engraved in the very depths of our being.

This evolution of mentalities could be achieved only if we perform the unification of Homo religious with Homo economicus.

Homo religiosus probably existed from the beginnings of the human species, at the moment when the human being tried to understand the meaning of his life. The sacred is his natural realm. He tried to capture the unseen from his observation of the visible world. His language is that of the imaginary, trying to penetrate higher levels of Reality—parables, symbols, myths, legends, revelation.

Homo economicus is a creation of modernity. He believes only in what is seen, observed, measured. The profane is his natural realm. His language is that of just one level of Reality, accessible through the analytic mind–hard and soft sciences, technology, theories and ideologies, mathematics, informatics.

Transdisciplinary methodology is able to identify the common germ of homo religiosus and of homo economicus—called homo sui transcendentalis in my *Manifesto of Transdisciplinarity.*[44] This identification could be done by taking into

[44]Nicolescu (1996a, 1996b), see this volume, p. XI ff.

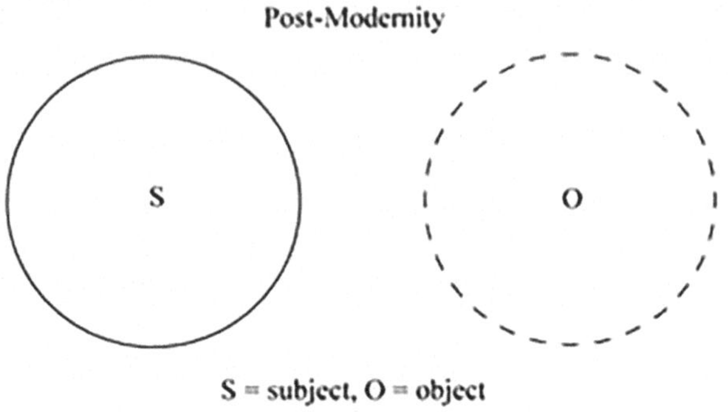

Fig. 3 Post-Modernity

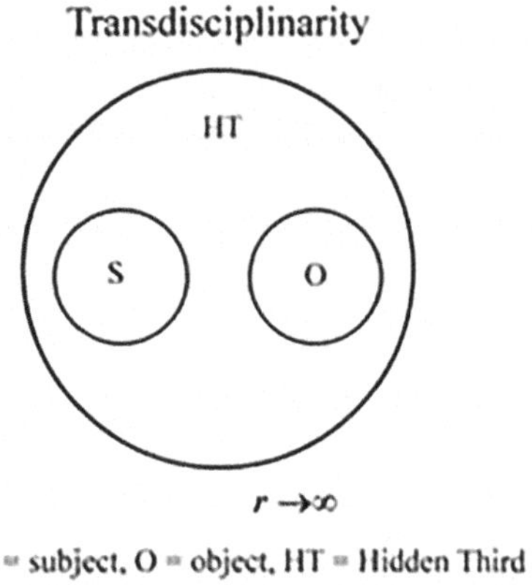

Fig. 4 Transdisciplinarity

account the new relation established by transdisciplinarity between Object and Subject.

In Pr — Modernity the Subject was immersed in the Object. Everything was trace, signature of a higher meaning. The world of the pre-modern human being was magical (see Fig. 1).

In Modernity, Subject and Object were totally separated (see Fig. 2) by a radical epistemological cut, allowing in such a way the development of modern science. The Object was just there, in order to be known, deciphered, dominated, and transformed.

In Post-Modernity (see Fig. 3) the roles of the Subject and Object are changed in comparison with Modernity and are reversed in comparison with Pre-Modernity: the Object, still considered as being outside the Subject, is nevertheless a social construction. It is not really "there". It looks more like an emanation of the Subject.

Transdisciplinarity leads to a new understanding of the relation between Subject and Object, which is illustrated in Fig. 4: The Subject and the Object are, like in Modernity, separated but they are unified by their immersion in the Hidden Third, whose ray of action is infinite.

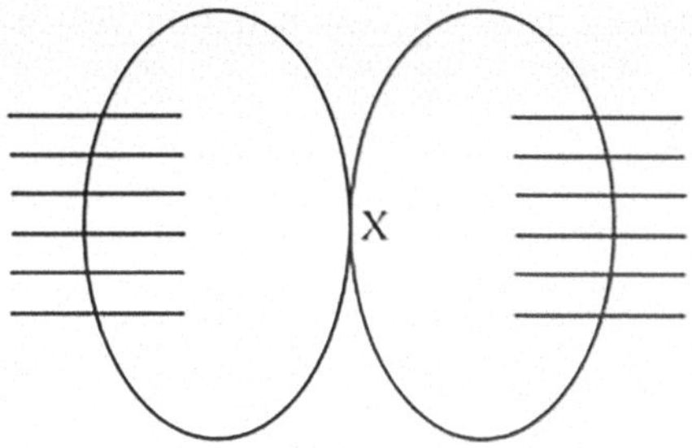

Fig. 5 Hidden Third

The transdisciplinary Object and its levels, the transdisciplinary Subject and its levels and the Hidden Third define the Transdisciplinary Reality or Trans-Reality[45] (see Fig. 5).

"What is Reality?"—asks Charles Sanders Peirce (1839–1914), a great philosopher, logician, mathematician of the beginning of the twentieth century.[46] He tells us that maybe there is nothing at all which corresponds to Reality. It may be just a working assumption in our desperate tentative in knowing. But if there is a Reality—tells us Peirce—it has to consist in the fact that the world lives, moves and has in itself a logic of events, which corresponds to our reason. Peirce's view on Reality totally corresponds to the transdisciplinary view on Reality.

The unified theory of levels of Reality is crucial in building sustainable development and sustainable futures. The considerations made till now in these matters are based upon reductionist and binary thinking: everything is reduced to society, economy and environment. The individual level of Reality, the spiritual level of Reality and the cosmic level of Reality are completely ignored. Sustainable futures, so necessary for our survival, can only be based on a unified theory of levels of Reality. We are part of the ordered movement of Reality. Our freedom consists in entering into the movement or perturbing it. Reality depends on us. Reality is plastic. We can respond to the movement or impose our will of power and domination. Our responsibility is to build sustainable futures in agreement with the overall movement of Reality.

References

Apostel, L., Berger, G., Briggs, A., & Michaud, G. (Eds.). (1972). *L'interdisciplinarité –Problèmes d'enseignement et de recherche*. Centre pour la Recherche et l'Innovation dans l'Enseignement, Organisation de Coopération et de développement économique.

[45] Nicolescu (2009).

[46] Peirce (1976, vol. IV, pp. 383–384); cf. id. (1931–1958).

Badescu, H. & Nicolescu, B. (Eds.) (1999) *Stéphane Lupasco - L'homme et l'œuvre*. Le Rocher, Monaco. [Translation in Portuguese: *Stéphane Lupasco - O Homem e a Obra*. TRIOM et University of Sa˜o Paulo, Sa˜o Paulo, 2001, translation by Lucia Pereira de Souza].
Camus, M., Magnin, T., Nicolescu, B., & Voss, K.-C. (1998). Levels of representation and levels of reality: Towards an ontology of science. In N. H. Gregersen, M. W. S. Parsons, & C. Wassermann (Eds.), *The concept of nature in science and theology (part II)* (pp. 94–103). Editions Labor et Fides.
Cicovacki, P. (2003). Transdisciplinarity as an Interactive Method: A Critical Reflection on the Three Pillars of Transdisciplinarity, in: Communication at the congress *The Unifying Aspects of Cultures*, Vienna, November, pp. 7–9.
d'Espagnat, B. (1981). *A la recherche du réel*. Gauthier Villars.
Descombes, V. (2004). *Le complément du sujet*. Gallimard.
Duguet, Pierre (1972). "L'approche des problèmes", in: Apostel et al. (eds.; 1972).
Freire, P. (1968). *Pedagogy of the oppressed*. The Seabury Press.
Gablik, S. (2004). *Has modernism failed?* Thames & Hudson.
Galileo, G. (1992). *Dialogue sur les deux grands systémes du monde*. Seuil, Paris, translated from the Italian by René Fréreux with the collaboration of Francois de Gandt, pp. 128–130; [*Dialogue on the Great World Systems*, University of Chicago Press, Chicago, 1956, with an introduction by Giorgio de Santillana].
Gibbons, M., Limoges, C., Nowotny, H., Schwartzman, S., Scott, P., & Trow, M. (Eds.). (1994). *The new production of knowledge*. SAGE.
Guerre (1997). Guerre et paix *entre les sciences: disciplinarité et transdisciplinarité*. Revue du MAUSS, No 10, Paris.
Heisenberg, W. (1998). *Philosophie – Le manuscrit de 1942*. Seuil, Paris, translation from German and introduction by Catherine Chevalley [German original edition: Ordnung der Wirklichkeit, R. Piper, München, 1989. Published first in: W. Blum, H. P. Dürr, and H. Rechenberg (Ed.), W. Heisenberg Gesammelte Werke, Vol. C-I: Physik und Erkenntnis, 1927–1955. R. Piper, Munich, 1984, pp. 218–306. A translation in English of this book can be found on the Internet page http://werner-heisenberg.unh.edu/t-OdW english.htm#seg01].
Heisenberg, W. (1971). *Physics and beyond: Encounters and conversations*. Harper & Row.
Husserl, E. (1966). *Méditations cartésiennes*. Vrin. Translated from German by Gabrielle Peiffer and Emmanuel Levinas.
Jantsch, E. (1972a). Vers l'interdisciplinarité et la transdisciplinarité dans l'enseignement et l'innovation, in: Léo Apostel et al. (1972).
Jantsch, E. (1972b). *Technological planning and social futures*. Cassell, Associated Business Programes, New York University Press.
Lichnerowicz, A. (1972). "Mathématique et transdisciplinarité", in Léo Apostel et al.
Lupasco, St. (1951). Le principe d'antagonisme et la logique de l'énergie - Prolégoménes à une science de la contradiction, Hermann & Cie, Coll. Actualités scientifiques et industrielles, no. 1133, Paris; 2nd ed. Le Rocher, Monaco 1987, foreword by Basarab Nicolescu. Rocher, Monaco, 1987, foreword by Basarab Nicolescu.
Morin, E. (1977). La méthode I - La nature de la nature. Paris : Seuil.
Morin, E. (1980). La méthode II - La vie de la vie. Paris : Seuil.
Morin, E. (1986). La méthode III - La connaissance de la connaissance. Paris : Seuil.
Morin, E. (1991). La méthode IV - Les idées, leur habitat, leur vie, leurs moeurs, leur organisation. Paris: Seuil.
Morin, E. (1999). Seven Complex Lessons in Education (translated from the French by Nidra Poller). Paris: UNESCO.
Morin, E. (2001). La méthode V - L'humanité de l'humanité. Paris: Seuil.
Morin, E. (2004). La méthode VI - Ethique. Paris: Seuil.
Nagel, E., & Newman, J. R. (1958). *Gödel's Proof*. New York Univsersity Press.
Nicolescu, B. (1982). Sociologie et m'ecanique quantique, 3e Milnaire, no 1, Paris, March–April.

Nicolescu, B. (1986). Science as Testimony. In Proceedings of the Symposium *Science and the Boundaries of Knowlege: the Prologue of Our Cultural Past*, organised by UNESCO in collaboration with the Cini Foundation (Venice, March 3–7, 1986), Paris, UNESCO, pp. 9–30.

Nicolescu, B. (1996a). *La transdisciplinarité – manifeste*. Le Rocher, Monaco, Transdisciplinarité Series, foreword by Joscelyn Godwin, afterword by Antoine Faivre. [Englisch translation: *Manifesto of Transdisciplinarity*. SUNY Press, 2002, translation from the French by Karen-Claire Voss. Translation in Portuguese: O Manifesto da Transdisciplinaridade, Triom, Sao Paulo, 1999, translation by Lucia Pereira de Souza; 2nd edition: 2001].

Nicolescu, B. (1996b). Levels of Complexity and Levels of Reality. In Bernard Pullman (ed.), *The Emergence of Complexity in Mathematics, Physics, Chemistry, and Biology*. Pontificia Academia Scientiarum, Vatican City, pp. 393–417. Distributed by Princeton University Press; Proceedings of the Plenary Session of the Pontifical Academy of Sciences, 27–31 October 1992, Casina Pio IV, Vatican.

Nicolescu, B. (1998a). The Transdisciplinary Evolution of the University, Condition for Sustainable Development. In *Rencontres Transdisciplinaires*, No 12, February; http://perso.club-internet.fr/nicol/ciret/bulletin/b12/b12.c8.htm.

Nicolescu, B. (2002). Nous, la particule et le monde. Le Mail, Paris, 1985; 2nd edition: Le Rocher, Monaco, "Transdisciplinarité" Series.

Nicolescu, B. (1991). *Science, meaning and evolution – The cosmology of Jacob Boehme, with selected texts by Jacob Boehme*. Parabola Books. translated from the French by Rob Baker.

Nicolescu, B. (1998b). Gödelian aspects of nature and knowledge. In G. Altmann & W. A. Koch (Eds.), *Systems - new paradigms for the human sciences* (pp. 385–403). Walter de Gruyter.

Nicolescu, B. (2000). Hylemorphism, quantum physics and levels of reality. In D. Sfendoni-Mentzou (Ed.), *Aristotle and contemporary science* (Vol. I, pp. 173–184). Peter Lang. Introduction by Hilary Putnam.

Nicolescu, B. (2004). Toward a methodological Foundation of the Dialogue between the Technoscientific and spiritual cultures. In L. Moreva (Ed.), *Differentiation and integration of worldviews*. Eidos.

Nicolescu, B. (2005). Toward transdisciplinary education. *The Journal for Transdisciplinary Research in Southern Africa, 1*(1), 5–15. North-West University, Wanderbijlpark, South Africa.

Nicolescu, Basarab (2007): Transdisciplinarity as a methodological framework for going beyond the science – Religion debate, Transdisciplinarity in Science and Religion, 2, Curtea Veche Publ., , pp. 35–60.

Nicolescu, B. (2008a). Spiritual dimension of democracy: Utopia or necessity? In A. Marga, T. Bercheim, & J. Sadlak (Eds.), *Living in truth* (pp. 509–516). Cluj University Press, Babes-Bolyai University.

Nicolescu, Basarab (2008b): The idea of levels of reality – Its relevance for non-reduction and personhood, in: Transdisciplinarity in Science and Religion, 4, Curtea Veche Publ., , pp. 11–25.

Nicolescu, B. (Ed.). (2008c). *Transdisciplinarity–theory and practice*. Hampton Press.

Nicolescu, B. (2009). *Qu'est-ce que la réalité?* Liber.

Nowotny, H. (1994). The potential of transdisciplinarity. In: http://www.interdisciplines.org/interdisciplinaritypapers/5.

Pauli, W. (1999). *Physique moderne et philosophie*. Albin Michel. translated from the German by Claude Maillard.

Peirce, C. S. (1931–1958): *Collected Papers of Charles Sanders Peirce*, 8 volumes, In C. Hartshorne, P. Weiss, & A. Burks (Eds.), Harvard University Press.

Peirce, C. S. (1966). *Selected writings (values in a universe of chance)*. Dover Publications. edited with an introduction and notes by Philip P. Wiener.

Peirce, C. S. (1976). *The new elements of mathematics*. (4 volumes, ed. by C. Eisele). Mouton Humanities Press.

Piaget, J. (1972). Lépistémologie des relations interdisciplinaires. In Léo Apostel et al. (1972).

The Charter. (1994). The Charter of Transdisciplinarity (in French, Spanish, English, Portuguese, Turkish, Italian, Arab and Romanian): http://basarab.nicolescu.perso.sfr.fr/ciret/; also in this volume.

Thomson Klein, J., Grossenbacher-Mansuy, W., Häberli, R., Bill, A., Scholz, R. W., & Welti, M. (Eds.). (2001). *Transdisciplinarity: Joint problem solving among science, technology, and society—An effective way for managing complexity*. Birkhäuser Verlag.

Basarab Nicolescu is a theoretical physicist in elementary particle physics at CNRS, France and professor of philosophy at the Babes-Bolyai University, Cluj-Napoca, Romania. He is member of the Romanian Academy and president-founder of the International Center for Transdisciplinary Research and Studies (CIRET). Basarab Nicolescu is founder and director of the Transdisciplinarity series, Rocher, Monaco, of the Romanians in Paris series, Oxus, Paris and of the Science and Religion and Science, Spirituality, Society Series, Curtea Veche, Bucharest. His books include: Manifesto of Transdisciplinarity, State University of New York (SUNY) Press, New York, 2002; Nous, la particule et le monde, E. M. E. Inter Communications, Brussels, 2012 (3rd edition); Science, Meaning and Evolution—The Cosmology of Jacob Boehme, Parabola Books, New York, 1991. He edited Transdisciplinarity—Theory and Practice, Hampton Press, Cresskill, New Jersey, 2008. Forthcoming: Science, Culture and Spirituality From Modernity to Cosmodernity, State University of New York (SUNY) Press, New York (to be published). A complete bio-bibliography of Basarab Nicolescu can be found on the page http://www.basarab-nicolescu.fr/index.php.

Open Access This chapter is licensed under the terms of the Creative Commons Attribution 4.0 International License (http://creativecommons.org/licenses/by/4.0/), which permits use, sharing, adaptation, distribution and reproduction in any medium or format, as long as you give appropriate credit to the original author(s) and the source, provide a link to the Creative Commons license and indicate if changes were made.

The images or other third party material in this chapter are included in the chapter's Creative Commons license, unless indicated otherwise in a credit line to the material. If material is not included in the chapter's Creative Commons license and your intended use is not permitted by statutory regulation or exceeds the permitted use, you will need to obtain permission directly from the copyright holder.

History of Science in a Transdisciplinary Perspective: From Transdisciplinarity to Transcultural Science

Marco Bischof

1 Introduction

In my book *"Tachyonen—Orgonenergie—Skalarwellen—Feinstoffliche Felder zwischen Mythos und Wissenschaft"* [1]I advocated that the history of science should treat scientific disciplines as holistic-cultural phenomena. The history of a discipline must not only be described from the view of the actually dominating standpoint, but must incorporate the undercurrents and subcultures of science, which are currently marginalized or completely excluded. A discipline consists not only of the currently dominating doctrine, but also of its repressed notions and unconscious, i.e., those conceptions that at a given moment have been considered not to be relevant, or not scientific enough, and have therefore been excluded. It must also include Non-Hegemonic Science, i.e., science that is currently not dominating, either within the Western world or in Non-Western ex-colonized countries.

Non-hegemonic sciences are for example feminist science,[2] Romantic science,[3] Rosicrucian optics and alchemy[4], esoteric currents in science, gnostic[5] and hermetic[6]

[1]Bischof (2002).

[2]Crasnow (2020).

[3]Cunningham & Jardine (1990).

[4]Schrödter (1954).

[5]Hassan (1995); Wegener (2015); DeConick (2016).

[6]Hannak (2013); Hanegraaff (2022).

M. Bischof (✉)
Institute for Transcultural Health Studies, Viadrina University at Frankfurt (Oder), Frankfurt, Germany
e-mail: mb@marcobischof.com

© The Author(s), under exclusive license to Springer Nature Switzerland AG 2025

M. Schüz (ed.), *Transdisciplinary Thinking and Acting*, Transdisciplinary Management of Ecological and Social Crises,
https://doi.org/10.1007/978-3-031-70680-6_3

currents, the Pythagorean-Platonic tradition[7], New Age Science[8], Transpersonal Psychology[9], parapsychology[10], paraphysics,[11] Shamanic Sciences,[12] Complementary and Alternative Medicine (CAM),[13] holistic approaches in biology and medicine.[14]

Additionally, science is subject to fashions like any other cultural manifestations, and therefore what is taught at the universities is not all there is to say on the subject in question. The notions that are not in fashion or repressed, however, have at some other time definitely been present in science and may someday emerge again from the "underground" and may again be considered as significant. In other words: also the parascientific "superstition" and "pseudoscience" aspects of the academic discipline in question must be included. Without their consideration, I postulated, a complete and integral comprehension of a research and scientific area is not possible at all.

Such a broad comprehension is especially important for the further development of a discipline, particularly in times in which the old paradigm is used up and doesn't bite anymore, and the necessity of a more comprehensive view becomes apparent.[15]

In another place I wrote in the same book:

> Human concepts and theories, including the scientific ones, are as a rule at the beginning of their development more alive, richer, deeper, and more complete than in their later implementations. They are of necessity in the course their further development simplified and reduced to certain aspects, in order to become more manageable and more applicable, but also comprehensible to a wider circle of people; in this process, however, they lose their depth, richness of aspects and interconnections, but also their inner tensions and contradictions. After some time are the concepts in this development away from the origin then so rigidified and used-up, that they rather hem instead of promote progress. Then it becomes necessary to return to their origins in order to renew them and make them again fertile. Fresh inspirations may then emerge before all from the previously not heeded and not developed aspects of those concepts.[16]

---

7 Bischof (2008).

8 Capra (1975).

9 Tart (1978); Bischof (2003).

10 Radin (1997).

11 Bischof (2002).

12 Gottwald & Rätsch (1998); von Stuckrad (2003).

13 Cf. Freeman & Lawlis (2008).

14 Bischof (1998).

15 Bischof (2002: 157).

16 Bischof (2002: 16)

2 A Short Overview of the History of Science

Since the enlightenment in the eighteenth century until about 1970, the historiography of science has been dominated by narratives of progress and the central position of the Scientific Revolution.[17] The renowned English historian and philosopher Herbert Butterfield (1900–1979), professor of history at the University of Cambridge, criticized in his 1931 book *The Whig Interpretation of History* any interpretation of history that *"studies the past with reference to the present"*[18] *"Past science is judged,"* writes David Alvarogonzalez about Butterworth, *"according to its contribution to theorems held as true in the present; the past is interpreted through current values, with a consequent dismissal of the problems and ideas of earlier scientists."*[19] Butterfield argued that historians should attempt to understand the past in terms of the past. Science is seen as emerging from a series of victories over pre-scientific thinking.[20] The term "whiggish history of science" is now also used to refer to historiography that focuses on the successful chains of theories and experiments that led to present-day theories while ignoring failed theories and dead ends.[21] Alvarogonzalez writes that the history of science should also recognize and study failed theories, closed paths, and dead ends, such as phlogiston theory in chemistry and Lamarckian evolution in biology.[22] The renowned historian of biology Ernst Mayr remarked: *"Inevitably this includes a treatment of false starts and of competing theories."*[23]

Stimulated largely by the publication of Thomas S. Kuhn's *The Structure of Scientific Revolutions*[24] the history of science since about 1970 has broken decisively with the teleological historiography, as well as with the view of scientific knowledge as sharply demarcated from its context.[25] The American physicist and historian of science Thomas S. Kuhn (1922–1996) is known by his division of the development of science as a series of phases of "Normal Science," interrupted by "Scientific Revolutions":

1. *Pre-paradigmatic science (or protoscience):* For the researcher exists here a large freedom in the choice of his experiments, so that scientists can investigate very different aspects of their subject area. In this manner are often many competing and incompatible views generated among scientists:

[17] Daston (2001: 6842).
[18] Butterfield (1931).
[19] Alvargonzález (2013: 85)
[20] Burrow (2008).
[21] Mayr (1990).
[22] Alvargonzález (2013).
[23] Mayr (1990).
[24] Kuhn (1962).
[25] Daston (2001).

2. *Normal science:* Characteristic for normal science is the acceptance of a "paradigm" by the scientific community, on whose basis research is undertaken. A paradigm is the ensemble of basic concepts, which at a certain time constitute a scientific discipline. On the one hand, the scope of relevant problems by the paradigm is greatly restricted; on the other hand, this same fact implies the possibility to undertake research in depth. The task of the scientist in normal science is the solution of problems, whose rules of solution are implied by the paradigm.
3. *Extraordinary or revolutionary phase of science:* Not until there are emerging problems over a considerable time at central places or surprising discoveries made, the phase of "extraordinary science" begins. In this stage there is also discussion on the foundations themselves. Such a "crisis" may lead to a "paradigm change" in which the paradigm of the discipline is rejected and replaced by another. The accumulation of knowledge is in contrast to normal science not cumulative, as important parts of the old paradigm are given up. In scientific revolutions change after Kuhn not only the theories, but also the general worldview and the scientific practice.

Kuhn's account of the development of science held that science enjoys periods of stable growth punctuated by revisionary revolutions.[26] In these revolutions, so-called *"anomalies"* play a key role, i.e., phenomena that have no place in the actual theoretical considerations of science. They become the inducement for new concepts that revolutionize science.

3 Societal Background

The American historian of science Val Dusek, emeritus professor at the University of New Hampshire, writes:

> Two major periods when the views of the excluded and oppressed outsiders entered into the literate discourse of the social establishments were the Renaissance and the French Revolution. During both of these periods the dominant structures of society broke down, the relation of the social classes became more fluid, and the thought of women and peasants was mediated by declassé intellectuals.[27]

Antonio Gramsci (1891–1937), an Italian Marxist, discusses the Marxist social movement in relation to such earlier movements as the Reformation and highlights the role of "organic intellectuals."[28] Gramsci's term "hegemony," used with respect to the dominant ruling-class belief structure, became popular in postmodern circles. One scholar who earlier dealt with themes of illiterate peasant and lower-class

---

[26] Bird (2018).

[27] Dusek (1999: 5–6).

[28] Gramsci (1978).

cultures in their relation to "higher" intellectual culture was Friedrich Heer (1916–1983), Austrian professor of the history of ideas. Heer, with his notion of the "underground" of history, gave an account of the mediation between the thought of excluded groups (such as peasants, the lower classes, and women) and the high culture of a period, sometimes subverting and always energizing the latter, but often simultaneously being partially co-opted by it. Heer's use of the term "underground" has a surprisingly contemporary ring, probably stemming both from Dostoyevsky's "underground man" and from the European Resistance movements during World War II.

By the "underground" Heer means the currents of the popular culture that are not under the hegemony of the elite and high culture.

> The underground includes not only the archaic, peasant, non-lettered culture, but also the culture of the bohemians, those intellectuals who are not part of the institutions of the official intellectual culture and who are in contact with the lower-class elements. This lumpen-intelligentsia acts as a mediator between the lower-class culture and the literate intellectual culture.[29]

Throughout the Middle Ages and the early Modern period these economically unfortunate intellectuals mixed with the oppressed classes and produced eschatological, utopian, and revolutionary theories, which were not generally the direct recording of the ideas, myths, and programs for action of the oppressed, of peasants and proletarians. Instead, the genesis of underground culture was the result of the mediating individual who had somehow received an education in the high culture of the day as well as been exposed to the moods, situations, and problems of the oppressed.

> In one sense, the underground culture was a reflection of lower-class culture, but primarily in terms of mood and attitude. The particular theoretical contents of the countercultural system were not, however, a simple borrowing of peasant myths or proletarian politics. The countertheories were an expression of problems and pre-reflective orientations as modified and restructured in terms of concepts and ideas borrowed from the literate ruling culture[30]

What Heer shows, however, is the tremendous innovation produced by the displaced intelligentsia forced into solidarity with the oppressed productive classes.

The era of the late Roman Empire and Early Christianity was one such time of innovation, a period of social disruption in which dispossessed and discouraged intellectuals were thrown into contact with slaves and proletarians in various mystery cults, Christian sects, and heresies. It was also a period of occultism and massive superstition, but also a time of great ethical and religious innovation.

The late Middle Ages, Renaissance, and Reformation also saw tremendous cultural innovation. These were, like the late Roman Empire, eras of occultism and superstition, but also one of ethical and religious innovation. Antonio Gramsci was

[29] Heer (1953).

[30] Dusek (1999: 7).

interested in the differences between Renaissance humanism, as a movement of elite intellectuals, and the Reformation, involving peasant religions and political movements as well as changes in the high culture.[31]

The period from the early nineteenth to the early twentieth centuries is another era of fusion between the lower-class popular culture and the intellectual culture. The Romantic movement around 1800 and the Bohemian artistic movements of the late nineteenth and early twentieth centuries marked periods when mediating declassé intellectuals assumed successively greater importance and the underground began to burst through into both the high culture and dominant political movements.

Certainly Heer's underground notions are applicable to the counterculture or Bohemian movements, including the Beatniks of the 1960s, hippies of the 1960s, and the New Age culture of the late twentieth century.

The late Roman Empire (with the rise of Christianity) and the Renaissance were also high waters of the occult, astrology, and various superstitions. The intellectual-artisan culture of the Renaissance that gave birth to modern science was steeped in alchemy, astrology, and magic. In some respects, the popularity of the occult among the cultured classes during the Renaissance was greater than it was in the Dark Ages, and was surpassed only by that in the later Roman Empire during the transition to Christianity and feudalism. If Kepler's astronomy and Paracelsus's chemistry and medicine emerged from their occultist preoccupations, similar phenomena may be possible with our own counterculture. Ecological consciousness may supply the cultural milieu that could give rise to new scientific concepts. (…).

Romantic science and Romantic philosophy of nature had fewer direct ties with peasant culture than did either early Chinese Taoism or Renaissance occultism. However, Romanticism in its early phases was energized and often instigated by the events of the French Revolution, with its upsurge of ordinary people against the aristocracy and clergy, even if many Romantics were conservatives or became such after disillusionment with the Terror, the excesses of the Revolution, and the military dictatorship of Napoleon.

The Romantics revived the doctrines and views of Renaissance occultist and hermeticist figures, including Giordano Bruno and Paracelsus, whose doctrines and personalities were ridiculed and discredited during the height of the enlightenment."[32]

Independently of Friedrich Heer historian of philosophy Richard H. Popkin identified a third force in the thought of the seventeenth century (Popkin, 1992, 2008; see also Coudert, 1999). In this time developed a third fashion of thought besides the Cartesian rationalism and the British empiricism. This movement is to be understood as a reaction to the increasing skepticism of the time. This new line of thought achieved in the course of the seventeenth century some strange alliances between the new science and theology, mainly among the organizers of England's ambitious scientific efforts, and had finally its climax in the thought of Sir Isaac

[31] Gramsci (1977, 1978).

[32] Dusek (1999: 5–11).

Newton. Richard Popkin called it the "third force."[33] It is characterized by its combination of the elements of empirical as well as rationalistic thought with theosophical speculation and millenaristic bible exegesis. All this was used in order to counter the skepticist challenge.

The leading theoretician of the "third force" was Joseph Mede (1586–1639), a biblical scholar, Egyptologist and Hebraist, and also a naturalist. His disciples John Dury (1596–1680), Samuel Hartlib (1600–1662), and Henry More (1614–1687) each describe the epistemological, skeptical crisis they had personally undergone, which they each resolved by finding certainty in Biblical materials, prophecies, texts, or spiritual ideas. All of them were enthusiastic about the ideas of Johann Valentin Andreae (1586–1654), author of the Rosicrucian manifestos, on the just government and the reformation of the world.[34] They found also support for their positive views in the philosophical theosophies of Jan Amos Comenius (1592–1670) and Jacob Boehme (1575–1624), so that their finished statements verged far away from what we are used to in the Montaigne-Descartes tradition. Not only did these people and their followers work out a different answer to philosophical skepticism; they also contributed greatly in terms of their solution, to the formulation of modern science, especially in the work of Henry More and Isaac Newton.[35]

Hartlib, who was of German origin, began to recognize more and more the virtue of technological development as a means to improving people's lot and enlightening them on the powers of the mind. He saw his task as the recording of all human knowledge to make it universally available for the education of all humankind.[36]

Hartlib therefore entertained a correspondence network all over Western and Central Europe, an association of personal friends, now called the "Hartlib Circle." Hartlieb and Dury were the main figures. Elias Ashmole (1617–1692), Thomas Vaughn (1621–1666), Robert Hooke (1635–1703), and Robert Boyle (1626–1691) were some of the members. Boyle was an Irish naturalist who worked in England. Boyle was a practicing alchemist, but simultaneously he became a co-founder of modern science, which was based on detailed and published experiments, mainly of physics and chemistry. The influence of Comenius on the Hartlieb Circle cannot be overestimated.

> Comenius was the most important theoretician of the "new science". Comenius contended *"that we must conjoin Divine Revelation, Reason and Sensory Perception in order to reach faith, understanding, and certainty. The senses provide evidence. Reason gains certainty the more it is based on the senses. Reason however, also corrects the senses when the senses judge incorrectly about apparently illusory data. In the same way, Revelation corrects reason when the latter reaches false views about invisible matter. When these corrections occur, reason is not contrary to sense, nor faith to reason"*.[37]

[33] Popkin (1992, 2008).

[34] Churton (2009).

[35] Popkin (1992).

[36] Churton (2009: 290).

[37] Comenius (1651), preface, cited by Popkin (2008: 205).

4 The Truth about the History of Science

We can conclude from these assessments that neither the *time of the Scientific Revolution* nor the time of the Enlightenment of the eighteenth century *have never been purely rational times*, but have been mixtures of rational discourse with esotericism, spiritualism, and illuminism.[38] Thus, scientific concepts have never been developed from a purely rational discourse, but from the *disputes between rational discourse and esotericism, spiritualism, and illuminism*. We need in the history of science to know all the sides of this dispute. Dan Edelstein writes in *The Super-Enlightenment*:

> The portrait of the Enlightenment that emerges from the 'Super' perspective is less one of a homogeneous movement, (…) than a movement continually oscillating between opposing poles.[39]

5 The Cultural Milieu of the Upbringing of Students

In the popular culture in which many scientists-to-be grow up, before the professionalization as a scientist sets in, there is not yet discrimination between orthodox science and myth; it consists of fairy tales, comics, science fiction, and popular science (Hauser, 2005–2016). This connects it with Friedrich Heer's "underground." This popular culture is informed by scientific fictions, i.e., *"long-lasting intellectual games in the world of science, from the world of science and for the world of science."*[40] The milieu of upbringing of students is molded by these scientific fictions and forms the longest-lasting impression on the future scientist and provides the motivation for him/her to become a scientist.

6 What Students of the History of Science Must Learn

They must learn of the scientific disputes at a given moment of history and of the cultural and philosophical background of these disputes. In these disputes, not all lost positions are forever worthless, but can be of value in the future. Science is also subject to fashion like all manifestations of culture, and not all the subjects of these disputes are exhaustive; some of them are just not in fashion at the time (Crane 1969; Rothman, 1989; Turner, 1992; Barber, 1993; Reville, 1999). As Erwin Schrödinger as early as 1932 said, in a lecture to the Prussian Academy of Sciences, that *"yes,*

---

[38] As Gabay (2005), Edelstein (2010), Fleming (2013), Monod (2013), Neugebauer-Wölk (1999, 2008a, 2008b, 2013), and Neugebauer-Wölk et al. (2013) have shown.

[39] Edelstein (2010: 31)

[40] Hauser (2016: 17), introduction.

science can be (and often is) a Fashion of the Times, a consequence of the Temper of the Age."[41]

7 From Transdisciplinary Science to Transcultural Science

The inclusion of the mentioned fields into the history of science is, however, not enough to obtain a global science that can be accepted by all peoples of the world. Science must also be urgently decolonized.[42]

A first step in this task has been the acceptance that other cultures than the European have also developed science; we have to leave behind the Eurocentric view of science. Until now this has been done concerning Chinese science. The renowned British biologist and historian of science Joseph Needham (1900–1995) started in the year 1954 to publish the multi-volume work *Science and Civilization in China.*[43] To date there are 7 volumes in 27 books. By asking his grand question: why did modern science not develop in China, even though China was technologically superior to the West prior to the sixteenth century? Needham's investigation is recognized as one of the most influential works in stimulating the discourse on the multicultural roots of modern science (Wikipedia).

Concerning Islamic science the pioneering work was Seyyed Hossein Nasr's *Science and Civilization in Islam*, named after Needham's work, published in 1968, followed by *An Annotated Bibliography of Islamic Science*[44] and *Islamic Science: An Illustrated Study.*[45]

Indian science was first treated by Debiprasad Chattopadhyaya in 1977 in the book *History of Science and Technology in Ancient India*, with a foreword by Joseph Needham.[46]

"Indigenous science" is a notion developed as a consequence of varied critiques of the dominance of Western ways of knowing.[47] It is defined as *"knowledge which is spatially and/or culturally context-specific, collective, holistic and adaptive. It is communicated orally and passed down from generation to generation."*[48] Although it was previously ignored in academic circles, it is now increasingly acknowledged and put to practice in the fields of development and conservation, environmental practice and education, agriculture, and other areas.

To summarize the growing number of works published since these initial efforts, the book of American historian of science Helaine Selin, *Encyclopedia of the History*

---

[41] Cited from Lukacs (2002: 98).

[42] Alvares (1991).

[43] Needham (1954)-present.

[44] Nasr (1975).

[45] Nasr (1976).

[46] Chattopadhyaya (1977).

[47] Le Grange (2004).

[48] Collective (2009).

of Science, Technology, and Medicine in Non-Western Cultures, was first published in 1997 in one volume; the third, greatly enlarged edition of this comprehensive work was issued in 2016 in five volumes and 4706 pages.[49]

Even "Western science" turns out, according to the latest findings, to be not so Western, after all, as the British historian of science James Poskett has shown this year in his book *Horizons*.[50] In the middle of the twentieth century almost all the historians of science were still convinced that modern science—and with it modern civilization—originated in Europe, sometime around the sixteenth century. Herbert Butterfield for instance declared: "*The scientific revolution we must regard … as a creative product of the West.*"[51] Students at Yale University in the 1950s were taught that "*the West generated the natural sciences… the East did not*," while readers of the journal *Science*—one of the most prestigious scientific magazines in the world—were informed that "*a small circle of Western European nations provided the original home for modern science.*" This Eurocentric view is simply wrong, as Poskett showed in many examples. Science is not a product of a unique European culture. Rather, modern science has always depended upon bringing together people and ideas from different cultures around the world. The development of modern science depended upon global cultural exchange. Whatever period we look at, the history of science cannot be told as a story, which focuses solely on Europe.[52]

8 Transcultural Science

The notion of transculturality was first proposed by the Cuban sociologist Fernando Ortiz (1881–1969) in the 1940s and then, from the 1990s onward, taken up and adapted by the German philosopher and theorist of postmodernity, Wolfgang Welsch (born 1946).[53] Many authors assume that transdisciplinarity will by necessity develop into transculturality.[54] According to these authors, monodisciplinarity will consequentially develop into pluridisciplinarity, pluri- or multidisciplinarity into interdisciplinarity, interdisciplinarity into transdisciplinarity, transdisciplinarity into transculturality; in ascending order these are scientific practices with increasing complexity and research expenditure. While multi-or pluridisciplinarity stands for a coexistence of professional perspectives on the same research topic, interdisciplinarity involves the scientific exchange of representatives of different disciplines on this research topic and the attempt to arrive at collaborative results. Transdisciplinarity represents once again a progression as also the practical

[49] Selin (1997).

[50] Poskett (2022).

[51] Butterfield (1949: 191).

[52] Poskett (2022).

[53] Reichardt (2018).

[54] Franz-Balsen (2012).

perspective (stakeholders, aggrieved parties) temporarily is included in the research process. That is to say that the problems or the research questions possess the character of a "living environment" (Lebenswelt). Transdisciplinary research provides knowledge for a practical solution geared to the common welfare. The new paradigm "transculturality" integrates from my point of view into this context. It complements and makes more precise the transdisciplinary approach by questioning the conventional view of the cultural dimension of the living environment (Lebenswelt). Transcultural research acknowledges the complexity, the constant change, and the hybridity of cultural constructions and dispossesses one of the conventional securities.[55]

9 Conclusion

In order to bring these approaches together in a veritable transcultural science, we must first overcome some difficulties inherent in such an attempt.

> Transdisciplinarity signifies "*a research strategy that crosses many disciplinary boundaries to create a holistic approach.*"[56]

This includes non-hegemonic sciences, the underground of sciences, as well as the paranormal, esoteric, gnostic, and hermetic currents, feminist science, ecocritical thought, black criticism, etc. Also non-Western science must be incorporated, including indigenous, orally transmitted knowledge. Moreover all this knowledge must be fused into one, albeit pluralist science.[57] In order to accomplish this, science must change, not a little bit, but greatly. It must change in order to accommodate alien concepts from other cultures that are incompatible with actual Western science.

A glimpse on the difficulties that occur if science encounters foreign cultures is the problem of Chinese "Qi" or the acupuncture meridians. Qi is the "energy" that is supposed to flow in the meridians of Chinese medicine.[58] Decades have passed since the Western science encountered Chinese medicine in the wake of US president Richard M. Nixon's visit to China in February 1972. Over this time, Western science tried to reduce Qi to electromagnetism or radioactivity, but didn't succeed.[59] Today most scientists agree that science must develop new concepts to accommodate Qi or the Indian "prana." In this attempt, science can fall back to earlier developments from the time when vitalist concepts were still part of European science until the nineteenth century, in the form of pneuma, animal spirits, and nerve fluids.[60]

---

[55] Ibid.

[56] Wikipedia (2023).

[57] Chang (2017a, 2017b, 2020); Ludwig & Ruphy (2021).

[58] Kubny (1995).

[59] Cf. e.g. Lu (1997).

[60] Bischof (2008: 62–79).

The Chinese scientist Yuk Hui, professor of philosophy at the City University of Hong Kong, provides the philosophical framework for such a non-reducing concept of Qi in his book *The Question Concerning Technology in China. An Essay in Cosmotechnics*.[61] He turns against the erroneous idea of a universal science and technology, which came out of Western modernity, resulting in the techno-cultural uniformization or homogenization of the planet. Instead, he proposes to cultivate different conceptions of technology through diverse forms of thinking and practice ("technodiversity") in order to invent alternatives.[62] Hui wants to reopen the question concerning technology in favor of plural "cosmotechnics" based on different cosmologies. In this, he turns against Heidegger who only believed in a singular technology based on the origin in Greek philosophy. This Western concept of technics is in the process of totally overpowering Chinese culture (as well as practically all other cultures on the planet) to the point of completely obliterating its own cultural origins. The Chinese, according to Hui, should reconnect with their own techno-cultural origin, which has been very different from that of the West and which he aims to uncover precisely in terms of a (Chinese) cosmotechnics. Non-Western cultural spheres should inquire into their own cultural archives and so recuperate their own cosmotechnical resources, in order to effectively confront their own danger of being engulfed and anesthetized by Western values, instead of relying exclusively on the Western archive.[63] Hui has tried in his book to do so by reconstructing a Chinese cosmotechnics based on Qi and Dao, which is the guiding principle of Daoism (Taoism) where Yang (Light) and Yin (Shadow) form a dynamic balance in nature.

Best seems such a transcultural science so far to be realized in the *"Superhumanities"* project of Jeffrey J. Kripal. Kripal occupies the Chair in Philosophy and Religious Thought at Rice University in Houston (Texas), a private research university, and is an internationally known historian of religion. According to Kripal, the "superhumanities" are the humanities (Geisteswissenschaften), but in contrast to the present state of things under full acknowledgment of the *"Superfaculties"* which make Man into Superman: the paranormal faculties of telepathy, clairvoyance, precognition and telekinesis, etc.[64] Superhumanities are a conscious and intended extension of humanism. At their origins, the humanities were born in the Italian Renaissance. This was a broad spiritual and intellectual movement understood by its founders to be a rebirth (hence *renaissance*) and recovery of Ancient Egyptian, Latin, Greek, and Hebrew Esoteric Wisdom.

Perhaps no one did more toward such a rebirth than Marsilio Ficino (1433–1499), head of the Platonic Academy founded in 1462 in Florence. He started translating and interpreting the *"Corpus Hermeticum,"* scriptures that have their origin in the first few centuries of the Christian era, and have established a whole spiritual current

---

[61] Hui (2016).

[62] Hui (2017).

[63] Lemmens (2020).

[64] Kripal (2022).

of great importance. It involved salvational knowledge or *gnosis,* that is, "privatized experiential-oriented religion." April DeConick, historian of early Jewish and Christian thought and fellow professor of Kripal at Rice University, writes that *such revolutionary gnostic spiritualities are structurally 'countercultural' and that they stand very much against, or above, the 'servant spiritualities' of the public religions (in which the gods are imagined to be kings, lords, or slave masters demanding submission, servitude and sacrifice).*[65]

Later the origins of the humanities in the superhumanities were practically forgotten. Today the situation is as follows. Kripal writes on this topic:

> I think the humanities are so important because the humanities are really the superhumanities. I think there is something cosmic or superhuman smoldering in the human, something that seems ever ready to burst into flames, and sometimes does. A few fortunate souls intuit this superhuman smoldering in particularly inspired books, themselves about to burst into flames—yes, books, of all things—and decide to give their lives to the pursuit und nurturing of that, whatever 'that' is; usually, of course, such souls end up pursuing the (super)human in terms that are more respectable and more culturally bound, which is to say: in ways that can get (a very few of them) paying jobs. The superhuman becomes the simply human again. These proper humanists will talk endlessly now about anything but the super and reduce it to something else, something always bad and sad. That's how we forget. Or are made to forget, But, deep down, what at least some of these individuals really want, or at least what they once wanted, is to become superhumans, like the authors who they first read and loved. That's because they really are. So are you.[66]

Kripal asks the question:

> Can you imagine what the humanities might become if they were seen anew and widely understood as the superhumanities? If we restored the anomalous and inexplicable to the heart of the inquiry in the humanities, if we would acknowledge the undercurrent of the fantastic within our humanistic disciplines? The core idea of the superhumanities is: pursue diversity among people until the end, make it more radical, render it ontological. In short: do not just decolonize history, religion, literature, or society, as academics like to say. Decolonize reality itself. It is not enough, after all, to question and challenge all the ways that our colonial and slave-owning pasts have shaped our cultures and our very selves. It is simply not enough to question and criticize these histories and then shove all these historical actors into a flat white European metaphysics of materialism, historicism, scientism, and moralism, as if these people were nothing more than biomedical bodies, historical actors, and social animals. What of their souls? Of their supernatural experiences? Of their encounters with god? Of their miraculous saints? Of their psychoactive mushrooms and plant medicines? Of the nonhuman agents and spectral entities, the demonic angels, with whom they routinely interact in terror, healing, and transformation? We must take the rogue and fantastic aspects of religious experience not just critically or historically (that is the white European metaphysics again) but also experientially or empirically—that is, as telling us something potentially true not only about ourselves, but about the actual structure of space, time, and matter itself.[67]

---

[65] DeConick (2016: 12).

[66] Kripal (2022: 8–9).

[67] Kripal (2022: 14–15).

Such an understanding of human diversity Kripal found in the works of the anthropologist and cultural psychologist Richard A. Shweder (born 1945), professor of Human Development at the University of Chicago. Shweder was writing on the "reality-posits" of other cultures. He was asking readers to take such realities seriously, indeed, potentially as their own.[68] In this way, we can see the expressions and institutionalization of religious diversity as a privileged path to more and more insight about reality, more and more value, truth, and being—in short, more of the real. Shweder writes:

> There are different 'reality-posits' expressed or experienced by different people in different states of mind, and the anthropologist or historian of religions is a kind of intellectual shaman called to travel through these other worlds and, by virtue of this constant traveling back and forth, provide new comparative insights for his or her own community. Different peoples actualize different truths and values, even different realities, to know and experience as 'true'. Truth is relative, then, not because there is no truth but because there is too much truth.[69]

Also part of Kripal's understanding of the superhumanities is his notion of the *"Human as Two"*:

> He means *"the double vision of the human that on the one hand fully recognizes and affirms all that can be said and understood with our present modes of knowledge (humanistic, social scientific, and scientific) but on the other hand also recognizes and affirms a supernature that can never be fully said, much less understood and defined."*[70] He continues to explain: *"The key to what I am calling the superhuman is that the 'Human as Two' includes and embraces both levels: both the subject/object representations of our ordinary, constructed, and determined individual historical and embodied experience and that which is beyond all representation and embodiment and so not fully constructed (and which we thus fear)."*[71]

Kripal wants to express that the conventional humanities, with their strict focus on the constructions and differences of human subjectivity and embodiment, and the sciences, with their exclusive focus on material objects and mathematical processes, are both finally inadequate to the fullness of the living world.

> :*If one cannot appreciate and understand paranormal phenomena, then one cannot appreciate and understand what a human being is—a superhuman, that is, a human being who is (and is not) a human being."*[72] And he specifies elsewhere:). *"One has to consider the psychical abilities like telepathy and precognition as evolutionary buds of newly emerging human faculties.*[73]

[68] Shweder (1991).

[69] Kripal (2022: 15 f).

[70] Ibid: 120.

[71] Ibid: 122.

[72] Ibid: 122

[73] Ibid: 63.

References

Alvares, C. A. (1991). *Decolonizing history: Technology and culture in India, China and the west 1492 to the present day*. The Other India Press/Apex Press.

Alvargonzález, D. (2013). Is the history of science essentially whiggish? *History of Science, 51*, 85–99.

Barber, B. (1993). Fashion in science. In *Constructing the social system*. Routledge.

Bird, A. (2018). Thomas Kuhn. *Stanford Encyclopedia of Philosophy*. Available: https://plato.stanford.edu/entries/thomas-kuhn/

Bischof, M. (1998). Holism and field theories in biology–non-molecular approaches and their relevance to biophysics. In J. J. Chang, J. Fisch, & F. A. Popp (Eds.), *Biophotons* (pp. 375–394). Kluwer Academic Publishers.

Bischof, M. (2002). Tachyonen–Orgonenergie–Skalarwellen. In *Feinstoffliche Felder zwischen Mythos und Wissenschaft*. AT Verlag.

Bischof, M. (2003). Vom integralen Weltbild zu einer neuen Wissenschaft. Der Wandel In Gesellschaft und Wissenschaft und die neue Religiosität. In M. Utsch & J. Fischer (Eds.), *Im Dialog über die Seele—Transpersonal Psychologie und christlicher Glaube* (pp. 3–51). LIT Verlag.

Bischof, M. (2008). *Der Kristallplanet. Globale Netze, platonische Körper und die Musik der Erde*. Drachen Verlag.

Burrow, J. (2008). *A history of histories*. Allen Lane.

Butterfield, H. (1931). *The Whig interpretation of history*. G. Bell.

Butterfield, H. (1949). *The origins of modern science*. G. Bell and Sons.

Capra, F. (1975). *The Tao of physics*. Shambala Publications.

Chang, H. (2017a). Is pluralism compatible with scientific realism? In J. Saatsi (Ed.), *The Routledge handbook of scientific realism*. Routledge.

Chang, H. (2017b). Presentist history for pluralist science. *Journal of General Philosophy of Science, 52*, 97. https://doi.org/10.1007/s10838-020-09512-8

Chattopadhyaya, D. (1977). *History of science and technology in ancient India*. Research India Publications.

Churton, T. (2009). *The invisible history of the Rosicrucians*. Inner Traditions.

Collective. (2009). Indigenous knowledge. From: *International Encyclopedia of Human Geography*. Retrieved Dec 24, 2022, from http://www.sciencedirect.com/topics/earth-and-planetary-sciences/

Comenius, J. A. (1651). *Natural philosophy reformed by divine light*.

Coudert, A. P. (1999). Newton and the Rosicrucian enlightenment. In J. E. Force & R. H. Popkin (Eds.), *Newton and religion: Context, nature and influence* (pp. 17–43). Springer.

Crasnow, S. (2020). Feminist perspectives on science. *Stanford encyclopedia of philosophy*. Accessed Oct 24, 2022, from https://plato.stanford.edu/entries/feminist-science/

Cunningham, A., & Jardine, N. (Eds.). (1990). *Romanticism and the sciences*. Cambridge University Press.

Daston, L. (2001). History of science. In N. J. Smelser & P. B. Baltes (Eds.), *International encyclopedia of the social & behavioral sciences* (pp. 6842–6848). Elsevier.

DeConick, A. D. (2016). *The gnostic new age: How a countercultural spirituality revolutionized religion from antiquity to today*. Columbia University Press.

Dusek, V. (1999). *The holistic inspirations of physics. The underground history of electromagnetic theory*. Rutgers University Press.

Edelstein, D. (Ed.). (2010). *The super-enlightenment: Daring to know too much*. Voltaire Foundation.

Fleming, J. V. (2013). *The dark side of the enlightenment. Wizards, alchemists, and spiritual seekers in the age of reason*. W. W. Norton.

Franz-Balsen, A. (2012). Higher Education for Sustainable Development meets Gender Mainstreaming. In C. Cremer-Renz & B. Jansen-Schulz (Eds.), *Von der Internationalisierung*

der Hochschule zur Transkulturellen Wissenschaft. Wissenschaftliche Konferenz 2010 an der Leuphana Universität Lüneburg (pp. 125–142). Nomos Verlagsgesellschaft.
Freeman, L. W., & Lawlis, G. F. (2008). *Mosby's complementary and alternative medicine—A research-based approach* (3rd ed.). Mosby Elsevier.
Gabay, A. J. (2005). *The Covert Enlightenment–Eighteenth Century Counterculture and its Aftermath.* Swedenborg Foundation Publishers.
Gottwald, F.-T., & Rätsch, C. (Eds.). (1998). *Schamanische Wissenschaft. Ökologie, Naturwissenschaft und Kunst.* Eugen Diederichs Verlag.
Gramsci, A. (1977). *Selections from the prison notebooks of Antonio Gramsci.* International Publishers.
Gramsci, A. (1978). *Selections from the political writings, 1921–1926.* International Publishers.
Hannak, K. (2013). *Geistreiche Kritik: Hermetik, Mystik und das Werden der Aufklärung in spiritualistischer Literatur der frühen Neuzeit.* Walter De Gruyter.
Hanegraaff, W. J. (2022). *Hermetic spirituality and the historical imagination. Altered states of knowledge in later antiquity.* Cambridge University Press.
Hassan, I. (1995). The new gnosticism–speculations on an aspect of the postmodern mind. In P. A. Bové (Ed.), *Early postmodernism. Foundational essays* (pp. 77–99). Duke University Press.
Hauser, Linus (2005, 2009, 2016). *Kritik der neomythischen Vernunft.* Schöningh. 3 volumes.
Heer, F. (1953). *Europäische Geistesgeschichte.* W. Kohlhammer Verlag. English Translation: *The Intellectual History of Europe.* Translated by Jonathan Steinberg. Weidenfeld & Nicolson, 1966.
Hui, Y. (2016). *The question concerning Technology in China. An essay in Cosmotechnics.* Urbanomic Media.
Hui, Y. (2017, November). Cosmotechnics as Cosmopolitics. *E-flux Journal*, No. 86. Available: http://www.e-flux.com/journal/86/161887/cosmotechnics-as-cosmopolitics/
Kripal, J. J. (2022). *The Superhumanities. Historical precedents, moral objections, and new realities.* Chicago University Press.
Kubny, M. (1995). *Qi–Lebenskraftkonzepte in China–Definitionen, Theorien und Grundlagen.* Karl F. Haug Verlag.
Kuhn, T. S. (1962). *The structure of scientific revolution.* University of Chicago Press.
Le Grange, L. (2004). Western science and indigenous knowledge: Competing perspectives or complementary frameworks? *South African Journal of Higher Education, 18*(3), 82–91.
Lemmens, P. (2020). Other turnings. *Angelaki–Journal of the Theoretical Humanities, 25*(4), 9–25.
Lu, Z. (1997). *Scientific qigong exploration–the wonders and mysteries of Qi.* Amber Leaf Press.
Ludwig, D., & Ruphy, S. (2021). Scientific Pluralism. *Stanford Encyclopedia of Philosophy.*
Lukacs, J. (2002). *At the end of an age.* Yale University Press.
Mayr, E. (1990). When is historiography whiggish? *Journal of the History of Ideas, 51*(2), 301–309.
Monod, P. K. (2013). *Solomon's secret arts–the occult in the age of enlightenment.* Yale University Press.
Nasr, S. H. (1975). *An Annotated Bibliography of Islamic Science (with W. Chittick)* (Vol. 1). Imperial Academy of Philosophy.
Nasr, S. H. (1976). *Islamic science: An illustrated study.* World of Islam Festival Trust.
Needham, J. (1954). *Science and civilization in China.* Cambridge University Press.
Neugebauer-Wölk, M. (Ed.). (1999). *Aufklärung und Esoterik.* Felix Meiner Verlag.
Neugebauer-Wölk, M. (Ed.). (2008a). *Aufklärung und Esoterik. Rezeption–Integration–Konfrontation.* Max Niemeyer Verlag.
Neugebauer-Wölk, M. (2008b). Esoterik im 18. Jahrhundert–Aufklärung und Esoterik. Eine Einleitung. In M. Neugebauer-Wölk (Ed.), *Aufklärung und Esoterik. Rezeption–integration–Konfrontation* (pp. 1–37). Max Niemeyer Verlag.
Neugebauer-Wölk, M., Geffahrt, R., & Meumann, M. (Eds.). (2013). *Aufklärung und Esoterik: Wege in die Moderne.* Walter De Gruyter.

Neugebauer-Wölk, M. (2013). Aufklärung–Esoterik–Moderne. Konzeptionelle Überlegungen zur Einführung. In M. Neugebauer-Wölk, R. Geffahrt, & M. Meumann (Eds.), *Aufklärung und Esoterik: Wege in die Moderne* (pp. 1–33). Walter De Gruyter.

Popkin, R. H. (1992). *The third Force in Seventeenth-century thought*. E. J. Brill.

Popkin, R. H. (2008). Die dritte Kraft im Denken des 17. Jahrhunderts. In *Mit allen Makeln. Erinnerungen eines Philosophiehistorikers* (pp. 181–224). Felix Meiner.

Poskett, J. (2022). *Horizons—A global history of science*. Viking/Penguin Books.

Radin, D. J. (1997). *The Conscious Universe–The Scientific Truth of Psychic Phenomena*. HarperCollins.

Reichardt, D. (2018). Creating notions of Transculturality. The work of Fernando Ortiz and his impact on Europe. In *Komparatistik–Jahrbuch der Deutschen Gesellschaft für Allgemeine und Vergleichende Literaturwissenschaft 2017* (pp. 67–81). Aisthesis Verlag.

Reville, William (1999, May 31). Science is not immune to fashion's trends. in: *The Irish Times*. Retrieved Aug 6, 2023, from https://www.irishtimes.com/news/science-is-not-immune-to-fashion-s-trends-1.190694

Rothman, T. (1989). *Science a la mode: Physical fashions and fictions*. Princeton University Press.

Schrödter, W. (1954). *Die Geheimkünste der Rosenkreuzer*. Baumgartner Verlag.

Selin, Helaine (ed.) (1997): *Encyclopedia of the history of science, technology, and medicine in non-Western cultures*. : Third Springer, Kluwer Academic Publishers, 2016.

Shweder, R. A. (1991). *Thinking through cultures: Expeditions in cultural psychology*. Harvard University Press.

von Stuckrad, K. (2003). *Schamanismus und Esoterik. Kultur- und wissenschaftsgeschichtliche Betrachtungen*. Peeters.

Turner, S. (1992). Sperber's "fashions in science". *Social Epistemology, 6*(1), 77–90.

Wegener, F. (2015). *Gnosis in high tech und science fiction*. Createspace Independent Publishers.

Wikipedia (2023). *Transdisciplinarity*. Available in: https://en.wikipedia.org/wiki/Transdisciplinarity

Bischof, Marco, Dr. phil., made a training in Breathing Therapy at the Institute for Breathing Therapy and Breathing Education (Prof. Ilse Middendorf) at Berlin (Germany) from 1974-76. He then studied Cultural Anthropology, History of Religions and European Anthropology at University of Zurich (Switzerland) from 1981-85. He was active in journalism, specializing in Alternative and Complementary Medicine, Border Areas of Science, Biophysics and new Health Concepts and as a science consultant for customers in business, science, politics and the press from 1985 onwards. 1992 he was a Visiting Scholar at the Center for Frontier Sciences of Temple University at Philadelphia (USA), 1994-95 he was Managing Director of the International Institute of Biophysics (Prof. Dr. F. A. Popp) and from 1995-2010 he was General Secretary and member of the Board of Trustees of the International Institute of Biophysics in Kaiserslautern and then in Neuss (Germany). From 2000-2002 he was assistant professor and member of the Working Group for Agrarian Culture and Social Ecology at Humboldt University at Berlin. He earned a Ph.D. at the Faculty of Cultural Studies of the Viadrina University at Frankfurt (Oder) in 2009, with a dissertation on salutogenetic health concepts and then was lecturer at the Institute for Transcultural Health Studies at the same university from 2009-2011. He published numerous articles in the popular and scientific press and is also known for his books, e.g. "Biophotons" (1995), "Tachyons, Orgone Energy and Scalar Waves—Subtle Fields between Science and Myth" (2002) and "Salutogenesis" (2010).

Levels of Reflection in Transdisciplinarity: The Paradigm of Philosophy as a Theory of Reflection or "Science Teaching"

Johannes Heinrichs

1 Foresight

Many contributions in this book demonstrate the necessity of inter- and transdisciplinary work and understanding.[1] Important topics are the still barely bridged gap between the humanities and the natural sciences, the claim of philosophy to be the foundation of all sciences, but also the gap between theory and practice. My two basic questions are what "transdisciplinary" means in contrast to the older term "interdisciplinary," and especially what claims arise from a clearer understanding of "transdisciplinary."

In the first part I deal with known forms and stages of interdisciplinary practice in the scientific field. This short inventory is already based on a systematic grid, which I relate in the second part to the stages of interpersonal relations as "lived reflection." In the third part, I examine the lived forms of reflection of human experiential spaces in general in comparison with scientifically objectifying (theoretical) reflection. Interdisciplinarity and transdisciplinarity are concerned with *theoretical reflection* even as they target practice. However, in order to build bridges between theory and practice, it is crucially important to understand forms of practice themselves as those of *implicitly lived reflection*, a notion that has not been common so far because "reflection" is always understood only as objectifying after the fact. In the fourth part, I will consider the future role of philosophy as a comprehensive theoretical discipline of reflection: both for bridging to practically lived reflection and for understanding with and between the empirical individual sciences.

[1] See, for example, contribution by Schüz chapter 1 in this volume.

J. Heinrichs (✉)
Social Ecology, Humboldt University Berlin, Berlin/Duisburg, Germany

© The Author(s), under exclusive license to Springer Nature Switzerland AG 2025

M. Schüz (ed.), *Transdisciplinary Thinking and Acting*, Transdisciplinary Management of Ecological and Social Crises,
https://doi.org/10.1007/978-3-031-70680-6_4

As far as the designation of philosophy as a discipline of reflection or "theory of reflection" is concerned, I understand it as an important, leaping further development of the Kantian expression "transcendental philosophy" as at the same time of the Hegelian understanding of "dialectic" (Heinrichs, 2020). Kant declared, "I call all cognition transcendental, which is not concerned both with objects, but is concerned with our mode of cognition of objects [. . .] in general"[2], i.e., a reflection on *how* the cognizing subject can gain objective cognition at all, i.e., on conditions that lie in the self-consciousness of the knowing, evaluating, willing human being. He was aware of the importance of this reflexive turn and proudly declared it as "Copernican turn":

> It is herewith just as with the first thoughts of Copernicus, who, after it did not want to go well with the explanation of the celestial movements, if he assumed, the whole star army turns around the spectator, tried, whether it would not succeed better, if he let the spectator turn and on the other hand the stars in rest.[3]

Besides this reflexive turn of Kant, the term "reflection theory" is inspired by the logician Gotthard Günther. It states that not only the form (method), but also the contents of philosophy are primarily the reflexive, i.e. self-referential relations, which are characterized by consciousness.[4] To what extent these reflexive, logically "multi-valued" relations—beyond two-valued either-or relations—not only play a paramount role in the cultural and social sciences, but also become important in the natural sciences, is a challenging question. A central question of inter- and transdisciplinarity, which we now want to approach step by step.

2 From Interdisciplinarity to Transdisciplinarity (First Phenomenology)

In the following gradation into 1.1 mere cognition, 1.2 inner confrontation, 1.3 intersubjective discussion, and 1.4 transsubjective space of cognition it is always presupposed that it is about the encounter of different disciplines of cognition in the sense of subject areas or methods. This interdisciplinary constellation is combined with the *reflective stages of interpersonality* that the author has presented in numerous publications: 1. mere objective cognition of the Other (objective stage), 2. reckoning with the Other as one's own source of activity (strategic stage), 3. entering into a commonality with the Other, e.g., in the opposition of the gaze (communicative stage, 4. making agreements and norms with the Other (metacommunicative stage).

[2] Kant (1956: B 25).

[3] Ibid: XVI f.

[4] Günther (1976: esp. 24–74); however, Günther himself did not clearly recognize the fundamental difference between implicitly lived and theoretical reflection, as evidenced by his futile attempts at formalization; cf. Heinrichs (2005: 58–64) as well as Heinrichs 2018: 184–200: "Reflexions-Systemtheorie als potentielle Sozialkybernetik und die philosophiegeschichtliche Stellung Gotthard Günther.

2.1 *The Objective and Most Common Level of Interdisciplinarity Today: The Mere Awareness of the Diversity of Other Disciplinary Approaches*

Prototypical for this kind of interdisciplinarity are, for example, *lecture series* or readers in which a topic is treated from very different perspectives, with very different methods and backgrounds. The attraction and value of such interdisciplinarity lies precisely in the perception of this *diversity* and *multiperspectivity*. One would hate to admit, however, that there is generally no real understanding among the active participants, the various professional representatives, still less among most of the audience or readers. It remains essentially to note the diversity of access, often combined with positive *amazement,* but more often with negatively tinged *non-understanding*, to the point of covert or overt rejection of foreign access. Mostly, however, one enjoys mutual *tolerance* in a kind of taster attitude and assures oneself and others that the prevailing darkness belongs precisely to the constitutional, unavoidable limitation of our cognitive capacity, into which one must send oneself if one does not want to fall back into dogmatic attitudes of the pre-Enlightenment era.

2.2 *The Subjective Level: Inner Confrontation with Other Points of View*

Competing views, disputes, and rejections, even if they are only tacit, go beyond mere indifferent notice. Inner rejection can already be the beginning of a sustained active engagement with the views and methods of the other discipline. In the best case, this inner critical discussion leads not only to a better understanding of the matter, but also to a self-correction of the recipient—or even to the next step:

2.3 *The Intersubjective Level: Discussion and Joint Interdisciplinary Research*

A prime example of the step out of the subjective confrontation with the other may be the famous encounter between Schiller and Goethe, in which in the latter at first "the old resentment" rose against Schiller's principled thinking at the expense of the unbiased perception of the phenomena, before a breakthrough and exchange, in the further course a most fruitful synthesis of the knowledge attitudes took place.[5]

[5] According to Goethe in "Glückliches Ereignis," Werke Vol. 10, 541.

Only the actually intersubjective, not merely inner-subjective-concealed discussion leads beyond the respective subjective, purely internal confrontation, also beyond the mere, seemingly tolerant recognition of diversity of step 1.1. The decisive step here is that one recognizes the right of the other's point of view and thus the deficit in one's own point of view. However, this by no means always has to be abandoned. It can be corrected and supplemented. Interdisciplinarity in this productive sense seems to be practiced much less often today. It presupposes a truer and also more optimistic relationship to scientific truth and its recognizability, which was probably more present in the humanities at the time of the German classics than it is today.

2.4 The Transsubjective Level: The Medium (the In-Between) of a Transdisciplinary Thinking Space

The last discussed active, intersubjective interdisciplinarity can open a space of transdisciplinarity or presupposes it, that is: a *common knowledge space* beyond the different methods and subjective attitudes like talents. Without this precondition and the research for a common, albeit partly hidden truth, there can be no talk of transdisciplinarity. In my opinion one should stop to speak of transdisciplinarity, if only that pseudo-tolerant, basically revealing, and relativistic juxtaposition of 1.1 is meant! Externally, both forms, the mere knowledge of other access (1.1) and the certainty of a more or less hidden, common truth (1.4.), are often confusingly similar. Precisely for this reason it is all the more important to distinguish whether, despite all differences, one is researching together in a *space of truth*—or in a space of pseudo-tolerant indifference to scientific truth, in which the considerable efforts of passionate research are satirized by status poses or silenced by statements of power.

There are different paradigms of approach to truth, to speak with Thomas Kuhn, e.g. once the geocentric or heliocentric system. These are not always compatible, as in this example. Often they are compatible as different ways of approaching truth. Whether and how the paradigm of relativity (limitation of effects by the speed of light) is compatible with that of quantum physics (instantaneous effects and placelessness of entangled events) is still argued and researched today. The precondition is that the two paradigms must be compatible in a deeper truth, that both paradigms are presumably incomplete.

In truth, true transdisciplinarity in the aforementioned sense does not work without passion and sacrifice, not least in the renunciation of status posturing. (Einstein's outstretched tongue might indicate this.) Even the mere juxtaposition of formulated statements in collections of articles such as this one cannot so quickly generate the transdisciplinary spirit, namely *commonly articulated foundations, i.e. presuppositions of knowledge, and the resulting common questions,* even in the face of differences in the presupposed or articulated paradigms.

One could argue that an articulated commonality of foundations and questions already limits transdisciplinarity and that the previous levels of interdisciplinarity granted more freedom in the sense of undogmatic openness. This openness is indeed greatest at level 1 above: mere juxtaposition of contributions, disciplines, and methods. There is no denying that this openness has its appeal and temporary justification. Nevertheless, it must be admitted that this openness of juxtaposition is only a kind of collection of materials, a mere attempt at interdisciplinary understanding and debate, let alone transdisciplinary research. This attitude reminds me of Hegel's mockery of knowing before knowing, "whereby that which calls itself fear of error rather reveals itself as fear of truth."[6]. Regarding transdisciplinarity: the mere coexistence of disciplines for fear of dogmatic one-sidedness is rather a cult of the respective one-sidedness for fear of a deeper commonality. Transdisciplinarity thus requires more than mere "respectful" coexistence of approaches and methods. It requires *agreement on fundamental commonalities and* thus opens up a common research space.

The German Wikipedia states: "In the German-speaking world, transdisciplinarity mostly refers to a 'principle of integrative research.'" This designation contrasts with the view of transdisciplinarity as a "universal theoretical unity principle." In other words, an attitude in the research process rather than a theoretical commonality. In this way, an alternative is posited between a *pragmatic* or merely procedural commonality of research on the one hand, and a theoretical commonality on the other, while suspecting the latter as dogmatic. I would not object to a procedural commonality in the scientific search for truth, but I would object to a frivolous renunciation of theoretical commonality. In my opinion, such a renunciation nullifies the difference between interdisciplinarity and transdisciplinarity.

A definition of "transdisciplinarity" in contrast to "interdisciplinarity" is rarely found in the relevant literature. Rather, a "conceptual ambiguity of interdisciplinarity and transdisciplinarity" is noted and lamented.[7] An older work by R. Stichweh is cited there, wherein transdisciplinary terms are defined as those "which are located from the outset at a level at which their meaning does not refer to specific problems of individual disciplines."[8] As far as this aims not only at pragmatic (directly action-related), but also at theoretical commonality, this corresponds to the opening of a *common theoretical research space* claimed here, which *is at least potentially also formulable.* In my view, this task falls to philosophy as a theory of reflection in the sense already briefly sketched of a transcendental philosophy, which is continued as "Wissenschaftslehre" (J.G. Fichte), but not dogmatically, but critically in the Kantian sense and therefore capable of dialogue. I will come back to this.

---

[6] Hegel (1988: 3), Introduction.

[7] Bogner et al., (2010: 13).

[8] Stichweh (1994: 36 f).

3 The Stages of Social Reflection

The gradation of levels of reflection outlined in the foregoing finds its justification in stages of intersubjectivity as social (not only subjective) reflection. When we speak of interdisciplinarity and transdisciplinarity, it is not only about the object reference. Science is always at the same time a social process. Its theoretically objectifying reflection (which is often, but too superficially, called "reflection" alone) is embedded in that already mentioned implicitly lived reflexivity, both subjective and social. Even when a scientist is doing research very much alone in borderline cases, he or she is already in a social context through language and prior cultural knowledge. Levels of sociality can be identified in this process, even better and more clearly than those of subjectivity, of individual self-consciousness alone. I speak of *social reflection* as at once the most essential building principle of society: namely, how social commonality emerges from individual consciousness and action (how, conversely, the social shapes the individual). The stages of this lived, i.e. not only theoretical but practical, social reflection are of fundamental importance. Society as a whole emerges from the *reciprocal dynamic reflection of* acting individuals, i.e. self-conscious beings. This is neither a purely theoretical reflection nor a mere "assumption of perspective"[9] which again remains subjective, but a *practical reflection, i.e. one that changes the interpersonal situation.*

The four stages of this practical-social and at the same time ontological (lived) reflection have already been listed at the beginning of sect. 1. They can be reconstructed in more detail:

1. I can treat the other instrumentally, as, for example, a doctor treats his patient in relation to the body, or people in the economy interact with each other solely under the objective aspect of goods and money. Such a relationship is unreflective: I view the other as an object, or in the context of objectivity, in which all that is personal is initially blanked out.
2. I can strategically consider the other and his freedom for my own interests, as, e.g., road users expect the (in)attention of the other with respect to their own driving. Such a relationship is one-way-simply-reflective: I "reflect" on your freedom in terms of my purpose. I regard the other as someone who can see and act for himself, who can pursue goals.
3. We can take the freedom of the other, our expectations and desires seriously and respond to them communicatively, e.g. in friendship. Such a relationship is mutually doubly reflective: I look at the other as he sees me as the one who sees him. We swing into a reciprocity of the gaze.
4. I take a position on this factual reciprocity (of the gaze) (embarrassed, friendly, dismissive, etc.) and at the same time take a position on the other's position: renewed reflection of reciprocity or concluding reflection. If this metacommunication succeeds, we agree, mostly tacitly and implicitly, on

[9] Geulen (1982: 24–72).

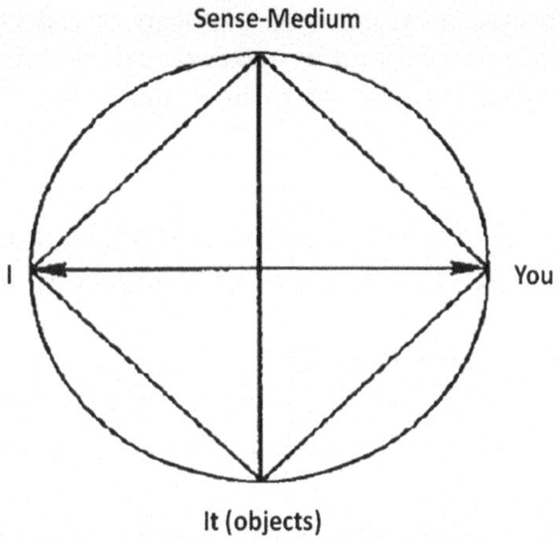

Fig. 1 The four elements of the human system of meaning and action

norms of behavior, in explicit form, e.g., quite elementary on temporal agreement. From the lived metacommunication on (which as a completed, lived reflection is much more than words about words!) it cannot structurally go further, but communication can be resumed under the new communication condition. Likewise, the other levels of reflection can be repeated and infinitely enriched with content.

These levels of reflection of social action, presented in more detail elsewhere, correspond to the following four relational poles of the human system of meaning and action, which I call "sense elements"[10] (Fig. 1):

Because obviously correspond

- the instrumental treatment (1) to the object in Fig. 1
- the strategic attitude and action (2) to the subjective subject (I),
- the communication (3) to the other objective subject (you).
- the metacommunication (4) to the common sense medium.[11]

This "crash course" of a social philosophy as well as of a philosophy of reflection *theory in general*, i.e. of such a reflection discipline, which starts from the lived self-reflection of the I in relation to the other elements, shall on the one hand justify the reflection stages of interdisciplinarity, which have been so harmlessly enumerated so far. On the other hand, however, it is to serve as preparation and justification for what has already been announced, for going beyond *intra-scientific* interdisciplinarity and transdisciplinarity into the general human forms of encounter. However, these are

[10] The expressions "instrumental, strategic, communicative" but not "metacommunicative" are inspired by Jürgen Habermas, who in turn was inspired by Max Weber. Habermas, however, from a philosophical and social theoretical point of view, lacks the reflexive logical clarification and gradation of those expressions. See my critical, unanswered open letter to Habermas (Heinrichs (2007: 469–482).

[11] Heinrichs (2005): passim

themselves understood as forms of reflection, so that a mediation of theory and practice, i.e. of forms of theoretical-objectifying reflection on the one hand and lived reflection on the other can occur.

4 Forms of Existentially Lived Reflection from the Perspective of Reflection Theory

4.1 The Big Picture

As the preparatory sessions for this publication have shown, two needs and currents of the participants could be clearly distinguished: One part of the group aimed far beyond intra-scientific and thus beyond conceptual thinking at the need and desire for interpersonal and universal understanding. Not only interpersonal love and friendship, but also the artistic and spiritual aspects and last but not least the relationship of human beings to nature were to be included in the transdisciplinary view and attitude toward life. Now, however, this completely justified desire for understanding beyond all "disciplinary" boundaries of the experience of being human cannot simply stand *next to the* equally justified postulate of scientific and intellectual discipline. The scientific-intellectual view must build bridges to the whole-human "transdisciplinarity" in a broader sense and vice versa, namely bridges that can be recognized as insightful and helpful from an existential (whole-human) point of view.

The short sketch in this third part is meant to serve this bridge-building. From my philosophical-reflection-theoretical point of view the general bridge consists in the realization that existential or *spontaneously lived reflection* on the one hand and *explicit-theoretical* reflection on the other hand are not alien to each other, but consist of the same "material," namely that *theory* wants to be nothing else than the "digital," i.e. objectivizing, representation of lived, existential reflection happening in pictorial object-images and linguistic signs. The latter is the active mirror in which all objectivity appears to us and with which we construct it. These two types of reflection cannot be "cleanly" separated, but under no circumstances should they be confused in principle and in linguistic usage, if one is not to indulge in a completely undisciplined, unstructured, and theoretically sterile sense of "unity."

In order to illustrate what "lived reflection" means as the content of objectifying reflection, the following thesis of the four levels of meaning or "semiotic" levels will now serve: Semiotics is the study of meaning processes by means of signs, which I understand not only as a single-scientific-technical, but as a philosophical discipline.

1. The level of human *action* and the corresponding, passively accompanying experience.
2. The level of *language*, which is essentially a self-regulating meta-action through signs.

Table 1 Classification of actions

1. Object-related action	*3. Social action*
1.1 Object change	3.1 Treating the Other as an Object
1.2 Movement action	3.2 Strategic action
1.3 Social object reference: work	3.3 Communicative action
1.4 Movement of goods	3.4 Norm-related action (social roles)
2. Intrasubjective action	*4. Expressive action*
2.1 Body-related action	4.1 Expression objects
2.2 Self-determination (decisions)	4.2 Facial expressions and gestures
2.3 Provisional decisions on social measures	4.3 Community expression
2.4 Sense designs	4.4 Sign action

3. The level of the *arts*, which are to be understood as meta-languages beyond everyday languages.
4. The level of *mystical-spiritual experience*, which can be understood as a kind of meta-art, in which the one-sidedness of active action is transcended to mere "active receptivity" to the universal medium of meaning.

This very comprehensive overall view of semiotic levels may seem bold. However, it is briefly outlined below, with references to more detailed accounts.[12] The view enables the *lived transdisciplinarity of human experience to be* supported *by a bridging* theory. As a result, the recurring dispute between holistic experience and what is perceived as fragmenting theory is to be settled. This is itself done on the side of a theory, albeit one that draws its legitimacy from our existential experience and must transcend the boundaries of the disciplines—without, however, abandoning them to irrationality.

4.2 *Sketch of the Theory of Action*

Human action differs from inner consciousness activity and also from diffuse outward activity in that a piece of the world is changed by it. Action is a change of the world, and the respective part of the world that is consciously changed leads to the main classification of action according to its respective intentions (Table 1):

Each of the main titles is further subdivided according to the principle of dialectical subsumption (i.e. repetition of the overall structure in the subdivisions). The great, also philosophical-historical importance of this type of subdivision by dialectical subsumption, which leads to a fractal subdivision, cannot be discussed here in more detail.

[12] Already summarized in Heinrichs (2014, chaps. 4–7).

4.3 Language Theoretical Sketch

The division is continued in sign action and intensified by the reflection theory of language as meta-action, i.e., as self-regulating action through its own, execution-immanent (i.e., syntactic) signs.

The so-called semiotic *dimensions of* language, first distinguished, albeit incompletely, by Charles W. Morris under this name[13], read in completed form, quite parallel to the levels of action mentioned:

- the sigmatic dimension (the primary object reference of speech signs)[14].
- the semantic dimension (subjectively internalized meanings).
- the pragmatic dimension (language as intersubjective action, speech act theory).
- the syntactic dimension (the self-regulating and system-building dimension).

4.4 Art Theoretical Sketch

The conception of the arts as lived meta-languages (in a different sense than linguistics distinguishes between object language and a meta-language discussing it!) can be understood according to the pattern of pantomime: Why does mime omit words? Or from the point of view of poetry and stylistics, which deals with the material of everyday language, but lifts this material into new stylistic contexts: Stylistics is a higher reflected syntax of language. However, poetry with all its genres as a whole already represents the third level of the arts, which can be divided into:

1. objective fine arts (e.g. painting, architecture).
2. subjective, performing arts (e.g. drama, dance).
3. poetry (language art).
4. music (based on syntactically designed space-time relationships of tones).

Of course there are mixed forms (e.g. between language arts, drama, and music) as well as many subdivisions. What counts in this context is merely the reflective-logical, higher rationality of the seemingly irrational of the arts.

The reflection-logical, higher rationality is closely connected with the unconscious in all its forms, which are essentially different: as physical unconscious, as psychic preconscious, as superconscious openness to the universal Logos, the medium of meaning.

[13] Morris (1975: 24 ff) and passim.

[14] The sigmatic dimension was rightly added to the three dimensions distinguished by Morris by the GDR philosopher Georg Klaus. See Heinrichs (2011).

4.5 *Mysticism as the Essence of Religions and Spiritualities*

The apparent irrationality of the arts as lived, performed, thus not only theoretical meta-languages is structurally transcended once more into a "meta-art." There, the one-sided subject control of human action is completely suspended in a reciprocity with the self-active medium of meaning. The empirically hardly deniable phenomenon of mysticism on its different levels can be understood reflectively and structurally as an activity of the medium of meaning itself. The subjective-human part ideally consists of pure receptivity. This is the semiotic and reflection-theoretical concept of mysticism, which is thus at least conceptually withdrawn from irrationality. Instead of delving into the rich history of religious (communal) and spiritual (more individual) manifestations of mysticism, only the levels of manifestation or expression of mystical experience are again differentiated structurally and generally according to the elements of meaning:[15]

1. "Objective" nature mysticism (emotion of natural phenomena)
2. "Subjective" inwardness mysticism (meditative contemplation)
3. Social mysticism (experience of the divine in the religious community).
4. Sign mysticism (interpretation and recording of signs as media of mystical experience)

This sketch of theoretical transdisciplinarity from the idea of lived reflexivity as self-relation-in-external-relation (self-relation-in-external-relation) wanted to make clear that philosophy as a "transcendental," overarching theory of reflection is able, at least in the field of the humanities or cultural sciences, to create that *common space of thought which* makes real transdisciplinarity possible in contrast to merely "pluralistic" interdisciplinarity of juxtaposition. Nothing is taken away from the individual disciplines by such a commonality of thought space. Of course, they need their own insights, methods, and principles. These are by no means simply deducible, at best *reconstructable*, in a dialogue of concept and experience.

On the whole, the gap between existentially lived reflection of man (in action and experience, language, art, and spiritual experience) and theoretical reflection remains an irrevocable feature of theory. The claims of the latter must not be ignored with pathos alien to theory and claiming "wholeness" of human activities and their respective theories. But theories themselves differ greatly in the extent to which they are able to close the gap between experience and theory in a life-enhancing way. Reflection-theoretical philosophy claims this to a great extent, i.e., to be able to put the two fundamentally different kinds of reflection (existentially lived and theoretically objectifying reflection) into fruitful relation.

[15] Heinrichs (2014, chap.4–7).

5 Outlook for Further Tasks

At this point, the aim was not to dive deeper into the many individual topics, but to grasp a transdisciplinary connection between philosophy as a born discipline of reflection and the many individual disciplines. Only as "transcendental philosophy" (Kant, Fichte, after the impulse of Descartes) it recognized itself as such, as a discipline of self-reflection, which was already with Kant the exact meaning of "transcendental" and "criticism," even if the latter did not yet grasp reflexivity as the *content of* the philosophical *form* of reflection. In all sciences dealing with the living, this part of lived, ontological reflexivity is undeniable. Not only the conscious life of humans and animals, but also the life of organisms in general is essentially based on self-reference and self-organization, in analogical gradation. Already the medieval thinker Thomas Aquinas recognized this analogically graded reflexivity[16] without immediately becoming a modern, post-Cartesian reflection theorist.

Reflexivity or inner self-reference is not only decisive for objectifying reflection in the humanities or cultural sciences. For these as well as for the foundations of the natural sciences, philosophy forms an indispensable theory of reflection. It stimulates empirical questions like the following: *By what is self-reflection, which undoubtedly constitutes the essence of human self-consciousness, physically made possible?* In my opinion, a *transcendental (reflection-logical) grid* can be developed from the sense elements listed above, which structures a common thinking space also for the primarily mathematical natural sciences. At the same time, it can enter into a close alliance with the still mythological element theory of Vedic philosophy[17]. The mythological thing about it is the literal reference to the ancient (also Greek) natural elements fire, water, air, earth, and ether. These are analogous to the sense elements listed here and their crossing point in the middle (see above graphic). M.E. one-sided "materialistic," today: "naturalistic" positions, as well as one-sided "idealistic" positions, which e.g. reduce everything material only to information (sense), can be avoided by this. By the way, Carl Friedrich von Weizsäcker also tried to avoid such one-sidedness with his "onto-epistemological"[18] interpretation of the natural sciences.[19]

6 Conclusions

Let us summarize: In the structure of the fourfoldness of the sense elements can be distinguished:

[16] Thomas Aquinas (2013: IV, 11).

[17] Schmieke (1995, 1997).

[18] Schüz (1986: 142, 183 f) characterizes it this way.

[19] Cf. Schüz (1986); Görnitz (2016).

1. materiality of the extended,
2. the individuality of the mental and animate, which is distinguished from it, due to reflexivity,
3. sociality as a product of reciprocal, double reflection as well as
4. the universal mediality of the spiritual, which is not to be confused with the soul-individual, but is presupposed to it as a condition of possibility.

This fourfold structure seems to structure all interpenetrating fields of experience and sciences as a transcendental (reflection-theoretical) fractal grid. Thus, it also serves as a *transdisciplinary theoretical framework* that is formulated here as a basis for further discussion.

Also the "consciousness of machines"[20], as cybernetics tries to capture it, is based on the imitation of reflexive relations. To what extent even the inorganic world of physics and chemistry is characterized by reflective structures analogous to human consciousness should be the content of further "transdisciplinary" investigations. How, for example, can the known four basic physical forces (strong and weak interaction, electromagnetism and gravitation) be reconstructed in terms of reflection-logic? Through which series of intermediate elements does the play of sense elements, which we assumed above, show up in them, too? To answer such questions, according to the present state of empirical knowledge, is only possible transdisciplinary and only with the help of a philosophy disciplined by reflection-logic and further developed on its part.

Hegel, who is underestimated in this respect, already thought in this direction of an ontological logicality of natural phenomena in his philosophy of nature[21], and he did so with greater logical precision than his childhood friend Schelling, who is generally regarded rather as a romanticizing natural philosopher[22].

In the meantime, however, a good two centuries of scientific research and technological development have taken place, which are unique in the history of mankind. In their disciplinary diversity, they require reflexive reappraisal, even unification. We do not do well either to look snobbishly at the earlier intuitions or to deal with them in a merely historicizing way. The meant unification makes, as said at the beginning, the difference of the transdisciplinary in the demanding sense to the mere interdisciplinary coexistence.

However, the old claim of philosophy as an integral universal science[23] par excellence must not simply be placed *above* that of the "individual sciences," as was the case before the separation of the individual sciences from the "philosophical faculty." The natural sciences as well as, e.g., psychology, sociology, semiotics, and linguistics claim today an "eye level" with philosophy. Yes, they merely look down

[20] Günther (1963).

[21] Wandschneider (2008).

[22] Schelling (1975).

[23] Scholastically formulated: "scientia universalis naturali ratione comparanda", in contrast to the self-understanding of ecclesiastical theology as "scientia universalis superrationali ratione comparanda." On this, for example, de Vries (1964: 12).

on philosophy, if it sees itself only as a history of itself rather than as a systematic basic discipline. It should therefore develop further into a strictly scientific, logical basic discipline.

Philosophy, as it is understood here as a theory of reflection, means a discipline of ontological structural reflection. Such a discipline opens a transdisciplinary space of dialogue with the other sciences. On closer examination, as humanities, they are not alone in being characterized not only by epistemological but also by ontological structures of reflexivity. As outlined in section 3, however, the ever-growing, possibly itself mutable and unfinishable insight into these universal structures cannot remain a merely intra-scientific matter. Rather, it concerns a large part of practice, of the reality of human action and experience. Philosophy is by no means practical only as prescriptive ethics moralizing the individual. Its structural insights are always at the same time insights into qualities of value.

References

Bogner, A., Kastenhofer, K., & Torgersen, H. (2010). *Inter- and transdisciplinarity in transition*. No.

de Vries, J. (1964). *Logica*. Barcinone. Friburgi, Brisgoviae, Romae: Herder.

Geulen, D. (Ed.). (1982). *Perspective-taking and social action. Texts on social-cognitive development*. Frankfurt/M.

Görnitz, T. (2016). *From quantum physics to consciousness–cosmos, mind, and matter*. Springer.

Günther, G. (1963). *The consciousness of machines. A metaphysics of cybernetics*. Agis.

Günther, G. (1976). *Beiträge zur Grundlegung einer operationsfähigen Dialektik* (Vol. 1). Meiner.

Hegel, G. W. F. (1988). *Phenomenologie des Geistes [EA Bamberg u. Würzburg 1807]*. Meiner.

Heinrichs, J. (2005). *Logic of the social*. Academia.

Heinrichs, J. (2007). *Actions. The periodic system of action types*. Academia.

Heinrichs, J. (2011). Sigmatics as a fourth semiotic dimension in Georg Klaus–a critical appraisal. *Journal of Semiotics, 33*(3–4), 237–225.

Heinrichs, J. (2014). *Integral Philosophy*. Academia.

Heinrichs, J. (2018). *Gelebte Reflexion. Writings on reflection systems theory*. Academia.

Heinrichs, J. (2020). *Dialectics. Dialectics beyond Hegel and Corona. Integral structural logic as Hegel's mission for a philosophy of the future*. Academia.

Kant, I. (1956). *Critique of pure reason [EA Riga 1781]*. Meiner.

Morris, C. W. (1975). *Foundations of sign theory. Aesthetics and the theory of signs*. Hanser.

Schelling, F. W. J. (1975). Ideen zu einer Philosophie der Natur, (EA Leipzig 1797). In *Ausgewählte Werke, Schriften von 1794–1798*. Wiss. Buchgesellschaft.

Schmieke, M. (1995). *Natural science and consciousness*. Synergia.

Schmieke, M. (1997). *The field of life–scientific foundations of a spiritual conception of life*. Tattva Viveka Schriftenreihe.

Schüz, M. (1986). *Die Einheit des Wirklichen–Carl Friedrich von Weizsäckers Denkweg*. Neske.

Stichweh, R. (1994). *Wissenschaft, Universität, Profession: Soziologische Analysen*. Suhrkamp.

Aquinas, T. (2013). *Summa contra gentiles, [first printed Strasbourg 1472], Lat.-dt. ed.* L. Schneider.

Wandschneider, D. (2008). *Philosophy of nature*. Buchners.

Heinrichs, Johannes, Prof. Dr., studied philosophy, theology, and social sciences at various German universities and in Paris. He lectured—initially as a Jesuit—philosophy in Frankfurt and Bonn, and Social Ecology in Berlin. He developed the social and political system of "Value-Levels-Democracy" (short English version Auroville 2019). His social theory of ontological reflection as well as his semiotics of the reflexive levels action, language, art and mystics are parts of a general reflection-system theory which is a creative up-to-date development of "German idealism", essentially understood as a multi-valued logic which departs from human self-reflection in relation to matter, to the other self-reflected beings and to the common Medium of Sense. He spent many months in India, esp. in the international city Auroville. An English summary of some 40 books by him appeared 2018 as "Integral Philosophy" (ibidem Stuttgart/Columbia University Press, New York), and his autobiographie in 2023. www.johannesheinrichs.de.

The Thought Pattern of Constellatory Self-unfolding and its Role for Understanding our Autogentic Universe

Albrecht v. Müller

"And the world bursts into singing, once you hit the magic word."[1]

1 The Essential Points in a Few Words

In the patterns of thought currently available to us in science and everyday life, a crucial conceptual figure is missing: that of "constellatory self-unfolding."

Comparable to the blooming of a flower, this conceptual figure describes how something can unfold out of itself, within itself, and toward itself.

The thesis posits that without this conceptual figure, we cannot comprehend the self-configuration of matter, life, and consciousness, nor can we grasp certain defining features of human existence, such as beauty, freedom, or goodness.

At the same time, thought patterns of constellatory self-unfolding might be crucial for mastering the challenges of our epoch, i.e., for shaping the further self-unfolding of our world.

The thesis is, thus, that this hitherto missing concept can enrich and deepen our understanding of the world, as well as our interaction with it—just as insinuated in the verse of Eichendorff cited in the title.

In order to clarify and elaborate this concept we need to unearth the "categorial apparatus" (pl.) that lie metaphorically speaking still one level below our most fundamental scientific theories. In a way one could characterize this as making a further philosophical step "with Kant beyond Kant."

[1] *Verses from: Joseph von Eichendorff (1788–1857): Wünschelrute* (Joseph von Eichendorffs Gedicht „Wünschelrute" (planetlyrik.de), *translated by the author.*

A. v. Müller (✉)
Parmenides Foundation, Pöcking, Germany
e-mail: avm@parmenides-foundation.org

© The Author(s), under exclusive license to Springer Nature Switzerland AG 2025
M. Schüz (ed.), *Transdisciplinary Thinking and Acting*, Transdisciplinary Management of Ecological and Social Crises,
https://doi.org/10.1007/978-3-031-70680-6_5

2 Painted with a Broad Brush

Our universe likely emerged from an initial singularity, the so-called "Big Bang." Thus, it brought itself into existence without external cause and, it is argued, this "arche" (in the dual ancient Greek sense of "origin" and "underlying principle") characterizes the unfolding of reality until today.

This means that self-constitution and self-unfolding are characteristic and cross-cutting aspects of all of reality and that this "autogenesis" continues until today.

Autogenesis is more than self-organization and also more than autopoiesis. In the former, a structure or even an entire system of structures forms more or less spontaneously. In the latter, existing material is organized in such a way that a self-maintaining and possibly even self-reproducing system emerges. Characteristic of autogenesis is that even the "material" on which this process draws and the framework in which it takes place arise only during the process itself. In this sense, all that comes into being actually unfolds out of, within, and toward itself.

In the case of our autogenetic universe, even space and time, within which its emergence and unfolding occur, are not a pre-given framework. They constitute themselves much more in the event and as a part of it.

It is argued that this character of the autogenetic permeates the entire reality of our universe, from the original singularity to today. What we call "reduction of the wave function" in quantum physics can be seen as a "footprint" of this ongoing self-constitution, especially also with regard to its characteristic "acausal" nature, i.e., the constitutive absence of external causes.

Because one didn't have the conceptual figure of constellatory self-unfolding available, it was extremely difficult to even name, let alone understand, what happens in quantum physics. Exactly the same applies to the subsequent self-constitution of the living, and as the third and fourth levels of this pyramid—the self-constitution of consciousness (as observable already in animals), and subsequently the emergence of explicit self-awareness or mind (that goes hand-in-hand with certain concept-based, self-referential thinking operations and the experience to exist within a present).

Thus, matter, life, consciousness, and mind can be seen as four forms or levels of constellatory self-unfolding that build upon each other or iteratively emerge from one another.

Once we have the thought pattern of constellatory self-unfolding at our disposal, descriptions of the four iterative unfolding stages no longer require separate ontological domains. Instead, a conceptually coherent description that could be characterized as a "modest ToE" becomes possible. (This "Theory of Everything" is "modest" in the sense that it explicitly points to its own incompleteness and does not claim to predict all future events.)

Obviously, traditional, causal accounts of reality do not become obsolete. They remain very valuable and powerful portraits of certain aspects of reality. But they become in a way "derivative aspects" of the taking place of reality and they are no longer seen as the most fundamental features of the unfolding of reality. Due to their precision and rigidity causal account has great advantages and science should continue to draw on them wherever possible. We should just give up the illusion that they constitute the only valid, respectively, the most fundamental framework of description.

Summing up: We find ourselves in a universe that essentially unfolds a) out of itself (thus without external causation), b) within itself (thus in a framework that is simultaneously emerging), and c) toward itself (thus toward an emergent self-perception, as present in human consciousness). For a brief elaboration of the conceptual prerequisites of this portrait of reality, see below paragraph 5.

But, obviously, the novel account doesn't end here. At least equally important as the theoretical considerations are the practical ones, i.e. in what all of the above arguments result regarding our interaction with an autogenetic universe. Also for this part of the argument, a quick, broad-brush preview can be provided.

The most sensible basic attitude in and toward such a universe could be characterized as "grateful attentiveness." Hand in hand with this comes an understanding of the world in which explainability and wonderfulness do not exclude each other, but even deepen mutually.

At the same time, with this new form of appreciating the world around us and the process of reality, we arrive at a fundamentally new paradigm of action. This could be described as "dialogic shaping."

Comparable to a potter, the agent repeatedly steps back and considers what has "resulted" in the dialogue between their own intervention and what has emerged. They then respond to it, and so on. . . Overall, this creates a kind of dialogic relation between the agent and the self-unfolding of reality that he or she tries to shape. It should just be mentioned that usually also the goals and ambitions of the agent continue to modify and further unfold during this dialogic interaction.

Overall, the addition of this pattern of thought of constellatory self-unfolding results in a fundamentally new form of being-in-the-world.

As the last point in this brief preview, I would like to mention the difference between the present state of human cognitive skills and what could be needed for surviving the present spiral of increasingly dangerous threats that modern civilization poses to itself. At present, we should characterize ourselves more appropriately as a "homo instrumentalis," and not yet a real "homo sapiens."

The instrumentalis is already very good in understanding causal relations and in local optimization. Due to the modular structure of modern science and technology, this has led to a self-accelerating increase of our instrumental power. But, at the same time the instrumentalis is still very weak in appreciating complex issues in their entirety and in acting with according responsibility. The instrumentalis is already very strong in (point-to-point) rationality, but still rather weak in reason (understood as appreciating entire, and in the above defined sense self-unfolding constellations).

If the above argument is valid holds water, learning to appreciate and shape processes the underlying dynamics of constellatory self-unfolding could be an important step on our way toward a real homo sapiens.

3 A Preliminary Working Definition of the Challenge and Its Solution

Something that unfolds itself maintains its identity; it remains itself, even though it simultaneously becomes richer and deeper. To imagine how this might fit together, the development of a personality might be a quite good metaphor: We continuously unfold, despite remaining the same person.

In ancient philosophy, there was once the conceptual figure of "causa sui," which pointed very much in the direction of self-constitution and self-unfolding. However, it was sorted out in modern science because it could not be thought consistently with the "cognitive frameworks" that had been available.

To be able to think consistently about the conceptual figure of constellatory self-unfolding, we must take a considerable epistemological detour. For example, we must no longer reduce time to its linear-sequential aspect. We need a second, complementary aspect of time, a non-local "time-space of the present." Only if we have this complement to the sequential schematization of time is there a place where this highly self-referential event can take place.

Strong self-referentiality, according to which something refers to itself as a whole, always bursts the bounds of a purely sequential description of time. The notion that something remains itself while becoming richer and deeper can only be thought if we have "temporal nonlocality" at our disposal (whose existence has now been conclusively demonstrated in quantum physics).

But, before we take a closer look at the prerequisites of the new conceptual figure, however, let us briefly address the causes and effects of the previous deficits.

4 The Emergence of the Deficits and Their Extent

From an evolutionary perspective, human thinking is an extremely recent development, akin to an early "beta release." Overall, we are probably still more of the already mentioned homo instrumentalis, adept at crafting, and far from a fully-fledged homo sapiens in the sense of true wisdom.

Characteristic of the homo instrumentalis is the ability to create very powerful artifacts of considerable complexity, which have greatly expanded their instrumental power, reaching from the splitting or fusion of atomic nuclei to the targeted manipulation of their own genetic codes.

While this form of "instrumental rationality" is already highly developed, the complementary capacity for reason—in the sense of the ability to perceive "a whole as a whole" (Carl Friedrich von Weizsäcker)—lags far behind. The modern homo instrumentalis thus resembles the sorcerer's apprentice from Goethe's poem of the same name. He is already capable of unleashing incredible forces, but often still lacks the ability to shape them in a meaningful way. (However, the unfortunate difference from the situation described in Goethe's poem is that in our case, there probably is no master waiting in the wings who could come to the aid of the overwhelmed apprentice.)

Either we manage a rapid further development of our cognitive abilities—that is, we succeed in bringing our capacity for reason and responsibility up to par with the already achieved instrumental power—or the so far very one-sided cognitive development of the homo instrumentalis could also lead very quickly to its self-destruction.

But let's first take a step back. In the proposed view of time and reality, facts are only the traces that the original event of reality leaves on the (simultaneously emerging) "canvas" of local space-time. Facts are precise, they can be strung like pearls along the linear-sequential aspect of time, and their emergence can be (at least in principle) causally reconstructed.

The phenomenon constellatory self-unfolding in the non-local time-space of the present is much more difficult to think about and describe in comparison.

Science has made very considerable progress in the last 200 years. At the same time, however, it must also be noted that we have not come all that much further in the really fundamental questions. There is still an unbridgeable gap between classical-relativistic physics and quantum physics. We still cannot precisely explain what makes the difference between inanimate matter and the phenomenon of the living. And we still cannot explain how consciousness emerges from matter and life, and how from consciousness than emerge the phenomena of explicit self- awareness and mind.

According to the thesis presented here, an answer to all these processes and interconnections can only be found when we expand the thesaurus of our basic thought patterns by the conceptual figure of constellatory self-unfolding.

However, this is not so simple, for it only becomes possible in a consistent form once one has uncovered the existence and significance of what could be called "categorial apparatus" (pl.). This refers to a level of cognitive processing of reality that lies, so to speak, "beneath" all our substantive scientific theories.

5 The Prerequisites of the New Conceptual Framework

Immanuel Kant pointed out that space and time are fundamental forms of appreciating reality, without which we could not meaningfully recognize or think anything. The somewhat exaggerated conclusion drawn from this observation—that space and time, therefore, have nothing to do with "actual reality"—has been rightly contested. The core argument is that our cognition is also a product of an evolutionary process and that these fundamental forms of intuition would hardly have survived if they had no fit with the underlying reality.

Simplified: A monkey that, due to a visual impairment, always sees branches in the wrong place would quickly be a dead monkey—and thus could not reproduce. Massively dysfunctional features of our cognitive apparatus would be quickly sorted out in the course of evolution.

The thesis presented here is that our appreciation of reality is not only based on space and time, or as we would say today, the "local space-time," but that "down there", beneath our content-based thought processes and theories, an entire "categorial apparatus" (pl.) at work. (Please note in this context the important difference between "categorial" and "categorical." The latter refers to a most fundamental norm, like in Immanuel Kant's categorical imperative. "Categorial", instead, refers to categories in the sense of most basic structures in our appreciation of reality and time.

A categorial apparatus consists of 4 mutually presupposing and functionally complementary components. These are:

1. The structure of a predication space, or a specific logic,
2. a basic pattern of how reality occurs and how events can be connected,
3. a specific aspect of time, and finally,
4. an epistemological basic structure, i.e., a certain relationship between observer and observed object, or between subject and object.

In the categorial apparatus underlying our typical, fact-oriented everyday consciousness and classical physics, these four slots are filled with the following patterns of thought:

1. Binary or Boolean logic.
2. The thought pattern of linear causality,
3. The linear-sequential aspect of time,
4. The epistemological scheme of a comprehensive separability of observer and observed object (what is referred to in physics as "Einstein objectivity," as laid out by Albert Einstein in the introduction to the famous EPR paper).

These four components form an "apparatus" in the sense that they presuppose, condition, and support each other. For example, linearly directed causality can only be consistently thought of if one assumes a linear-sequential (i.e. a neither looped nor non-local) structure of time. The same applies to all 12 bilateral relationships between the four components of the first or "classical" categorical apparatus. This corresponds to or constitutes the "classical" or "factual aspect of reality" that is so familiar to us and therefore hardly perceived in its specificity.

This aspect of the reality process is extremely important and powerful, but it is not the only aspect we need for a comprehensive appreciation of reality. For this, a second fundamental aspect (or "portrait") of time and reality is required. For this we have introduced the term "statu-nascendi aspect." This aspect or portrait corresponds to a second, complementary categorical apparatus.

This second categorial apparatus is needed for addressing the original event of reality, the actual or primordial (i.e. pre-factual) "taking place of reality—from which facts emerge as the traces that it leaves on the canvas of local space-time. The latter is co-emergent with the factual shape of reality and time. Without facts (that persist within it) no local space-time, and without the rigid framework of local spacetime non facts.

The statu-nascendi aspect deals with the ongoing self-constitution of reality. It describes reality as it actually unfolds itself—still in the non-local time-space of the present. Only one it has "taken (a) place," it has become a fact and has, hand- in-hand with this, become part of local space-time.

The temporal correlate of this reality that is just happening is the already mentioned non-local time-space of the present. Here there is not yet clearly separable in a "before" or "after"—and accordingly, there is no causality available. Nor are bivalent logic or the clear distinction between subject and object applicable. The second categorial apparatus consists also of four fully interdependent components that fill the same four niches: An aspect of time, a mode how reality materializes, the structure of a predication space, and a basic epistemological setting.

The four constituents of this second categorial apparatus are:

1. The non-local time-space of the present, which forms the temporal platform on which the primordial taking place of reality can occur,
2. the phenomenon of constellatory self-unfolding as the modality in which reality emerges here,
3. a paratactic predication space respectively a constellatory logic (in which the overall meaning only emerges from the configuration of all individual statements; such a predication space can tolerate contradictory statements because there don't exist compelling conclusions from which the well- known "ex falso quodlibet catastrophe" of Boolean predication spaces would result),
4. the basic structure of strong self-referentiality, according to which something can refer to itself in its entirety (and thus transcending any clear-cut separability of subject and object).

Readers who have already dealt with quantum physics may have noticed that all the properties of reality mentioned here occur in quantum physics in one form or another. There they constitute the much-lamented "incomprehensibility" of quantum physics.

This apparent incomprehensibility arises as soon as one—not yet knowing about the existence and significance of two different categorical apparatus (pl.)—quite naturally applies the classical categorial apparatus when appreciating and interpreting the phenomena of quantum physics.

And it is exactly this "categorial mistake" which is at the roots of the apparent incomprehensibility of quantum physics: One tries to describe the primordial state of reality (in the sense of "prior" to the orderedness that comes with the availability of local space-time) already within the categorial framework that applies only to the factual aspect of reality, i.e. when the original taking place of reality has settled down to facts. (In order to make this unfamiliar, but highly important distinction a bit clearer, one might also use for the prior state of reality the notion of "preality.") As a brief outlook, it should be noted that this new perspective also opens up to a novel interpretation of the relationship between classico-relativistic physics and quantum physics: The former describes the factual aspect of reality, while the second deals with the primordial self-unfolding and taking place of reality.

Both aspects of reality are valid, they cannot be reduced to one another, and all attempts to subordinate one to the other (e.g., the search for "hidden variables" or the quantization of gravity) are doomed to fail from the outset.

The fundamental complementarity of the two descriptions of reality claimed here is even evident already in their current formulation: The singularities of general relativity are the place where the background structure of local space-time (driven by the self-referentiality of gravity!) melts back into the statu nascendi, respectively "preality." Conversely, the reduction of the wave function marks the point where preality, i.e. reality in statu-nascendi transitions into the factual reality mode, describable in the rigid framework of local space-time.

Within the novel conceptual framework outlined here, the singularities of general relativity and the reduction of the wave function in quantum physics can be interpreted as *inverse transitions between the two complementary aspects of reality*.

In closing this part of the argumentation, I would like to emphasize once again that adding the thought pattern of constellatory self-unfolding only becomes consistently possible if we (a) uncover the existence and significance of categorical apparatus (pl.) and then realize (b) that there are two such, each internally consistent categorical apparatus (pl.), each corresponding to an independent aspect of reality.

6 Hat Becomes Possible Via the New Pattern of Thought: Beyond Science

Expanding our thesaurus of basic thought patterns resembles a cornucopia: There is an abundance of new possibilities for appreciating and addressing, conceptualizing, and explaining hitherto intractable phenomena.

In the sixth and final section of this brief essay, I will give a few examples for additional contexts in which the thought pattern of constellatory unfolding can be useful. Obviously, this can only be very superficial, but may be it can serve in some cases as a trigger for further, independent thinking on practical applications.

- The first example concerns our self-understanding as human beings in an autogenetic universe. Within the mentioned ToE, humans can be seen as the place (or at least as one place), where our universe opens its eyes and begins to become aware of itself. With this step, however, the entire universe acquires a fundamentally novel quality and dignity. To clarify this claim, one could imagine two universes—one in which this step occurs and one in which it doesn't. While the latter would probably be perceived as a rather dull, purely mechanistic event, self-awareness leads to a fundamentally unending cascade of infinitely richer and deeper new perspectives. This deep and wonderful transformation occurs in the self-perception of humans—and it happens anew in each individual human being. The emergence of this fundamentally novel quality and dignity of the entire universe in every human being is the reason for the infinite, inviolable, and incommensurable dignity of each single human being.

- With this step in our argumentation, however, the concept of human dignity is finally also removed from the whims of political despots. The notion of dignity of humans—as the place of the beginning of self-awareness of our entire universe—becomes an integral part of the most potent scientific explanation of the world to date. To not recognize this dignity, or to consciously disregard it, thus turns out to be as inappropriate or arbitrarily foolish as the assertion that 2 + 2 equals 5.
- Beyond expanding our understanding of matter, life, and mind, the thought pattern of constellatory self-unfolding also opens up novel perspectives for how to organize societies or the international order. It can be argued that rule of law, democracy, and probably even a social market economy are the organizational patterns in which the self-unfolding of humans is supported best. In this vain, the idea of a "co-evolutionary society" could serve as a novel leitmotiv, that combines individual freedom, fair participation for all societal groupings with the ambition to leave nobody behind. On the international level, the leitmotiv of a "participatory global domestic policy" could be the one that best supports the goal of optimally supporting the further constellatory self-unfolding of mankind.
- Another domain in which the novel thought pattern might be useful pertains to our experience of the world or our being-in-the-world. Here, the focus could be on the authentic experience of the ongoing self-unfolding of reality, and a kind of profound joy (in the sense of Eudaimonia) about it. Consciously experiencing the "concertus mundi" and the resulting tenor of thankful attentiveness might be one of the best ways of achieving wisdom. And, typically for this pre-factual aspect, the story can also be told in reverse: Wisdom might be, what allows us to appreciate the concertus mundi in the tenor of thankful attentiveness.
- Even more generally speaking, everything that makes life worth living, such as beauty, love, freedom, and meaning, is in one way or the other characterized by the phenomenon of constellatory self-unfolding—and can therefore be thought and experienced much more deeply and clearly if we have this figure of thought at our disposal.
- With this, I would like to turn again to the new paradigm of acting that has already been mentioned. In a self-unfolding world, the attempt to influence the course of events through purely mechanistic steering is at least suboptimal, often even doomed to failure. What matters much more is the dialogic shaping of the further self-unfolding, which is quite comparable to the activity of a gardener or the potter, also already mentioned. At its core, it is about a "dialogic shaping," in which what emerges from itself here and what is intended by the agent enter a dialogue—and within which the intended can also continue to develop further. This action paradigm of dialogic shaping corresponds to the basic attitude of grateful mindfulness in and toward an autogenetic universe. Both together lead to far better results in a self-unfolding world than attempts at purely mechanistic steering according to preconceived objectives.
- In this context, I would also like to mention the convergence of the ethically required with what is in the long-term interest of the actors themselves, which presumably exists only in an autogenetic universe. As an example, consider the successful upbringing of children, or more precisely, the careful and mindful

support of the unfolding of their possibilities and the self-unfolding of their personality. This naturally serves the children—but also the parents themselves, who thereby receive the gift of dealing with wonderful people in their children. By this convergence, it becomes clear that all forms of repression and tyranny ultimately lead to a structural impoverishment of the world of the oppressors. Obviously, despotism is not yet overcome by this observation. But at least it becomes recognizable as a pattern of behavior that massively also impairs the chances of the despots, or any other suppressors of human unfolding, to lead a life worth living.

- And it is precisely this effect that can be generalized in the sense of a new ethic of constellatory self-unfolding: By focusing the actors on promoting the successful self-unfolding of their counterpart, their own world immediately becomes richer and deeper.
- An overemphasis on power, possession, and control is the driving force behind the spiraling self-threats of modern civilization. This thought trap is caused by an overemphasis on the factual aspect of reality. When we reduce the reality to its factual aspect, zero-sum games become dominant in all social contexts. What a crucial magnanimity, kindness, generosity also contribute to a successful life for the actor themselves probably only becomes apparent only when we have the thought pattern of constellatory self-unfolding at our disposal. The good news in that is that within the conceptual framework outlined here, overcoming the ruthless pursuit of power, possession, and control no longer represents mainly an altruistic act: It lies in the well-understood self-interest of the actors themselves.
- Last but not least there seems to exist the possibility to formulate a modified categorial imperative for an autogenetic universe: "Act in such a way that you support the further self-unfolding of the world at every moment and to the best of your abilities!"

In concluding this little invitation to expand our thesaurus of thought patterns by constellatory self-unfolding, I would like to offer four summarizing theses for discussion:

I. Due to the so far very one-sided development of his cognitive abilities, today's homo instrumentalis has entered a spiral of increasing threats that modern civilization poses for itself. This one-sidedness affects both the theoretical description and understanding of our world, as well as our practical interactions with it.

II. In order to overcome this spiral we must complement our already well-developed abilities in the area of analytico-instrumental rationality and local optimization with a corresponding development of our capacity for synoptic, respectively, constellatory reason and responsibility.

III. We will only survive the explosive expansion of our instrumental power if we succeed in making a significant step toward a true homo sapiens that is capable of also appreciating the ongoing self-unfolding of reality.

IV. The addition of the thought pattern of constellatory self-unfolding to our repertoire of ways to describe and understand the world around us and ourselves could be an important step toward a real homo sapiens—and hand-in-hand with this, it should make our own lives richer and deeper.

Müller, Albrecht v., Prof. Dr., is founder of the Parmenides Foundation and director of the Parmenides Center for the Study of Thinking His two main fields of interest are the concept of time and the phenomenon of thinking. After a PhD on "Time and Logic" at the LMU in 1982 he worked for many years at the Max Planck Society and taught philosophy at the LMU in Munich, altogether for more than three decades. Still in the 80ies he got involved in international security and arms control, was director of EUCIS (European Center for International Security), and served as co-chairman of the Pugwash Study Group on Conventional Arms Control. In 1983 he developed the theory of structural defense dominance as a novel paradigm for conventional arms control which later became a leitmotiv for the CFE treaty. In order to support complex thinking and decision-making processes, von Müller developed the visual reasoning methodology "Eidos" with which he served as scientific advisor to several governments, supranational institutions and large corporations. After a brief excursion into the world of business at the end of the 90ies, he established in 2001 the Parmenides Foundation as a new interdisciplinary research center with the task "to study the foundations and frontiers of human thinking". Together with other faculty members of the foundation he developed and teaches the MCA (Magister Cogitationis Artium) and its executive version, the CEP (Cognitive Excellence Program). Both offer an alternative to traditional MBAs for times of exploding complexity and an always faster pace of change. Under these novel conditions which standard recipes no longer work and outstanding authentic thinking become the essential success factor. Von Müller was external member of two multi-disciplinary research centers at the University of Munich, the Human Science Center and the Munich Center for Neuroscience. He served as member of the Board of Trustees of the Max Planck Institutes of Neurobiology and Biochemistry. Together with Ernst Pöppel he edits the Springer book series "On Thinking ". For further information see: www.parmenides-foundation.org.

Coherence as a Central Concept of a Transdisciplinary Dialogue

Marcus Schmieke

1 Introduction

A large part of the ecological problems on our planet have arisen from the use of unilateral technologies whose negative effects on humans and nature were not recognized in time. Technology is generally based on disciplinary organized science, which due to the tunnel vision of specialized disciplines cannot grasp the impact on the whole. If the negative effects of a specialized technology are to be counteracted, one-sided specialized technologies and fields of knowledge are generally used, which also do not have a view of the whole. This is where the transdisciplinary approach comes in, which on the one hand tries to establish the unity of knowledge and its technological application (mode 1), and on the other hand wants to bring transdisciplinary science to bear in society in cooperation with the various stakeholders (mode 2). In contrast to the separately acting disciplines, which generally draw a fragmented picture of reality, transdisciplinary science strives for a holistic description of the real. The discovery of coherence as a central property of the living in several individual disciplines represents an opportunity to arrive at a holistic science of life that removes the boundaries of the disciplines via an understanding of this property of life and its mechanisms. The further destruction of life by the analytical sciences is to be counteracted by a transdisciplinary approach based on coherence.

M. Schmieke (✉)
ECR Institute, Deetz, Germany

© The Author(s), under exclusive license to Springer Nature Switzerland AG 2025
M. Schüz (ed.), *Transdisciplinary Thinking and Acting*, Transdisciplinary Management of Ecological and Social Crises,
https://doi.org/10.1007/978-3-031-70680-6_6

2 The Two-Stage Objective of Transdisciplinarity

Two modes can be distinguished in transdisciplinarity. In the first step (mode 1), the aim is to bring about the unity of multiple scientific disciplines,[1] so that in the second step (mode 2), this unity of knowledge can be brought into an overall societal process to solve central problems and challenges in collaboration with the various stakeholders.[2] In this science–society collaboration, transdisciplinary science is characterized by the fact that it is ideally committed only to the truth and the common good, while the various social groups, parties, and stakeholders have multiple objectives. In the second step, science is about taking a leading role and aligning the multiple interests to solve the respective problems.

3 The Bohmian Dialogue

To this end, it is necessary to establish a genuine dialogue between the disciplines involved and between the social groups involved. The term dialogue is used by physicist David Bohm in this context in its original meaning of letting the logos or sense shine through.[3] Meaning here would be understood as the common objective of finding a sustainable solution to the problems or paradoxes in question. Bohm distinguishes between problems that can be solved at the practical level and the underlying paradoxes that exist at the level of each party's deeper underlying assumptions and beliefs and lead to the conflicts on the outside. These paradoxes are generally not solvable at the practical level, but require deeper insight and inner clarification among the groups or individuals involved. According to Bohm, it is hidden basic assumptions and paradoxes that lead to the fact that the coherence of the social system is disturbed and thus its meaning cannot flow freely. In the particular case, the various parties work outwardly to solve the recognized problem, but due to divergent basic assumptions they do not arrive at uniform answers or solutions to many practical questions, so that no purposeful joint action emerges. In such a case, the objective of the transdisciplinary process is to establish the coherence of the system or to create the most comprehensive coherence possible.

4 Transdisciplinary Definition of Coherence

Coherence is generally defined as the cohesion of interrelated systems or parts of systems. In linguistics, coherence is the meaning-giving connection of individual words.[4] In physics, coherence is a property of waves with respect to their

[1] Piaget (1972).

[2] Scholz and Steiner (2015).

[3] Bohm (2013).

[4] Sanders and Pander Maat (2006).

interference ability. For this to occur, the respective waves must match in both frequency and phase to allow constructive interference, i.e., mutual amplification. In quantum physics, coherence is a basic condition for quantum entanglement. It is lost when a quantum system is perturbed by unmanageable influences such as thermal noise. In quantum physics, one speaks of decoherence when such disturbances dissolve existing quantum entanglements and destroy the unity of a quantum system. In cell biology, one also speaks of coherence when cell contacts lead to unified cell communication.[5] In foreign policy, there is the so-called coherence imperative, which is that the actions of individual nations should serve the common goal of a federation such as the European Union. A society is considered coherent in a social sense if the different individuals and groups that are bound together in it have a common understanding of values and contribute to the social context in their actions.[6]

Taken together, coherence can be defined as the connection of a system given by its internal dynamics, which leads to a coordinated behavior of the individual parts of the system based on common properties. Here, the whole shapes the coordinated behavior of its parts without the parts having to interact with each other. If all parts of a whole possess independently of each other the same information, by which their movement is coordinated, then they can behave coordinated on each other, without having to exchange mutually information, energy, or forces. Coherence is a necessary property of transdisciplinary dialogue, defined and necessary both globally and locally for each subsystem. Here, dialogue is both the result and the path of a transdisciplinary process.

5 Dialogue and Paradoxes in the Natural Sciences

The dialogue of the sciences, in particular the natural sciences or exact sciences, is leading the way here, since at present the entire social discourse is strongly oriented toward the findings and the world view of the natural sciences. In the natural sciences, there has been a comprehensive profound paradigm shift in the past 120 years, caused by the parallel discovery of the theories of relativity and quantum physics. In the resulting new natural science, the notion of coherence plays a particularly important role, as the insights of quantum theory, and in particular quantum field theory, make the importance of coherence essential to the emergence of more complex systems, and especially living systems.[7] In the biological and medical sciences, however, the human body is for the most part still viewed and described entirely from the perspective of classical physics. In this sense, the natural sciences themselves have not yet coalesced as a coherent system. Already in physics,

[5]Lloyd (2011).

[6]Manca (2014).

[7]Lloyd (2011).

general relativity and quantum theory confront each other as incompatible and even contradictory theories.[8]

Nobel Prize-winning physicist Roger Penrose, like many others, sees the solution to this problem as related to the question of consciousness in the natural sciences.[9] He assumes that the relationship between physical processes and consciousness will be clarified if a unification of quantum theory and relativity is achieved. This is another expression of the principle of coherence. As long as the two most important theories for the description of the material are contradictory to each other, they cannot reflect in their totality the unity and coherence of consciousness, because unity is the most important property of consciousness. Only when the scientific theories describing the material are coherent with each other can they function as the material equivalent of consciousness, since consciousness itself, according to Tononi, is an expression of maximum integration with maximum differentiation.[10]

6 The Implicit Order Behind the Disciplines

Unifying different scientific theories, especially when they are in contradictory relation to each other, can only succeed if the implicit assumptions (axioms) behind the theories are uncovered and connected and harmonized. The same is true if one wants to establish coherence between different scientific disciplines. Such an endeavor can likewise only succeed if one does not work at the level of explicit expression of the individual disciplines, but harmonizes the assumptions and basic premises behind them.[11]

A level change from the explicit of the individual disciplines to the implicit is required, again to stay in the language of David Bohm. Bohm postulates an implicate order that stands behind the explicate one and thus has not been objectified by the disciplinary process. Each individual scientific discipline has its own method of explicating and thus objectifying certain parameters and properties of the real. In doing so, an implicit remainder, which cannot be objectified by the applied method, remains, which Bohm calls the implicate order. This implicit remainder must be thought of as more comprehensive than the objectified reality itself.

Since all disciplines presuppose such a remainder, one can infer an implicit unitary reality consisting of the common remainder of all possible perspectives of objectification and out of which all disciplines objectify. It is to this unitary reality that the perspective of transdisciplinarity in mode 1 applies. In order to establish coherence between disciplines, there is a need for a dialogue between them that becomes aware of the methods of objectification used, reflects on them itself, and thus makes them aware of the assumptions behind them. Here, the dialogue itself is

[8] Maldacena and Susskind (2013).

[9] Hameroff and Penrose (2014).

[10] Tononi et al. (2016).

[11] Nicolescu (2010).

an expression of the effort to achieve coherence. It is intended to reveal the common meaning behind the disciplines and to lead to additional coherent insights into the underlying one reality.

An example of at least a partially successful dialogue between disciplines is biophysics, which can be understood as the application of physics to the matter of living systems. For example, biophysics has discovered that cell assemblies emit ultra-weak coherent electromagnetic radiation, called biophotons,[12] which may serve inter- and intracellular communication. For decades, physicists have demonstrated both theoretically and experimentally coherent behavior in living systems, suggesting the applicability of quantum theory and quantum field theory in biology.[13] On the other hand, the importance of quantum fields and macroscopic coherent behavior finds relatively little acceptance among biologists, since within biology the molecular paradigm still largely prevails, which considers biochemical and molecular interactions to be determinant and believes that quantum effects can largely be neglected in biology.[14] In this, the biologists are even confirmed on the part of a large number of quantum physicists who consider that the coherence of quantum effects is lost at the high temperature of biological systems due to Brownian molecular motion even at the smallest distances and in the shortest time (decoherence). However, the application of quantum field theory to living systems predicts exactly this coherent behavior also in macroscopic systems, if the basic conditions are fulfilled. Here, an open dialogue even within physics would lead to a better result. An example of such a transdisciplinary dialogue was the summer school of the International Institute of Biophysics in Neuss, which annually brought together representatives of a wide variety of natural sciences to come closer to describing the matter of living systems.[15]

7 Paradigm Shift in the Natural Sciences

Within the natural sciences themselves, coherence is an important property, especially when it comes to describing the living, as in biology. While quantum theory describes the coherence of simple physical systems primarily at the microscopic level, quantum field theory can be used to derive a comprehensive coherence of complex macroscopic systems, in which long-range correlations also occur.[16] This refers to correlations within physical systems whose range extends far beyond the range of individual interactions and thus cannot have arisen from the interaction of its parts. Understanding coherent effects and systems at this level is of particular importance because it refutes reductionist concepts that assume that a description of

[12] cf. Popp et al. (1992).

[13] Beloussov et al. (2007).

[14] McFadden and Al-Khalili (2018).

[15] Bischof (2008a).

[16] Vitiello (2012).

the living is possible simply by understanding the interactions of the smallest building blocks of life, atoms and molecules at the biochemical and biophysical level. This molecular paradigm of biology is so prevalent that many natural scientists still currently assume that the effects described by quantum theory are negligible in the human body and that it is therefore sufficient to describe molecular interactions.[17] This paradigm assumes that a sufficiently complex interaction of atoms and molecules can give rise to living systems. The indivisibility of the life process as a unified coherent system finds no place in this.

8 The Unity Reality Behind the Life Processes

The paradigm of coherence is based on the indivisibility of living systems. Every approach, which divides the unified life process into parts, loses essential aspects, which make up life. Already the division of the human being into body and psyche ignores essential characteristics of the system, which consequently can express themselves in unsolvable paradoxes.[18]

David Bohm assumes, as already mentioned, that the explicate order of the physical is based on a deeper implicate order, on which physical and psychic phenomena are not separated from each other.[19] C.G. Jung and Wolfgang Pauli also postulated such a unified reality in which the physical and the psyche must be thought together.[20] The strict separation of these realms was already formulated by Rene Descartes as the opposition of *res extensa* and *res cogitans*, which has shaped the thinking of Western philosophy and science ever since. The hereby lost unity of these two realms leads to paradoxes. The *hard problem* formulated by Chalmers, namely how a material system like the brain can produce subjective experiences, is an example of such a paradox, since naturally in a purely two-valued system of thought like Aristotelian logic, the interaction between the two levels as the *tertium non datur* has no place.[21]

9 Coherence in Quantum Field Theory

The unitary reality described by Pauli and Jung as well as by many other scientists is not amenable to direct formalization. Here it is hypothesized that macroscopically coherent behavior in living systems is the bridge between the fragmented material level and the implicate order described by Bohm behind it, in which the physical and the mental are not yet distinguishable from each other. Such macroscopic coherent

[17] Fleming et al. (2011).
[18] Mansfield and Marvin Spiegelman (1996).
[19] Bohm (2002).
[20] Atmanspacher (2020).
[21] Chalmers (2020).

processes, which can encompass larger spatial domains, are predicted by quantum field theory when applied to the matter underlying life. Quantum field theory arises from what is known as second quantization, in which interacting particles, as well as their creation and annihilation, can also be described. The quantum field theories arising from this are able to describe and formally represent the coherence effects lost in a pure particle description.[22] In quantum field theory, macroscopic systems such as cells or even larger biological entities can be described as coherent systems characterized by long-range internal correlations that appear as non-local connections. Dissipative systems, i.e., systems that exchange energy and information with their environment, generate a double in quantum vacuum in this way, which contains the informational cohesion and internal dynamics of the system in the form of complex field quantum condensations.[23] These so-called Bose-Einstein condensates of field interaction quanta contain the unity of the system as a coherently acting organism. Since these condensates are states of lowest energy (vacuum states) their nature is informational, which is consistent with the insight of quantum physics that the essence of nature consists of an interplay of energy and information.[24] Whenever living systems occur, they appear as complex macroscopic coherent systems whose internal cohesion (coherence) is due to dynamics at a deeper informational level of the quantum vacuum. The appearance of overarching coherence in material systems is therefore always an expression of a deeper dynamics that is self-referential and fractal, like properties characteristic of living systems.[25] Coherence is in this way an expression of the unity of the living and at the same time an indication of the connection of the energetic-physical to the informative dynamic holistic happening in the background. These complex biophysical connections can be clarified by a simple example. If a group of people meets, for example, for a scientific symposium in a certain location, it is not the interactions of the participants, such as the exchange of emails or appointments or even an influence by forces between the participants, which led to the meeting, but it is the organization of a higher level, which planned the meeting and, for example, by sending an email to all participants and communicated the place and time of the symposium. This organizational level works with the simultaneous dissemination of information to all parts of the system (participants), so that the participants themselves do not have to interact directly with each other to meet in one place. In quantum field theory, this level represents the dynamical level where organizational quanta of lowest energy (which are vacuum states) control the behavior of molecules or other parts of the system. The parts, such as the water molecules, therefore do not have to interact directly with each other and exchange local forces to vibrate in coherence with each other.

---

[22] Blasone et al. (2011).

[23] Freeman and Vitiello (2008).

[24] Barzegar et al. (2021).

[25] VITIELLO (2009).

10 Life, Coherence, and Consciousness

In its development, life strives collectively as well as individually for the highest possible coherence. Collectively, for example, coherent flight behavior can be observed in migratory birds, whose formations appear to be designed from a central whole.[26] Individually, coherence can be detected in ultra-weak cellular radiation, but also in physiological functions such as heartbeat, respiration, and brain waves. Among themselves, these functions also exhibit coordinated coherent behavioral patterns.[27] In the organizational form of life, coherence means the ability to integrate and control complex systems and to achieve maximum integration in the presence of high diversity or diversification. These are precisely the properties that a neural system must possess in order to have consciousness according to Tononis theory of integrated information IIT.[28] Consciousness appears as a goal of life that it achieves when it reaches high complexity with self-reference that is both mapped into internal states and leads to a mapping of the process of consciousness into itself, which can be labeled self-consciousness.

Complex, self-referential, fractal, and coherent systems thus form the interface between self-consciousness on the one hand and the living organism on the other. At the same time, self-consciousness ties the living being to a collective space that it shares with other living beings. Man creates his language and culture in this comprehensive mental space, which Heinrichs calls the medium of meaning.[29] The emergence of language and culture is in turn already the expression of a comprehensive coherence achieved in the social exchange between humans. Meaning or sense is the coherent expression of people's collective reflective achievements.

In the transdisciplinary process, the manifold disciplines of contemporary science now strive for coherence by discovering the hidden meaning and order behind it, thus in turn enriching and expanding the disciplinary process. At the same time, the transdisciplinary science that emerges from this process strives for a coherent dialogue with social groups and stakeholders, aligned with the goal of solving the major local and global problems and paradoxes of the present. This dialogue is also about finding the common hidden meaning-giving order that has the power to harmonize the different interests and approaches.

[26] Portugal et al. (2014).

[27] Bischof (2008b).

[28] Tononi and Balduzzi (2007).

[29] Heinrichs (2021).

11 The Logical Structure of Coherent Processes

The emergence of complex macroscopic coherent processes in material systems requires three stages of reflection to connect the fragmented conception of life (elementary particles, atoms, molecules, and their interactions) to patterns of meaning of life. This principle is well demonstrated in the complex dynamics of water in living systems, which consist of up to 99% water molecules.

Water has the ability in living matter to form macroscopic quantum properties in the form of long-range correlations and the structures described. A simple example of such long-range correlations is a crystal, whose comprehensive super spatial regular structure is mediated and built by just such correlations through order quanta. Water can become a liquid crystal through such long-range correlations. Since the water molecule is a dipole with broken symmetry (the oxygen atom forms the negative pole, while the hydrogen atoms are relatively positively charged), its vibrational quanta (dipole-wave quanta) could form complex Bose-Einstein condensations in the quantum vacuum.[30] To form a Bose-Einstein condensate, all particles must be in the energetic ground state. In this state, they are indistinguishable from each other and behave as a single particle. Such a state is extremely stable because it is at the lowest energetic level.[31] The underlying dynamical processes of these fractal and self-similar structures are infinitely iterative, as shown by the mathematics of self-similar fractals.[32] Here, the same mathematical or geometric operation is performed over and over again at all scales to result in self-similar fractal patterns that have a similar shape at any given magnification. The resulting macroscopic coherent water domains connect the external world consisting of molecules to the internal world of life as a unit of meaningful information.

The transition from external fragmentation to internal wholeness can be described by a four-valued reflective system, which contains the infinite iterative process as a third value.[33] While a two-valued system possesses only the fundamental value and its reflection, i.e., an ontological subject that is reflected, a four-valued system possesses two ontological subjects, namely, being and reflection, both of which are reflected, leading to four-valuedness. Classical Aristotelian logic reaches its limits in describing the process of life, since it can only distinguish between purely objective matter and purely subjective consciousness and does not understand the interaction between the two and the intermediate realm of mixed subjectivity that escapes as *tertium non datur.*[34]

The coherent structures emerging from such a fractal iterative process occur at multiple levels of an organism. Water forms coherent domains (spatial regions in which molecules vibrate coherently, i.e., at one frequency and with the same

[30] Abbott et al. (2008).

[31] Fröhlich (1968).

[32] Vitiello (2012).

[33] Schmieke (2021).

[34] Günther (1976).

Table 1 4 values of the life process up to the quantum vacuum

Reflection level	*Level*	*Structure*
0	1	Water molecules
1	2	Dipole-wave quanta of water
2	3	Dipole-wave quantum condensate (Bose-Einstein condensate)
3	4	Integrated information state of the quantum vacuum

Source: The author

phase)[35] on the nanometer scale to filamentous structures on the micrometer scale,[36] as is the case with microtubules in the cytoskeleton.[37] Here, the coherent properties of water are of particular importance because, according to the theory of Hameroff and Penrose, the orchestrated reduction of the wave function within the intercellular network of microtubules is a strong candidate for the physical correlate of consciousness.[38] The link between coherence and consciousness is again particularly apparent here. The following (Table 1) gradation of reflection levels emerges:

Here, the four-valued reflective structure is expressed in 3 reflection levels, beginning with the waves of the water molecules, which represent the 0th reflection level as irreflexive matter.[39] The dipole-wave quanta form the first reflection level, because they have as order pattern the water molecules of the 0th level as object. The second reflection level can be iterated infinitely and brings out in this way the wholeness of the coherent structure of the Bose-Einstein condensate, which arises from a spontaneous symmetry breaking, which can be caused by external or internal electromagnetic fields. It has the order quanta of the first reflection level as its object and therefore corresponds to a reflection level of higher order (quantization). In this a fractal self-similar structure is formed. The concluding third reflection level corresponds to the wholeness of the meaningful information which is contained in the quantum vacuum (see Fig. 1).

The final reflection on the whole self-similar fractal coherent process generates the fourth logical value. The infinite iterative processes, which are the basis of the macroscopic coherence, thus result in the area, which lies between outside and inside, which constitutes the life and also forms the basis for the consciousness. Self-consciousness can be described as the most important property of highly developed living beings, with the help of a four-valued system of self-reflective processes in the quantum structure of the brain and its double in the quantum vacuum (see Fig. 2). This formalism contains, besides the pure states of dual logic S^S and O^O, the states S^O and O^S corresponding to the mixed subjectivity. They are the forbidden third of the classical logic. The exchange of energy and information between brain and environment leads to a doubling of the line of freedom and thus to

[35] del Giudice et al. (2009).

[36] Brizhik et al. (2019).

[37] Lin et al. (2021).

[38] Hameroff and Penrose (1996).

[39] Günther (1991).

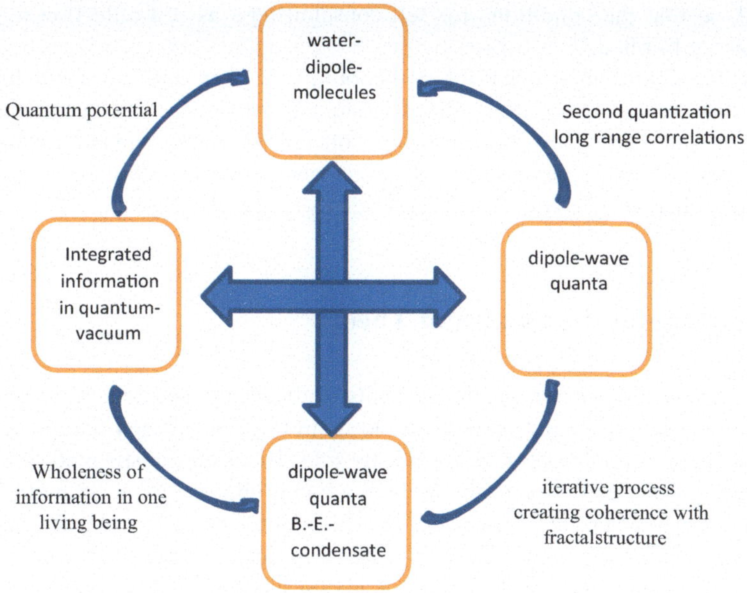

Fig. 1 Triple reflection from water to information. Source: The author

Fig. 2 Self-consciousness as communication between brain, environment, and double in the quantum vacuum. Source: The author

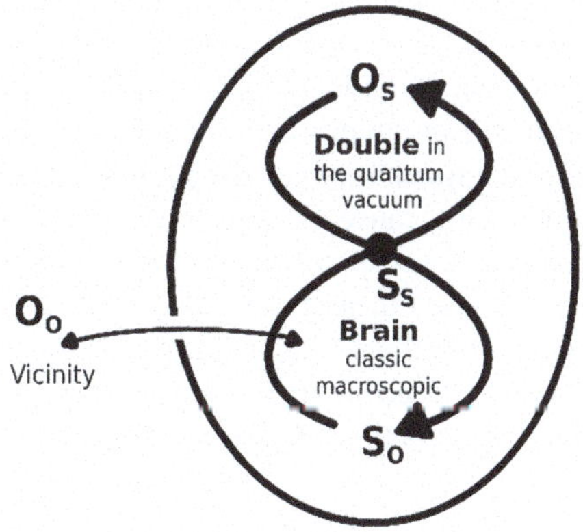

a double of the brain in the quantum vacuum. There the brain deposits meaningful images of the environment as Bose-Einstein condensates, which can be reactivated later as memory contents. Vitiello suggests that self-consciousness arises in the reflection between the brain and its double, which can be viewed as a self-reflective process of the brain.[40] In this self-reflection, the environment is simultaneously

[40] Vitiello (2020).

reflected, so that the conditions for self-consciousness as self-reflection in other-reflection are fulfilled.[41]

Thus, the Cartesian subject-object dichotomy is overcome by a quadrivalent process that creates a subjective-objective domain between consciousness and its material counterpart. The corresponding coherent dynamics generates coherent domains in the water in the brain,[42] in the pure coherent water within microtubules,[43] and in the coherent behavior of neuronal signals in the EEG.[44]

12 Splitting in the Quantum Theory

Already quantum physics exceeds the limits of dual logic by the complementarity principle between wave and particle and leads in the Copenhagen interpretation to two very different processes standing side by side. The deterministic mathematically describable unitary evolution of the wave function is opposed to the reduction of the wave function to one of its eigenstates. The cause of the non-deterministic reduction is not found within the theory itself, which is why many physicists regard it as the influence of consciousness in the course of an observation.[45] Others see it only as the result of the interaction with a classically described measuring device, which does not answer, however, in the end the question when and by what the reduction of the wave function takes place, with which, after all, according to Born's rule, chance enters the otherwise deterministic theory. More modern theories again pursue the approach that the reduction of the wave function must have objective reasons anchored in the physical theory itself, e.g. that certain parameters of the system are not superpositionable. Roger Penrose considers gravitational properties of space-time as the cause of objective reduction[46] while Chalmers takes the approach that physical correspondences of consciousness cannot be in states of superposition, since consciousness must always be unitary or one might say coherent. In all these approaches there is an attempt to make the transition from a quantum theoretical to a classical description of reality, where the unity of the quantum process passes into the fragmentation of classical objects.

[41] Vitiello (2007).
[42] Vitiello (2009).
[43] Lin et al. (2021).
[44] Freeman and Vitiello (2008).
[45] Stapp (1982).
[46] Penrose (1996).

13 Macroscopic Coherent Processes in Quantum Field Theory

Through the previously described dynamics of quantum field theory, this problem is solved for the life process in a particularly elegant way. From this point of view, life itself represents a completely coherent process in itself, which is also characterized by coherent quantum states at the macroscopic level. Due to the long-range entanglements corresponding to condensed vacuum states, life is an indivisible process that cannot be decomposed into body, mind, and spirit even at the fundamental level, and into smaller material units at the organic level, without losing essential aspects of the unity of life. For example, the body is considered to consist of cells. While this assumption is correct, there are quantum properties at the level of intercellular water,[47] microtubules, and even collective cell behavior[48] that must be described as unitary coherent quantum processes and that are lost when cells are described by themselves as purely classical biological entities. If one analyses individual cells by themselves using biological, biochemical, or biophysical methods, the experimental method chosen in each case impairs the coherence of the larger coherence through their physical interaction and in this way destroys the underlying quantum physical unity. Even more fundamentally, even treating a single living entity as a separate organism is an inadmissible simplification that ignores overarching coherences.

All living beings on Earth are part of a unified biosphere, which, according to some scientists, is permeated by a spiritual noosphere.[49] Only in their interplay, which is also based on coherence, can life be understood in its entirety. In this sense, Carl Friedrich von Weizsäcker's insight can be understood that the assumption of the existence of individual material objects is to be regarded as a classical approximation. Likewise, Weizsäcker concluded the existence of independent individual subjects would also be the result of classical approximation.[50]

14 Transdisciplinarity and Coherence

According to this, reality is to be understood as an all-encompassing unity, which is formed from an inner dynamic. However, with each perception and the subsequent thoughts, man objectifies and subjects his objectified cognitions to the laws of dual Aristotelian logic. Since objectification subjects the object of cognition to language and this itself is already no longer the coherent whole, the price for this is always the loss of aspects of wholeness and coherence. Essential aspects and properties of the real are necessarily lost. This is especially true of the specialized objectification undertaken by a single scientific discipline. Each discipline defines its objects in a

---

[47] del Giudice et al. (1988).

[48] del Giudice et al. (1985).

[49] Shushakov (2020); Zwart (2022).

[50] Görnitz (2014).

specific way, neglecting the coherences and entanglements corresponding to it. The result is extremely powerful tools for analyzing the parts and technically manipulating a system. However, herein also lie the limitation and dangers of the disciplinary approach. Because it serves to unilaterally understand and master a particular aspect of a system, it necessarily loses sight of the effect on the system as a whole and accepts collateral effects both epistemic and technical. No purely disciplinary approach does justice to the coherence of life. This is where the transdisciplinary approach comes in, by trying to restore the coherence of the whole system. To see the whole, one must turn one's attention to the larger interconnections, such as the long-range interconnections of quantum field theory and the ultra-weak coherent electromagnetic fields of biophysics.[51] In them, entanglements are expressed that go unnoticed in purely molecular considerations.

The quadrivalent process presented in this article introduces two levels of reflection of intermingled subjectivity to bridge the gap between subject and object. In terms of the Copenhagen interpretation of quantum physics, this means making the connection between the unitary evolution of the wave function and its spontaneous reduction subject to chance. Coherent systems emerging from the infinite iteration of the second reflection stage can afford such a transition, as they incorporate the microscopic constituents into the context of a macroscopic fractal and coherent state that dissipatively couples to the classical objects of its environment. Whereas the Copenhagen interpretation of quantum theory, as propagated by Bohr and Heisenberg, does not grant ontological character in the sense of physical reality to either the particle picture or the wave picture, the 1952 paper by the repeatedly cited David Bohm outlines an ontological realist quantum theory that grants physical ontological character to both particles and waves.[52]

15 David Bohm's Quantum Potential as Effectiveness of the Whole in the Part

The Bohmian model leads exactly to the same experimental descriptions and predictions as the Copenhagen interpretation, but considers both the physical objects like elementary particles and atoms, and the wave function as physical realities and at the same time gets along without the splitting brought about by the abrupt reduction of the wave function. For this purpose, Bohm introduces an additional quantum potential, which, as it were, like an internal energy, guides the particles through the information of the entire system under consideration.[53] The quantum potential represents the effectiveness of the whole in the part and in this way establishes the coherence of the whole system by guiding all its parts through the information of the whole. In this way, the individual parts or particles of a system do

[51] Popp (2003).

[52] Bohm (1952).

[53] Schmieke (2021).

not move independently of each other, but are coordinated by the quantum potential with the motion of all other particles. The more comprehensive the whole system considered by the wave function is, the more holistic information flows into the quantum potential and thus into the guidance of the parts of the system. Although the quantum potential enters the Hamiltonian operator of the Schrödinger equation as a kind of energy in addition to the classical potential, its effect is mainly characterized by its informative character. The quantum potential has negligible energy of its own compared to its potential effect, since it informs the energy of the system as a whole. Formally, the quantum potential is formed as the second spatial derivative of the real part of the wave function and contains the non-local influences of all other parts of a system on the particle under consideration.[54]

16 Self-Awareness and Coherence

The same principle can be extended to all coherent units of a system. Each subsystem forming a coherent unit is coherently related to the whole system by a quantum potential acting in it, which is the efficacy of the whole related to it. It is this transferred principle of generalized quantum potential that ensures the coherence of complex systems, from double-slit experiments to multibody systems to social systems. The difference between purely physical, biological, and human systems is that the parts have independence from the whole system to the extent that they are endowed with self-awareness. Self-awareness in this context is the ability for self-reflection with simultaneous reflection of the outside. Here, the third reflection stage of self-consciousness again corresponds to the total reflection on the infinite iterability of empirical self-consciousness, as described elsewhere. For human beings, this means being coherent with the whole to the extent that they are internally coherent and align their own perceptions and decisions with the guidance of the inner wholeness. Depth psychologically, this can be interpreted as an integration of conscious and unconscious psychological parts, as wholeness is expressed and acted upon in the unconscious.[55]

17 Coherence in Transdisciplinary Dialogue

In summary, coherence is a fundamental property of living systems, complex systems, and ultimately of the real itself. It is human perception with its objectifying reflection that separates out individual systems or objects from the coherent whole. This process is necessary for man's thinking and acting, but it must be completed by a subsequent recognition of the lost connections. The disciplinary splitting of scientific methodology leads to an extended loss of coherence, which should be

---

[54] Ibid.

[55] Zakai (2020).

compensated by a dialogue between disciplines aiming at coherence. This is about reflecting the hidden basic assumptions and at the same time letting the hidden meaning shine through. For this purpose, the concept of coherence should be defined as comprehensively as possible in an interdisciplinary way in order to track down and model the lost coherence, if possible, within as well as between the disciplines. Valuable insights should be derived from this, also for the solution of complex local and global problems. This could lead to a paradigm shift in the natural sciences, especially the sciences of life such as biology, by considering holistic views of life such as those of quantum field theory applied to biology and neuro-sciences.[56] A holistic natural science is more capable of producing a life-friendly technology that we desperately need to solve global ecological and related economic crises. A paradigm shift in the natural sciences that focuses on wholeness and coherence can simultaneously lead to a general shift in consciousness toward sustainability, as the findings of the natural sciences strongly shape people's consciousness.

References

Abbott, D., Gea-Banacloche, J., Davies, P. C. W., Hameroff, S., Zeilinger, A., Eisert, J., Wiseman, H., Bezrukov, S. M., & Frauenfelder, H. (2008). Plenary debate: Quantum effects in biology: Trivial or not? *Fluctuation and Noise Letters, 8*(1), C5.

Atmanspacher, H. (2020). The Pauli-Jung conjecture and its relatives: A formally augmented outline. *Open Philosophy, 3*(1). https://doi.org/10.1515/opphil-2020-0138

Barzegar, A., Taqavi, M., & Shafiee, A. (2021). Zeilinger on information and reality. *Foundations of Science, 26*(4), 1007. https://doi.org/10.1007/s10699-020-09696-8

Beloussov, L. V, Voeikov, V. L, & Martynyuk, V. S. (2007). *Biophotonics and coherent systems in biology.*

Bischof, M. (2008a). A tribute to Fritz-Albert Popp on the occasion of his 70th birthday. *Indian Journal of Experimental Biology, 46*(5), 267.

Bischof, M. (2008b). Synchronization and coherence as an organizing principle in the organism, social interaction, and consciousness. *NeuroQuantology, 6*(4). https://doi.org/10.14704/nq.2008.6.4.200

Blasone, M., Vitiello, G., & Jizba, P.. (2011). *Quantum field theory and its macroscopic manifestations.*

Bohm, D. (1952). A suggested interpretation of the quantum theory in terms of 'hidden' variables. II. *Physical Review, 85*(2). https://doi.org/10.1103/PhysRev.85.180

Bohm, D. (2002). *Wholeness and the implicate order*. p. 284.

Bohm, D. (2013). *On dialogue*. p. 137.

Brizhik, L., Chiappini, E., Stefanini, P., & Vitiello, G. (2019). Modeling meridians within the quantum field theory. *JAMS Journal of Acupuncture and Meridian Studies, 12*(1). https://doi.org/10.1016/j.jams.2018.06.009

Chalmers, D. J. (2020). Is the hard problem of consciousness universal? *Journal of Consciousness Studies, 27*(5–6).

Fleming, G. R., Scholes, G. D., & Cheng, Y. C. (2011). Quantum effects in biology. *Procedia Chemistry, 3*(1), 38–57. https://doi.org/10.1016/J.PROCHE.2011.08.011

[56] Hameroff et al. (2014).

Freeman, W. J., & Vitiello, G. (2008). Dissipation and spontaneous symmetry breaking in brain dynamics. *Journal of Physics A: Mathematical and Theoretical, 41*(30), 304042. https://doi.org/10.1088/1751-8113/41/30/304042

Fröhlich, H. (1968). Long-range coherence and energy storage in biological systems. *International Journal of Quantum Chemistry, 2*(5), 641. https://doi.org/10.1002/qua.560020505

del Giudice, E., Doglia, S., Milani, M., & Vitiello, G. (1985). A quantum field theoretical approach to the collective behaviour of biological systems. *Nuclear Physics B, 251*(C), 375–400. https://doi.org/10.1016/0550-3213(85)90267-6

del Giudice, E., Elia, V., & Tedeschi, A. (2009). The role of water in the living organisms. *Neural Network World, 19*(4).

del Giudice, E., Preparata, G., & Vitiello, G. (1988). Water as a free electric dipole laser. *Physical Review Letters, 61*(9), 1085. https://doi.org/10.1103/PhysRevLett.61.1085

Görnitz, T. (2014). Carl Friedrich von Weizsäckers Entwurf Einer Einheit Der Physik. *Acta Historica Leopoldina, 63*.

Günther, G. (1976). Cybernetic ontology and transjunctional operations. *Beiträge Zur Grundlegung Einer Operationsfähigen Dialektik.Erster Band, 1*(February).

Günther, G. (1991). *Idee Und Grundriß Einer Nicht-Aristotelischen Logik*.

Hameroff, S., & Penrose, R. (1996). Conscious events as orchestrated space-time selections. *Journal of Consciousness Studies, 3*(1). https://doi.org/10.14704/nq.2003.1.1.3

Hameroff, S., & Penrose, R. (2014). Consciousness in the universe: A review of the 'Orch OR' theory. *Physics of Life Reviews, 11*(1), 39–78. https://doi.org/10.1016/J.PLREV.2013.08.002

Hameroff, S. R., Craddock, T. J. A., & Tuszynski, J. A. (2014). Quantum effects in the understanding of consciousness. *Journal of Integrative Neuroscience, 13*(2), 229.

Heinrichs, J. (2021). *Action, language, art, and mysticism as reflection-levels of an alternative semiotics and the spiritual perspective of a value-levels-democracy*. in Pragmatism, Spirituality and Society.

Lin, C. D., Radu, C. M., Vitiello, G., Romano, P., Polcari, A., Iliceto, S., Simioni, P., & Tona, F. (2021). Sounds stimulation on in vitro Hl1 cells: A pilot study and a theoretical physical model. *International Journal of Molecular Sciences, 22*(1). https://doi.org/10.3390/ijms22010156

Lloyd, S. (2011). Quantum coherence in biological systems. *Journal of Physics: Conference Series., 302*.

Maldacena, J., & Susskind, L. (2013). Cool horizons for entangled black holes. *Fortschritte der Physik, 61*(9), 781–811. https://doi.org/10.1002/prop.201300020

Manca, A. R. (2014). Social cohesion. *Encyclopedia of Quality of Life and Well-Being Research, 6026–28*. https://doi.org/10.1007/978-94-007-0753-5_2739

Mansfield, V., & Marvin Spiegelman, J. (1996). On the physics and psychology of the transference as an interactive field. *Journal of Analytical Psychology, 41*(2), 179. https://doi.org/10.1111/j.1465-5922.1996.00179.x

McFadden, J., & Al-Khalili, J. (2018). The origins of quantum biology. *Proceedings of the Royal Society A: Mathematical, Physical and Engineering Sciences, 474*(2220).

Nicolescu, B. (2010). Methodology of transdisciplinarity–levels of reality, logic of the included middle and complexity. *Transdisciplinary Journal of Engineering & Science, 1*(1). https://doi.org/10.22545/2010/0009

Penrose, R. (1996). On Gravity's role in quantum state reduction. *General Relativity and Gravitation, 28*(5), 581. https://doi.org/10.1007/BF02105068

Piaget, J. (1972). *The epistemology of interdisciplinary relationships*. in Interdisciplinarity: Problems of Teaching and Research in Universities.

Popp, F. A., Li, K. H., & Gu, Q. (1992). Recent advances in biophoton research and its applications.

Popp, F. A. (2003). Properties of biophotons and their theoretical implications. *Indian Journal of Experimental Biology, 41*(5), 391.

Portugal, S. J., Hubel, T. Y., Fritz, J., Heese, S., Trobe, D., Voelkl, B., Hailes, S., Wilson, A. M., & Usherwood, J. R. (2014). Upwash exploitation and downwash avoidance by flap phasing in ibis formation flight. *Nature, 505*(7483), 399–402. https://doi.org/10.1038/nature12939

Sanders, T., and H. Pander Maat. 2006. *Cohesion and coherence: Linguistic approaches*. in Encyclopedia of Language & Linguistics.

Schmieke, M. (2021). Bohm's quantum potential approach to consciousness from the perspective of a four-valued logic. *Dev Sanskriti Interdisciplinary International Journal, 17*, 01. https://doi.org/10.36018/dsiij.v17i.206

Scholz, R. W., & Steiner, G. (2015). The real type and ideal type of transdisciplinary processes: Part I—Theoretical foundations. *Sustainability Science, 10*(4), 527–544. https://doi.org/10.1007/S11625-015-0326-4

Shushakov, E. (2020). *Genesis of the Term Noosphere and Its Use by P. Teilhard de Chardin and V. I. Vernadsky*. Vestnik Pravoslavnogo Sviato-Tikhonovskogo Gumanitarnogo Universiteta, Seria I. Bogoslovie, Filosofia, Religiovedenie (87).

Stapp, H. P. (1982). Mind, matter, and quantum mechanics. *Foundations of Physics, 12*(4), 363. https://doi.org/10.1007/BF00726783

Tononi, G., & Balduzzi, D.. 2007. *Measuring consciousness as integrated information*. in AAAI Fall Symposium–Technical Report. Vol. FS-07-01.

Tononi, G., Boly, M., Massimini, M., & Koch, C. (2016). Integrated information theory: From consciousness to its physical substrate. *Nature Reviews Neuroscience, 17*(7), 450–461. https://doi.org/10.1038/nrn.2016.44

Vitiello, G. (2007). Quantum dissipation and information: A route to consciousness modeling. *NeuroQuantology, 1*(2). https://doi.org/10.14704/nq.2003.1.2.15

Vitiello, G. (2009). Coherent States, Fractals And Brain Waves. *New Mathematics and Natural Computation, 05*(01), 245. https://doi.org/10.1142/s1793005709001271

Vitiello, G. (2012). Fractals, coherent states and self-similarity induced noncommutative geometry. *Physics Letters, Section A: General, Atomic and Solid State Physics, 376*(37), 2527. https://doi.org/10.1016/j.physleta.2012.06.035

Vitiello, G. (2020). Matter, mind and consciousness: From information to meaning. *Journal of Integrative Neuroscience, 19*(4), 701.

Zakai, A. (2020). Erich Neumann and the crisis of Western ethics. *Society, 57*(3), 332. https://doi.org/10.1007/s12115-020-00487-2

Zwart, H. (2022). Pierre Teilhard de Chardin's Phenomenology of the Noosphere. *Philosophy of Engineering and Technology, 38*.

Schmieke, Marcus, was born in 1966 in Oldenburg. During his studies of physics and philosophy in Heidelberg and Hanover, he developed the vision of exploring the interaction between matter and consciousness. This synthesis between science and philosophy is a recurrent pattern in his life. Among other things, he has also studied Vedic philosophy and architecture during his stays in various Indian monasteries. His encounter with the physicist Burkhard Heim, as well as his work on his own books about science and consciousness—The Last Secret and the Field of Life—laid the theoretical foundation for the development of Information Field technology. Through an intensive examination of Heim's model of the 12 dimensional structure of the universe, Marcus Schmieke gained a deep understanding of the physical information field. In 2007, based on the work of Burkhard Heim, the Russian physicist Nikolai Kozyrev and other scientists, Marcus Schmieke developed the first TimeWaver system and thus brought information field technology into practical application for the first time. Within a very short period it proved to be a success in complementary medicine. He is also the inventor of the Healy wearable frequency technology and the founder of Healy World. He also wrote more than 20 books about the connection of mind and matter and other spiritual and scientific topics, which have been translated into several languages. He is also the co-founder of the Tattva Viveka magazine, a forum for philosophical and scientific exchanges. Marcus Schmieke is the founder of the Vedic Academy e.V (now ConScience e.V.), co-founder of the Institute for Applied Consciousness Research (IACR) near Berlin, where he is investigating the interaction between matter and consciousness with several international physicists, and founder of the Existential Consciousness Research Institute ECR, which organizes the international "Worlds of Consciousness" events. He is an adjunct faculty member at the Dev Sanskriti University in Haridwar, India. He gives lectures on scientific, philosophical and medical topics worldwide.

The Wondering I as a Threshold Toward a Transdisciplinary Experience of Goodness

Salvatore Lavecchia

1 Which Notion of I Is Consonant with Transdisciplinarity?

The usual notion of "I"[1] is consonant with the image of the I which Ludwig Wittgenstein in his *Tractatus Logico-Philosophicus* connects with solipsism:

> The I in solipsism shrinks to an extensionless point and there remains the reality co-ordinated with it" (5.64); this notion is in turn coincident with the representation of a subject constituting "the limit—not a part of the world (5.641).[2]

In the encounter with the world we habitually tend, in fact, to represent the perceiving, thinking, understanding, knowing I as a point localizable within our body, whose interiority we intend as separated from the exteriority of the world we are perceiving. As a consequence, we are oriented to associate the modality of perception actualized by the I with the perspective of a tunnel vision,[3] which hopelessly splits the first person perspective from the second and third person perspectives, that is a postulated mere subjectivity of the perceiving consciousness from a postulated

---

[1] I prefer here the use of the term "I" instead of the term "self", since the second term necessarily involves frontality/objectivity. In consonance with the premises assumed in the present contribution, I do not accept the interpretation of the term "I" as exclusively deictic, paradigmatically proposed in De Gaynesford (2006). Basing on the considerations formulated in the present contribution, the "I" manifests itself as both transcending any modality of "self" and capable of integrating in its life many modalities of "self" and "selfness" (see also Di Francesco, 1998).

[2] See Wittgenstein (1922).

[3] Metzinger (2009) paradigmatically identifies this tunnel vision with the only possible modality of consciousness characterizing the I.

S. Lavecchia (✉)
University of Udine, Udine, Italy
e-mail: salvatore.lavecchia@uniud.it

© The Author(s), under exclusive license to Springer Nature Switzerland AG 2025
M. Schüz (ed.), *Transdisciplinary Thinking and Acting*, Transdisciplinary Management of Ecological and Social Crises,
https://doi.org/10.1007/978-3-031-70680-6_7

objective being. A contrastive dualism between the I and the world is thus generated, which immediately determines and configurates every field of life and knowledge.

Compared to the just described, dualistic notion of I, an unprejudiced observation concerning phenomenal experience of perception discloses a radically different possibility of characterizing the perceiving I. Actually, the image of a localizable point does contradict even the most elementary consideration regarding perception: in every perception we are, in reality, unable to localize the perceiving I at any place in space or moment in time, neither within a supposed merely subjective interiority nor in a supposed objective exteriority.[4] Moreover, both modalities of localization would involve some absurdities: in the case of identifying the I with a merely subjective interiority—with a point shrunk in itself—perception would be a bare projection of a subjectivity, without any authentic encounter with the perceived being; in the case of reducing the perceiving I to external determining factors, the perceiving I would be a more or less illusionary no-one,[5] and we would have no reason to be certain of any evidence manifested by this no-one. With respect to an unprejudiced observation of perception, the most consistent image of the perceiving I is hence the image of a nonlocalizable, nonphysical point or, more precisely, *center of attention,*[6] which, transcending any first, second, third person perspective—both singular and plural —instantaneously generates an unlimited *sphericity* illuminating the perceived being, offering it the possibility of manifesting itself, whereas the point and the sphere constitute an instant unity, not a temporal succession.[7] The aforesaid image preserves in fact both the indispensable openness toward the perceived being and the equally indispensable remaining centered in itself which characterizes the perceiving I. In the absence of the first, the I would act as a solipsistic subject; without the second the I would dwindle till to the status of an inessential no-one.

Which of the two considered representations of the perceiving I—the localizable point and the nonlocalizable center/sphere—is more consonant with a consciousness aiming at authentic transdisciplinarity?

In effect, the tendency toward hyperspecialization[8] permeating our present time is essentially supported by a tunnel vision, through which each individual scientific discipline, more or less consciously, tends to solipsistically represent its own methodological constraints as untranscendible limits. The necessary result is a hopeless separation of the single discipline from other disciplines as well as from any possibility of attaining a knowledge which could be really significant for the wholeness of life. Maybe is the aforementioned tunnel vision simply a projection of the tunnel vision characterizing the I perceiving itself as a localizable point shrunk in

---

[4]For a first introduction to the concept of perception and knowledge as transcending the opposition of exteriority and interiority, see Scheurle, (2013).

[5]See the also in this respect paradigmatic considerations in Metzinger (2003). For critical comments, see Lavecchia (2014d).

[6]A deep analysis concerning the notion of attention can be found in Ganeri (2017).

[7]For a consequent explication of this image, see Lavecchia (2014c).

[8]A first introduction to this notion can be found in Malone et al. (2011).

itself? Both visions share namely the same eminently undialogical, dualistic gesture, which negatively contracts the perception of the thinking and knowing I into a singular perspective, rendering it incapable of an authentically unprejudiced encounter with phenomena in whatever dimension and perspective of life. On the contrary, the image of the perceiving I as nonlocalizable center/sphere instantaneously manifests an intrinsic, unrestricted *dialogicity*. Accordingly, the aforesaid image helps us to (re)discover the I as intrinsically *transperspectival*: as a center/sphere of consciousness which intrinsically acts *transperspectively*, therefore transcending any first, second, third person perspective, both singular and plural.

In consonance with the just formulated premises, if not conditioned through conceptual prejudices or preconceived representations—like the habitually internalized representation of the localizable point shrunk in itself—the I encountering the world is constitutionally capable of preserving both its own autonomous self-awareness/-transparence and the possibility for the perceived being of autonomously manifesting its own transparence. This constitutionally, eminently *dialogical* I, not the solipsistic I represented through a localizable point, can be assumed as a consonant fundament for a philosophical anthropology able to be in accordance with any epistemic as well as ethical approach oriented toward a consistent transdisciplinarity. The development of an anthropology manifesting this characteristic—which, of course, cannot even elementarily provided in this context—should be perceived as an urgent prerequisite, since it would decisively contribute to more and more deeply integrate the dialogical gesture generating transdisciplinarity into the self-awareness of the person encountering both the own interiority and the world.

2 Toward a Transdisciplinary Notion of Good

The Present of a Platonic Heritage

The just characterized notion of *dialogical I* can help to eliminate any dichotomy not only between interiority and exteriority, subjectivity and objectivity, self and world, individuality and relationality/universality, but also between perception and knowledge on the one side, ethics and morality on the other side. The described dialogical I can be namely associated with a conception of the good capable of bridging the gap separating the two terms involved in the last mentioned dichotomy.[9] The here intended conception can be developed basing on two key passages of Plato's work, *Timaeus* 29e-30a and *Respublica* 506d6-509c.[10] The great interest these two passages can wake is connected with the fact that they presuppose a not prioritarily

---

[9]More extensive considerations concerning the here characterized notion of good are offered in Lavecchia (2014a).

[10]The whole work of Plato can be in fact interpreted basing on the centrality of this conception: see Lavecchia (2012, 2015).

contrastive or teleological notion of the good: here the good is understood neither as arising from the contrast with evil nor as a goal (*telos*) perceived as merely objective and transcendent in relation to a knowing or acting self—that is to individual, personal consciousness—on the contrary, the good is here prioritarily characterized through the gesture of an unconditional generosity, which disposes for the most unrestricted self-givingness, that is for absolute openness toward the autonomous manifestation of another being.

In *Timaeus* 29e-30a—the only passage in which Plato explicitly characterizes the notion of good—is declared the reason because of which the divine Demiurge gives life and form to the cosmos, identifying this reason with his being good, meaning being unrestrictedly free from envy, and therefore inclined to the highest form of generosity. The unconditionality, the absolute freedom characterizing the good is here evidenced on the one side by the fact that the cosmogonic activity of the Demiurge is not determined—neither by any need or necessity nor by an impulse toward self-affirmation, nor by any norm, imperative, commandment, nor by preceding experiences or by calculations regarding the future—, on the other by the fact that, according to *Timaeus* 34b6–8, the cosmos generated by the Demiurge is characterized by complete autonomy—it is self-sufficient, that is independent from any external factor with regard to preserving its existence and life—and self-consciousness.

The Demiurge is eminently good because he is capable of opening an unrestricted space of manifestation for another free and autonomous being. In *Respublica* 506d6-509c Plato paradigmatically illustrates the corresponding modality of consciousness through the analogy between the origin of every being and the sun. This origin, transcending every form of being (*Respublica* 509b8–10), is identified with the supreme good because, in the same way that the sun is unconceivable without its manifestation through light, the supreme good is, beyond any dichotomy between transcendence and immanence, unconceivable as separated from its unconditional, unrestricted manifestation through being, that is from its absolutely free openness toward the transparent manifestation of *another* autonomous and free reality. In other words, in Plato's analogy the light of the sun is illustrating the unconfined transparence, unhiddenness of the supreme good, deriving from its unconditional self-givingness, and producing the manifestation not only of merely objective being, but also, with equal originality, of *intellect* (*noûs*, *Respublica* 508b12-c2), that is of a self-conscious "I" or self being capable of autonomously knowing and acting.

In the two just considered passages the notion of good is prioritarily characterizing a particular modality of consciousness. This modality of consciousness can be described as paradigmatically, unrestrictedly *transperspectival and dialogical*, that is both transcending any determination of being and connoted by an unconditional, generative openness toward the manifestation of and the dialogical encounter with another autonomous consciousness or whatever other phenomenon. This very same openness may be in turn intended as paradigm not only for any authentically productive epistemic approach, but also for any act which can be perceived as veritably ethical.

The here characterized *agathological*[11] openness reveals itself as fundament both of productive, creative intelligence and of true virtue, whereas virtue is here understood as capability of generating relations which are harmonious and consistent both for the perceiving/knowing I and for the phenomena the I is encountering or going to encounter. This agathological openness is completely consonant with the representation of the dialogical I described above: the perceiving and knowing I would not veritably perceive and know *another* being, if this dialogical openness would not be intrinsically present in any act of unbiased perception and knowledge. In consonance with these premises, the here presupposed notion of good could be a key notion in any reflection concerning the perceiving and knowing I as transcending any tunnel vision, regarding both its epistemic and ethical operativity. As a consequence, this transperspectival, eminently dialogical notion of good may be considered not only as transdisciplinary, but also as fundament of any authentic and productive transdisciplinarity, since it is able to most efficiently manifest the fundamental gesture of the perceiving and knowing I as intrinsically consonant with a transdisciplinary consciousness.

3 Confidence and Wonder as Fundaments of Transdisciplinarity

Knowledge as Experience of Birth

The above delineated *agathological*, and thus transperspectival and dialogical I can be further illustrated through a seminal passage of Plato's dialogues.

Theaetetus 149a-151d3 is the only source in which Socrates represents himself through the well-known analogy with midwives[12]: Socrates helps intellectually and spiritually pregnant persons to give birth to true knowledge; however, interestingly enough, giving birth to an own knowledge or wisdom is forbidden to him while acting as midwife (*Theaetetus* 150c4-d2), although he has to be capable of distinguishing true from false knowledge, in order to really help to achieve a consistent birth (150b2–3, c1–3, 151d2–3). In any case, the aforesaid prohibition does not involve that Socrates never gave birth to true knowledge; otherwise, the analogy with the midwives would be inconsistent, since according to Socrates midwives are required to have experience in giving birth (149b5-c2). In consonance with this analogy, Socrates *has* therefore given birth to true knowledge. The obvious reason why he is not allowed to do it while acting as a midwife is the fact that *other*

[11] *Agathological* is here not primarily related to the field of ethics or practical philosophy, but rather to a philosophical approach which, transcending any dichotomy between practical and theoretical life, perceives a supreme good (*agathón*), intended as a modality of consciousness characterized by unconditionally dialogical generosity, as both ultimate origin and reason of a free and autonomous thinking (*lógos*). See Lavecchia (2014a).

[12] I gave a more detailed consideration of Socrates' midwifery in Lavecchia (2013, 2022).

persons—not Socrates—have to give autonomously birth to true knowledge, not passively learning something from Socrates (150c6, d1–2, d6–7-e1). A deeper reason is nonetheless the fact that helping another person to autonomously generate true knowledge presupposes a transperspectival modality of consciousness in the helping person. This modality is namely the only one able to transcend in every perception and situation the conditioning, inappropriately standardizing factors emerging from previous experiences: the perceiving I makes itself constantly and generatively *void* with respect to any already attained knowledge, and thus opens an unlimited, nonconditioning space of transparence for the inimitable individuality of the perceived being as well as for the way through which that individuality can manifest itself most transparently. Exactly to this modality of consciousness points the analogy with midwifery related to Socrates. In fact it presupposes the same unrestricted, generatively generous *confidence* required from a midwife by the event of a birth, where, notwithstanding her expertise, if she wishes to avoid mistakes, a midwife is never allowed to merely *deduce* from any past experience the singularity and uniqueness of the newborn, of the mother, of the situation concomitant with the birth: an efficient midwife makes herself constantly void with regard to her previous experience, exclusively concentrating on opening an unlimited space for the manifestation of the individuality presently going to be born.

Basing on these premises, Plato's representation of Socrates' midwifery can be considered perfectly consonant with the transperspectival and transdisciplinary notion of good deduced above from the analogy between the sun and the good. The unconditional confidence, to which the dialogical gesture of Socrates' midwifery points, appears namely as the ultimate ground of genuinely *good* choices, whereas *good* transcends here any dichotomy between epistemic and ethical domains, simply referring to a consciousness, a perceiving and knowing I, able to generate an unconditioned space for the unrestricted manifestation of any individual phenomenon in every dimension and field of life. In other words, the modality of consciousness presupposed by Socrates' midwifery can be perceived as paradigm for any person wishing to attain productive perception and authentically innovative knowledge in any conceivable field and discipline: harmonious perception and true knowledge cannot be generated through exclusive instantiation of exterior or interior norms, standards, methods, that is through formalized procedures and disciplines, since, no matter how refined and how supposedly anticipatory they may be, they are always a result deduced from past experiences, and therefore valid only in limited perspectives.

Socrates' transperspectival, dialogical voidness with regard to any preceding knowledge and experience implicitly reveals that a monarchy of formalized procedures and methods biases the encounter with the concrete individuality of phenomena. Socrates reveals that, on the contrary, both epistemic and ethical virtue involves the willingness of being unrestrictedly confident in the generative power of the *present* phenomenon and situation: confident in the fact that authentic manifestation of the good, that is genuine virtue, no matter in which dimension of life, presupposes the willingness to generate an unlimited space for the manifestation of a *new birth* (Plato, *Symposium* 212a2–5). In few words, Socrates points to a modality of consciousness in which the above considered, constitutionally transperspectival, dialogical gesture of our perceiving and knowing I becomes a constant, more and more conscious practice and operativity. This modality of consciousness can be additionally specified through a further passage of Plato's *Theaetetus*, in which Socrates considers the act of wondering as the origin of philosophy (155d1–5), whereas philosophy is here intended as striving for true knowledge and wisdom (*sophía*). Aristotle gives the best commentary to this passage in *Metaphysics* 982a30-983a20: wonder is origin of philosophy—of striving for *sophía*, with the same meaning as in Plato's *Theaetetus*—because it is generated in the encounter with phenomena which are incommensurable with preceding experiences, that is are not deducible from any already acquired experience, discipline, method, procedure, standard, and therefore stimulate to transcend the state of aporia with which they are associated, encouraging to generate authentically *new* dimensions and perspectives of knowledge.[13]

Aristotle's consideration regarding wonder can be perceived as additional declaration of the fact that only a transperspectival, eminently dialogical, and therefore transdisciplinary modality of consciousness is able to attain true knowledge, that is an experience of authentic *newness*. In this context wonder can be considered another element indicating the unrestricted openness which intrinsically characterizes the perceiving, thinking, understanding I, not imprisoning it in the limited perspectives, in the tunnel vision of any discipline, thus revealing it as an eminently dialogical and agathological reality, through whose transperspectival nature every discipline can receive an authentic sense and significance.

4 The Dialogical, Wondering I and the Present Crises

Consciousness Determines Being!

In the present time too many approaches which aim at developing solutions for crises are focused on the premise that only through implementing worldwide, objective evaluable control measures human beings can be oriented toward the good.[14] Such approaches presuppose, in other words, that transformations regarding consciousness and, consequently, acting in the world, can only emerge from objectively determining and determinable factors, since it is assumed that consciousness and the I are to be intended as merely emergent, and therefore eventually illusory phenomena. As a consequence, if present crises are generated by, for example, lacking sense of community, the just mentioned approaches presuppose that solely by implementing worldwide standardizing measures and programs, which aim at

[13] A more detailed interpretation of the here emphasized notion of wonder cab be found in Lavecchia (2014b).

[14] In Lavecchia (2021a) I formulated critical considerations regarding these approaches concerning the field of academic education.

generating and fostering the aforesaid quality, crises might be resolved. Additionally, if transdisciplinarity is perceived as an essential factor regarding solution of crises, such approaches trust only in supposedly objective, possibly worldwide control standards, procedures, evaluations, plans, incentives, and so on, in order to produce an effective transdisciplinarity.

The aforesaid trend, which reveals an exclusive credence in centralized planning strategies, as well as an equally radical distrust in the positive power and creativity of concrete individuals and communities, leads to the emergence of some paradoxes. The most conspicuous conceptual paradox emerges, for instance, from the presupposition that creating a permanently traceable and trackable digital self for every human being in the world would foster sense of community. In fact this assumption pretends to stimulate communitarian values basing exactly on the representation of the I as localizable point shrunk in itself, which, as we saw, Wittgenstein—rightly—characterized as consonant with solipsism, that is with the most unproductive form of individualism! Is the assumption really unquestionable that identifying human beings with digital selves will let emerge sense of community, and not the contrary?[15]

Interestingly, predominant approaches to present crises do not take seriously the productive implications a consistent philosophical anthropology could generate in coping with crises. According to the preceding considerations, consistency would be manifest in an anthropology able to develop an image of the—both inwardly and outwardly—perceiving I which would be consonant with the intrinsic transperspectival and dialogical nature of the I. This anthropology—which would necessarily understand the human being as triarticulated in body, soul and spirit[16]—would not lead to a new form of negative anthropocentrism; on the contrary, it would eliminate from the self-perception of human beings those representations and images which have generated the existing negative anthropocentrisms, manifesting the human being as constitutionally characterized by a positive orientation toward dialogical self-transcendence, that is by unprejudicial, generative openness toward harmonious relations with the world. In sum, the here intended anthropology would reveal that the fundament for sense of community—community not only with other human beings, but with every being in the world—is intrinsically present in *every* human being, since every human being *is* a transperspectival, agathological, and thus eminently dialogical I.

A consistent, unprejudicial philosophical anthropology would make manifest that the human modality of consciousness is, beyond every encounter with any being, intrinsically oriented toward dialogical community. This would constitute a not negligible premise for genuine creativity in coping with crises associated—as

[15] An introductory discussion concerning the so called *digital loneliness* is given in Brennan (2021).

[16] Introductory considerations concerning this approach to anthropology, which cannot be extensively described here, can be found in Lavecchia (2019), Weger & Wagemann (2021), Weger et al. (2021). For the possibility of integrating the notion of I discussed in the present contribution in the horizons of different spiritual traditions see Lavecchia (2020, 2021b).

every crisis actually is?—with missing sense of community, since it would encourage confidence in the fact that *every* human being is intrinsically able to autonomously generate sense of community, beyond any planning and standardizing intervention. This would also implicate that crises would be encountered much more efficiently if individual awareness of this premise as well as confidence in individual initiative would be fostered, instead of implementing more and more standardizing, collectivizing dynamics. Of course are predominant approaches to crises not based on the here presupposed confidence, since they have no confidence in the reality of the I. However, If taken seriously, this lack of confidence leads to manifest absurdities: Why should we namely trust in no matter which conceptual elaboration as well as both epistemic and ethical decision of a nonreal I, that is why should we trust in any theory and strategy, since only our I can decide to trust or not to trust? The absurdities which implies the elimination of trust in the reality of the perceiving, thinking, understanding, deciding, and acting I lead to the irreducibility of freedom,[17] that is to the irreducibility of phenomenal consciousness to any form of being,[18] and consequently to any form of standardization, algorithmization, computation.

The radicality of present crises is perhaps associated with mistrust in the just evoked irreducibility, that is in the fact that in a world governed by peace and harmony consciousness determines being, and not the contrary. Only seemingly has this provoking statement nothing to do with transdisciplinarity. In fact, would genuine transdisciplinarity be possible if we would not perceive the untranscendible irreducibility of consciousness and freedom? Is transdisciplinarity—together with all its impressive potentialities concerning the encounter with crises—possible without an intrinsically confident and wondering I, who in the encounter with all phenomena is constantly open for the birth of authentic newness, which is impossible without authentic freedom? Is transdisciplinarity possible in absence of an I never fearing before the transparence of phenomena and events transcending preceding experiences, theories, disciplinary perspectives and methods, as well as any dogmatic unilaterality with respect to research methods[19]? Maybe is transdisciplinarity genuinely realizable only if *I* am confident in the possibility of autonomously and constantly manifesting a *good* modality of consciousness, consisting in an unrestricted generosity: free not only from every being, but also from my*self*, that is from any *self*[20] and *selfness,*[21] and thus capable of transforming every perception in the experience of a wondering encounter with a new birth.

---

[17] On irreducibility of free will see Esfeld (2020).

[18] A defence of the assumption concerning the priority of phenomenal continuity in constructing a philosophy of self as well as of the originality and irreducibility of consciousness is given in Chalmers (1996), Bayne (2010), Dainton (2011), Bitbol (2014).

[19] A paradigmatic exhortation to transcending any unilaterality concerning scientific methods is contained in Heisenberg (2019).

[20] The transcendence of persons as beings with capacity of attention with respect to the notion of self is persuasively affirmed in Ganeri (2017).

[21] See Lavecchia (2020).

References

Bayne, T. (2010). *The unity of consciousness*. Oxford University Press.
Bitbol, M. (2014). *La conscience a-t-elle une origine? Des neurosciences à la pleine conscience: une nouvelle approche de l'esprit*. Flammarion.
Brennan, E. (2021). Digital loneliness. Asking too simple a question about a complex problem. In G. Balbi, N. Ribeiro, V. Schafer, & C. Schwarzenegger (Eds.), *Digital roots. Historicizing media and communication concepts of the digital age* (pp. 229–244). De Gruyter Oldenbourg.
Chalmers, D. J. (1996). *The conscious mind. In search of a fundamental theory*. Oxford University Press.
Dainton, B. (2011). *The phenomenal self*. Oxford University Press.
De Gaynesford, M. (2006). *I: The meaning of the first person term*. Oxford University Press.
Di Francesco, M. (1998). *L'io e i suoi sé. Identità personale e scienza della mente*. Raffaello Cortina Editore.
Esfeld, M. (2020). *Science and human freedom*. Palgrave Macmillan/Springer.
Ganeri, J. (2017). *Attention, not self*. Oxford University Press.
Heisenberg, W. (2019). *Ordnung der Wirklichkeit*. Mit einer Einleitung von Helmut Rechenberg und einem Kommentar von Ernst Peter Fischer. Springer.
Lavecchia, S. (2012). Agathologie. Denken als Wahrnehmung des Guten oder: Auf der Suche nach dem offenbarsten Geheimnis. *Perspektiven der Philosophie, 38*, 9–45.
Lavecchia, S. (2013). Vertrauen in die Ich-Geburt. Führung in Sokrates' Horizont. In G. Werner & P. Dellbrügger (Eds.), *Wozu Führung? Dimensionen einer Kunst* (pp. 73–81). Karlsruher Institut für Technologie Scientific Publishing.
Lavecchia, S. (2014a). Agathological realism: Searching for the good beyond subjectivity and objectivity or on the importance of being platonic. *Etica & Politica, 16*, 533–549. (https://www.openstarts.units.it/bitstream/10077/10742/1/LAVECCHIA.pdf), also in *Moral realism and political decisions*, ed. Gabriele De Anna and Riccardo Martinelli, pp. 29–50. Bamberg: University of Bamberg Press 2015.
Lavecchia, S. (2014b). Chi invidia non conosce meraviglia. Note su thaumazein ed essere aphthonos in Platone e Aristotele. *Thaumàzein, 2*, 169–194.
Lavecchia, S. (2014c). Das Ich und das Gute. Ansätze einer Licht-Philosophie in Anknüpfung an Novalis und Platon. *Perspektiven der Philosophie, 40*, 9–46.
Lavecchia, S. (2014d). Trasparenza o illusione dell'autocoscienza? Metafisica della luce e neurofilosofia. In F. Fabbro (Ed.), *Neuroscienze e spiritualità. Mente e coscienza nelle tradizioni religiose* (pp. 74–91). Astrolabio-Ubaldini Editore.
Lavecchia, S. (2015). *Generare la luce del bene. Incontrare veramente Platone*. Moretti&Vitali.
Lavecchia, S. (2019). Nella feconda luce dello Spirito. Per una antropologia triadica. *Mistica e Filosofia, 1-1*, 67–80.
Lavecchia, S. (2020). Libertà da sé e da altro. Comprendere come vuoto dell'io, oltre Oriente e Occidente. *Mistica e Filosofia, 2-1*, 29–45.
Lavecchia, S. (2021a). Im Namen der Un-Schuld. Philosophie für eine unakkreditierte Universität. *Allgemeine Zeitschrift für Philosophie, 46-1*, 81–101. (a reduced version is published in https://bildung-wissen.eu/fachbeitraege/im-namen-der-un-schuld-philosophie-fuer-eine-unakkreditierte-universitaet.html)
Lavecchia, S. (2021b). Towards a dialogical self. Metaphysics of light as a philosophical perspective for interreligious encounter. In A. A. Mosleh (Ed.), *Convergence and divergence. An intercultural-interreligious dialogue* (pp. 137–146). Verlag Traugott Bautz, Nordhausen.
Lavecchia, S. (2022). Erkenntnis und Wissenschaft als Geburt der Freiheit: Betrachtungen zu Sokrates' Hebammenkunst. *Perspektiven der Philosophie, 48*, 129–140.
Malone, T. W., Laubacher, R., & Johns, T. (2011). The big idea: The age of Hyperspecialization. *Harvard Business Review.*. Accessed Jan 4, 2023, from https://hbr.org/2011/07/the-big-idea-the-age-of-hyperspecialization, *82*, 106.
Metzinger, T. (2003). *Being no one. The self-model theory of subjectivity*. MIT Press.

Metzinger, T. (2009). *The ego tunnel: The science of mind and the myth of the self.* Basic Books.

Scheurle, H. J. (2013). Das Gehirn ist nicht einsam. Resonanzen zwischen Gehirn, Leib und Umwelt. Kohlhammer.

Weger, U., & Wagemann, J. (2021). Soul, Spirit, and consciousness in psychology and philosophy: Traditions, current views, perspectives. (editorial). *European Psychologist, 26-2*, 71–72.

Weger, U., Sparby, T., & Edelhäuser, F. (2021). Dualistic and Trichotomic approaches in psychological enquiry. *European Psychologist, 26-2*, 85–95.

Wittgenstein, L. (1922). *Tractatus Logico-Philosophicus.* Transl. Frank Plumpton Ramsey. Kegan Paul, Trench, Trubner &. Co.

Lavecchia, Salvatore, Prof. Dr., (born 1971), after study grant at the Scuola Normale Superiore (Pisa) both as undergraduate and PhD student (1990-1997), with supplementary research grants in Tübingen (1993), München (1996), Oxford (1997), in 1998 he achieved the PhD in Classics. From 1999-2001 he was research and teaching fellow for Classics at University College London, from 2002-2003 research fellow of the Alexander-von-Humboldt-Stiftung in Tübingen. From 2004-2009, financed through the programme "Rientro dei Cervelli" (Italian Ministry of Universities and Research), he has been lecturer for Classics and Ancient Philosophy, since 2009 professor for Ancient Philosophy at the University of Udine. From 2010-2015 he was teacher in the Master "Consulenza Filosofica di Trasformazione" (University of Verona); since 2018 he is teaching and member of the scientific commitee in the Master "Meditazione e Neuroscienze" (University of Udine), since 2022 in the Master "Contemplative Studies" (University of Padova). Main research interests: the socratic dialogue, also in its relationship to counselling; the notion of self and mind in Plato and Platonism; the metaphysics of light in Plato and Platonism; ancient aesthetics; the development of a philosophy of the self being capable of a productive dialogue both with neurosciences and spirituality. Since 2016 he is coordinating at the University of Udine an interdisciplinary seminar involving philosophy, mathematics, and physics.

The Epistemology of Vedantic Constructivism

Linking Eastern and Western Epistemological Approaches and Deriving Recommendations for Action

Elfriede Neuhold

1 Introduction

The world is currently facing great challenges. In particular, more and more resources are being consumed from available sources, while the sinks of the earth, i.e. the capacity to absorb pollutants and waste, are already heavily burdened or even overloaded[1]. These challenges will become even greater in connection with the population explosion casts.[2] Against this background, an attempt is made here to find new approaches to solutions on a philosophical basis, in order to find the path to a successful transformation. Under the concept or umbrella of "transformation" there are very different approaches, ranging from sociological analyses of social change to addressing technological trends and organizational issues to transformational aspects of daily life.[3] Just developing many concepts alone is not enough if the political will to implement them is lacking, or as Fred Luks puts it: "There is a gap between vehement commitments to sustainability and actually effective transformation steps."[4]

Transdisciplinarity can help to achieve transformation. The term transdisciplinarity, formed in the 80s of the twentieth century, belongs to the theories of science that are used to characterize forms of research that go beyond the subject-

[1]Cf. Meadows Donella (2020).

[2]Cf. Jorgen (2013).

[3]Cf. Luks (2019): 16.

[4]Luks (2019): 234.

E. Neuhold (✉)
University for Continuing Education Krems, Krems, Austria
e-mail: Elfriede.neuhold@donau-uni.ac.at

© The Author(s), under exclusive license to Springer Nature Switzerland AG 2025

M. Schüz (ed.), *Transdisciplinary Thinking and Acting*, Transdisciplinary Management of Ecological and Social Crises,
https://doi.org/10.1007/978-3-031-70680-6_8

specific and disciplinary constitution of science in a problem-related way.[5] Scholz and Steiner have expanded the concept to include the exchange of science with civil society and practitioners. In the journal "Sustainability Science," transdisciplinarity is defined as follows: "Transdisciplinarity integrates or relates different epistemics from science and practice... or from branches of disciplines if interdisciplinary integration is impossible...".[6]

In 1994, in the "Charter of Transdisciplinarity," Basarab Nicolescu started from a concept of transdisciplinarity based on mutual respect, on the recognition of cultures and religions and the promotion of the humanities. Even then, he warned against technocratic achievements of the sciences, against spiritual self-destruction and an increasingly impoverished inner identity. This article will build on Nicolescu by focusing on the dialogue or connection of East with West in relation to different cultures, religions, and philosophies.[7]

Philosophies are explanatory models and attempts to represent what the world is or how the world appears. From these arise world views, from which people derive their own positions with regard to values, attitudes, and relationships in dealing with nature and with their peers. In the present work, epistemological approaches of the West and the East are to be compared with each other, in order to derive from it extended world views and attitudes. For the problem described above is the consequence of a materialistic view of the world, which primarily allows profit-optimized economic activity based on pollution and exploitation of nature.

Therefore, this work deals with a comparison of Western epistemologies of modern times and an epistemology of the East, the Advaita Vedanta according to Śankara, in order to gain new approaches to solve our world problems through an expanded understanding of truth.

2 Epistemologies of the West and the East

2.1 Basics

Philosophy, as love (philo) of wisdom (sophia), is, like other sciences, in search of truth. It is divided into different areas, such as natural philosophy, ontology, metaphysics, and ethics as practical philosophy. Epistemology addresses the question of the conditions under which the knowledge of truth becomes possible. The most important epistemological approaches, especially of modern times and of ancient India, specifically Advaita Vedanta, are presented here.

The philosophy of Advaita Vedanta originated before 1000 BC in India in the Upanishads of the Vedas and was formulated as a philosophical system by Śaṅkara

[5]Cf. Mittelstraß, Encyclopaedia Philosophy and Philosophy of Science (2004a, 2004b): 329.

[6]Scholz and Steiner (2015): 1.

[7]Cf. Basarab Nicolescu (1994).

in around 900 AD. Śaṅkara's Vedanta philosophy is based on the Upanishads, on the one hand, and on Samkhya philosophy on the other.[8]

Samkhya philosophy is a school of Indian rationalism. It is dualistic and assumes that, on the one hand, there is the material world or nature, and, on the other hand, the primordial matter or the Self from which the material world emerged.[9]

The literature also discusses whether and to what extent Śaṅkara's philosophy incorporates elements from Buddhist Mahayana philosophy or the Vijñānavāda school (Yogacara). The latter was founded by Asanga and Vasubandhu in about the fourth century AD. Its core teaching is that all perceptible phenomena arise only on the basis of mind and as such are insubstantial. Consequently, all perceptions are understood as mental projections. Accordingly, the experience of the phenomenological world is an illusionary experience that only takes place in consciousness.[10]

While some say that Śaṅkara only adopted methodological elements from Buddhism,[11] others assume that philosophical ideas and arguments such as non-dualism were also adopted.[12]

In this respect, Śaṅkara based himself on the Upanishads, on the one hand, and, on the other hand, integrated influences from a dualistic Samkhya philosophy as well as of the non-dualistic Mahayana Buddhism and summarized them in many teaching poems as a systematic philosophy.

In verse 298 of his summary of the core thoughts of all the Upanishads, "*sarva-vedānta-siddhānta-sāra-saṅgraha,*" he writes about the concepts of reality and truth and about the relationship between the world and absolute reality:

> Reality means the Absolute Reality of the essence of Truth, Consciousness and Bliss. The universe is a projection onto the Absolute Reality, just as the blueness is a projection onto the sky.[13]

This results in the synonymous use of reality, truth, and consciousness. Of particular importance is the distinction between the impermanent and the eternal, whereby only the impermanent endures as really being and is thus true and real:

> The distinction between the eternal and the transient is defined as the knowledge that only the Absolute Reality is eternal, but everything else is transient.[14]

---

[8] Cf. Larson (1987): 41.

[9] Cf. Garbe (1894): 137.

[10] Cf. Tola and Dragonetti (2005).

[11] Cf. Reynolds (1975): 121.

[12] Cf. King (1998): 81.

[13] Sankaracarya, The Heart of Vedanta (2003): 143.

[14] ibid.: 31.

2.2 *Epistemologies of the Greeks and Modernity*

Theories of knowledge have existed since the time of the Greeks. Parmenides, one of the pre-Socratics from the sixth century BC, was one of the first philosophers to distinguish between experience mediated by the senses and knowledge opened up by the logos. Parmenides claims that only the logos allows true knowledge and that this would exclusively concern the imperishable being. However, when the logos, i.e. experience-free thinking, experiences that which exists, which transcends ordinary reality, at a particular moment, Parmenides encounters a reality that is beyond that which can be experienced with the senses and which thus transcends the world. He is considered the founder of the "doctrine of what is (greek On): the doctrine of being or ontology." Parmenides defines being as unbecome and imperishable, as one and uniform, and as unchanging and perfect, thus influencing Plato's theory of ideas and Aristotle's concept of God.[15]

Plato's theory of knowledge consists in the connection of the world of ideas with the world of the senses. The human soul has viewed the ideas before entering the human body. By allowing reason to remind the soul of the ideas it has seen before (anamnesis), knowledge is gained. It is the soul that forms the link between the world of ideas and the world of the senses. On the one hand, it belongs to the world of the senses through its desires and affects; on the other hand, it belongs to the world of ideas through reason.[16]

With his principles of cognition, Aristotle developed a doctrine that is not a generalization, but a cognition that carries out an abstraction based on perception by the intellect. Aristotle is neither an empiricist nor a rationalist per se.[17]

In modern times, the epistemologies of rationalism and empiricism in particular, as well as constructivism and realism, were developed as opposing approaches. Realism, however, never achieved significance. Kant then, as will be shown, derived the principles of cognition from pure reason with the Critique of Pure Reason, thus linking rationalism with empiricism.

2.3 *Rationalism*

While rationalism assumes that cognition arises primarily through thinking, the empiricism assumes that all cognition is based on experience gained through the senses.

René Descartes, mathematician and physicist, is considered the founder of rationalism. He seeks and finds the unshakable foundation of everything in thought:

[15]Cf. Höffe, Kleine Geschichte der Philosophie (2008a, 2008b): 27 ff.

[16]Cf. ibid.: 39.

[17]Cf. Höffe, Aristotle (2008a, 2008b): 55.

"'cogito ergo sum,' (I think, therefore I am)."[18] In his "Meditations on the Foundations of Philosophy" he tries to get to the foundations by doubting everything. In this first meditation, "What can be doubted," he explains that the senses had deceived him and that it is a precept of prudence never to trust completely those who have already deceived us once.[19]

In this essay, Descartes doubts everything he perceives. He compares perception to that in dreams, where, for example, the perception of the body and clothes can be identical to the perception of the waking state. He assumes that everything can be wrong except the colors with which our consciousness paints the pictures of things. Only in mathematics, more precisely in arithmetic and geometry, is there an undoubted certainty that, for example, $2 + 3 = 5$ or that a square always has four sides. Because this result is independent of whether one is awake or asleep.[20]

One could assume, according to Descartes, that there is a God who created everything. But it could also be: "...that there is no earth at all, no heaven, no extended thing, no shape, no size, no place, and yet that all this seems to be merely there just as it now appears to me."[21] In the second meditation "On the nature of the human mind" he deals with the question of what remains even if he doubts everything. Even if he were to be deceived by someone, a god or an evil spirit, it is undoubted that he is. Thus, the sentence "I am, I exist" is necessarily true.[22]

But knowing that I am (ego sum, ego existo) is not enough for Descartes. He wants to know exactly what he is. And after all that he is not, namely neither body nor sensation, thinking remains: for as long as he can think, he is a thinking being.[23] Consequently, we are a "res cogitans," a thinking thing or being. It should be mentioned here that cogitare is translated in some places as thinking, whereas cogitatio is translated in other places either as thinking or as consciousness. We do not find any further specification of this here. Descartes does not deal with the definition of spirit, soul, mind, and reason at this point.

2.4 Empiricism

According to the physician and philosopher John Locke, knowledge of reality is based on sensory perception. In his main work "An Essay concerning Human

[18] Cf. Mittelstraß, Encyclopaedia Philosophy and Philosophy of Science (2004a, 2004b), vol. 1: 453.

[19] Cf. Descartes (1992): 32.

[20] Cf. ibid.: 35.

[21] Ibid: 37.

[22] Cf. ibid.: 44.

[23] Cf. ibid.: 47.

Understanding," he deals with the epistemology of empiricism, which is based on observation and experience.[24]

He thus does not see thinking that is detached from experience: "Locke traces all the contents of thought back to simple ideas of external (sensation) and internal perception (reflection)."[25] Consequently, reflection per se is not possible without previous impressions and perceptions through the senses.

The two approaches of rationalism and empiricism can each be justified, but they are not compatible with each other.

2.5 *Critique of Pure Reason According to Kant*

Kant develops a new method of gaining knowledge.[26] He forms the synthesis of the contradictory theories of cognition, namely rationalism and empiricism. According to him, all cognition is a product of rational and empirical factors, but with one exception: he calls cognition derived from pure reason as cognition a priori, and cognition derived from sense perception as cognition a posteriori.

Those cognitions that "...take place absolutely independently of all experience..." or to which "nothing empirical is added at all..."[27] he calls cognitions a priori and thus pure. Kant discusses what he understands by "pure." Thus, he comes to the conclusion that a proposition is pure when two factors are fulfilled. First, the proposition's statement must be "necessary" (necessity); second, it must be so strictly general in thought that there can be no exception to it (strict generality).[28]

As examples of cognition a priori, which he derived from the concepts, he named mathematics in the first place, because they were necessary and could not be derived from experience.[29] Moreover, there are no exceptions to this. According to Kant, pure mathematics therefore contains no empirical basis, but only pure cognition a priori.

With this, he justifies the Critique of Pure Reason: "For reason is the faculty which provides the principles of cognition a priori. Therefore, pure reason is the one that contains the principles for knowing something a priori".[30]

---

[24] Cf. Höffe, Kleine Geschichte der Philosophie (2008a, 2008b): 182.

[25] Ibid: 183.

[26] Cf. ibid.: 211.

[27] Immanuel Kant (1978): 51.

[28] Cf. ibid.: 52.

[29] Cf. ibid.: 64.

[30] Ibid.: 73.

2.6 Constructivism

The basic thesis of constructivism assumes that humans do not have access to objective reality, but that reality is subjectively constructed. There are three founders of constructivism: the cyberneticist and physicist Heinz von Förster, the psychologist Ernst von Glasersfeld, and the psychotherapist and communication researcher Paul Watzlawick.[31] All three have somewhat different approaches to constructivism. Due to the importance of constructivism for this work, the theories of Watzlawick and Förster in particular will be examined in the following.

Radical Constructivism According to Paul Watzlawick

Paul Watzlawick sees truth as something that is experienced through the senses and interpreted on the basis of the experiences made. In this respect, objective truth is possible, but it is difficult for people to get it. The experienced truth is thus a reality constructed by man, which is subjective and therefore different from person to person. Thus, Watzlawick describes in a conversation with Bernhard Pörksen that our sense organs convey an image of reality via our central nervous system and that we create an individual second-order reality through the attribution of meaning, significance, and value.[32]

Here, a distinction is made between two realities. The first-order reality is therefore that which is conveyed to us by the sense organs. After that, there are attributions of sense, meaning, and value, which give rise to a second-order reality. And since this varies from person to person, there is no absolute truth, but only subjectively constructed realities, even if people come from the same cultural circle and family system, where there are similar value concepts and attributions of meaning.

Watzlawick even goes one step further when he says that all we know about real reality is that it is not. For only in the collapse of our constructions of reality, according to Watzlawick, do we understand that the world is not as we have designed it.[33]

The reason why this is so, is because of communication and language. In the book "How real is reality," Paul Watzlawick deals with reality, which is the result of communication. The belief that there is only one reality is the most dangerous of all self-deceptions. Rather, there are countless, even contradictory, perceptions of reality that are not the reflection of eternal, objective truths.[34]

---

[31] Cf. Pörksen and Watzlawick (1997): 383.

[32] Cf. ibid.: 389.

[33] Cf. ibid.: 389.

[34] Cf. Watzlawick (2005): 7.

He does not see the epistemology of radical constructivism as dogma, but questions it himself. He even claims that science does not serve the knowledge of truth, because absolute truth cannot be grasped.[35]

According to Watzlawick, there is no objective conception of reality.

Radical Constructivism According to Heinz von Förster

In his contribution "Constructing a Reality," Förster states that we have "blind spots" in the area of our senses. Consequently, he does not deal with thinking in general, but with the neuronal organization of our central nervous system and the interplay of the impulse of sense perception, the conduction of the impulse via electrical charges and the synaptic connections. In this context, the nervous system plays a far greater role than the sensory organs: "Since we only have about 100 million sensory cells, but our nervous system contains close to 10,000 billion synapses, we are 100,000 times more sensitive to changes in our internal environment than to changes in our external environment."[36]

After a detailed analysis of the processing of sensory impulses in the brain, he concludes: "The nervous system is organized, or organizes itself in such a way, that it calculates a stable reality." The organism therefore regulates itself autonomously.[37]

From this he deduces that reality is only constructed in the brain, but that this is dependent on a relationship to another. For if one were to assume that one only exists oneself, then all other people would only be in one's own head. This theory is solipsism, which Förster, however, rejects. He assumes that there are other organisms and that the image of the world develops through interaction: "It is the relationship between the YOU and the I, and this relationship is called identity: reality = community." From this, Förster concludes his ethical imperative: "Always act in such a way that further possibilities arise."[38]

Here, the five essential epistemologies are presented in overview (see Table 1):

Up to this point, five basic Western epistemologies have been presented. In the case of the theory of radical constructivism, more differentiation has been made between the two explanations because of the importance of this model. Overall, there is a tendency to depict cognition as a two-stage process.

---

[35] Cf. Pörksen and Watzlawick (1997): 391.

[36] Cf. Förster (1998): 51.

[37] Ibid.: 57.

[38] Ibid.: 60.

Table 1 Overview of western epistemologies

Theory	Author	Source of knowledge	Essence
Rationalism	Descartes	Mind	Everything starts from thinking
Empiricism	Locke	Experience	Everything starts from experience
Critique of pure reason	Kant	Principles of cognition 1. A priori 2. A posteriori	There are both, cognition through thinking and experience
Radical constructivism	Watzlawick	First order: senses Second order: attribution of value, meaning, and sense in the brain	Everyone constructs a different subjective idea of reality.
Radical constructivism	v.Förster	Stage 1: Sensory receptors Stage 2: Calculation of a stable reality in the brain	Brain autonomously constructs reality, only partly on the basis of sensory perception

Source: The author

2.7 *Finding Knowledge According to the Ancient Indian Philosophy of Advaita Vedanta*

The philosophy of Advaita Vedanta is recorded in the Upanishads of the Vedas, the ancient scriptures of India. The word "Veda" means knowledge and comes from the root "vid," to know. There are four major parts of the Vedas, Ṛg-Veda, Sāma-Veda, Yajur-Veda, and Atharva-Veda. Each of the four Vedas consists of four parts: the Samhitas as the original texts, the Brahmanas as commentaries, the meditations (āranyakas), and a philosophical part (upaniśads). The word Vedanta thus consists of the partial syllables "Veda" and "anta" (end), i.e. the Upanishads are the final part of the Vedas. These contain the philosophical writings that are the basis of the ancient Indian philosophy of Advaita Vedanta.[39]

Advaita consists of the word "Dvaita," which is duality, and the prefix "a," which is negation. Advaita means the non-duality. Thus, the philosophy of Advaita Vedanta is the doctrine of non-duality or non-separateness, whereby the world as duality is mere appearance (māyā).[40]

The conclusion from non-duality as the main statement of the Vedas is thus the non-duality or non-separateness of all being: everything is contained in Brahman, the eternally swelling Infinite. The infinite is also described in other words: Sat, Chit, Ananda, which means being, consciousness, and bliss.

This is reflected in the four principles of Vedanta (see Table 2):

All four statements describe the relationship of individual self (atman) with supreme self (brahman, the Infinite). Both from the direct perspective of the I and the You, and abstractly in the third person: there is no distinction: I, you, the Self,

[39] Cf. Sankaracarya (2003): 19.

[40] Mittelstraß (2004a, 2004b), vol. 1: 50.

Table 2 Main statements of the Upanishads

Principle	Translation	Upanishad	Part of the Vedas
prajñānam brahma	Brahman is consciousness	Aitareya upaniśad 5.3	ṛg-veda
Aham brahmāsmi	I am Brahman	brihadāranyaka upaniśad 1.4.10	Yajur-veda
Tat tvam asi	That is (also) you	Chandogya upaniśad 6, 6–7	sāma-veda
Ayam ātma brahma	The self is Brahman	mahadāranyaka upaniśad 2	Atharva- veda

Source: Sankaracarya (2002: 48)

and Brahman are one. And Brahman, the Infinite, is the consciousness in which and through which the endless multiplicity of the finite appears.

Thus, Advaita Vedanta is the doctrine of non-duality or non-separateness: non-separateness in the sense that **everything** exists **in and through the One,** and that is consciousness itself, or in other words, Brahman, the Infinite. The Infinite is the only thing that exists in reality and therefore consciousness is the actual reality or truth.

The reality of consciousness is to be explored and through meditation and other techniques it should be possible to connect with this reality and in this way to gain an intuitive, direct experience of the infinite. However, the direct experience or realization of the infinite is not part of this work.

Epistemological Approaches in Advaita Vedanta According to Śaṅkara

Śaṅkara wrote several works in which he summarized the Upanishads. His main work is the *"brahma-sūtra-bhāṣya,"* which is not only THE source of Advaita Vedanta, but also one of the most influential texts of the entire Indian tradition.[41]

In the various summaries by Śaṅkara, which this work is based on as presented in the first direct Sanskrit-German translations, Śaṅkara explains the connections to a disciple, where the context is of overcoming the fetters of the cycle of suffering through rebirth. Against this background, the knowledge of the philosophical part of the Vedas is explained and the non-separateness of all that exists is expounded. Thus, in the "Wreath of Verses on the Knowledge of the Absolute or Infinite" (*"brahma-jñāna-āvali-mālā"*), one of his shortest teaching poems, Sankara explains the essence of man, whether disciple or master, in verses 14 and 17:

> I am nothing but pure consciousness and pure knowledge. I do not act, do not experience. I am the 'I Am'. I am eternally imperishable. (14)

[41] Mittelstraß (2004a, 2004b), vol. 3: 667.

> The seer and the seen, these two categories are different from each other. The seer is the Absolute Reality, the seen is delusion (maya). This is the quintessence of Vedanta. (17)[42]

In verse 20, the concluding verse of this short text, Śaṅkara sums it up:

> The Absolute is real, the world is unreal. The soul is truly the Absolute Reality, nothing else. With this tenet is to be taught the science of truth. This is the quintessence of Vedanta.[43]

The Vedic philosophers—like all ancient masters—work with parables. One simile that is used quite often is the image of a rope, which is mistaken for a snake in the darkness, and the image of a jar made of clay. The jug is made of clay. When it breaks, the clay as the base material of the jar is still clay. In verse 19 of the above teaching poem, the parable is given:

> The whole earthy jar: floor, walls, etc., is always clay. In the same way, the whole world is the Absolute. That is the quintessence of Vedanta.[44]

Śaṅkara thus states that everything is made of the same material. Just as in the time before Christ most objects were probably made of clay or wood, he uses the material clay as an example from which many utilitarian objects, and even houses, are formed. Whatever the shape of that which is formed from it, the basis is and remains unchanged the basic material clay. And it is the same, according to Śaṅkara, with the forms and objects of the world, be they trees, people, or planets: everything is woven from the same material of consciousness.

The word **knowledge** also appears in these scriptures. However, since these scriptures are aimed at explaining what is behind this visible world and how to get out of the birth cycle, the knowledge to be gained does not fundamentally refer to this world, but to the reality of consciousness behind it. Thus, in verses 64 and 65, Śaṅkara writes about recognizing and knowing the truth:

> Whatever is seen or heard is nothing but the Absolute. When the truth is realised, one knows that everything is the non-dual Absolute Reality, Being, Consciousness and Bliss. (64)
>
> With the eye of knowledge, one sees the omnipresent being, consciousness and happiness. Without the eye of knowledge, one cannot see it, just as the blind man cannot see the shining sun. (65)[45]

Cognition according to Śaṅkara is a combination of the perception of the senses with cognition in the mind, because first one hears and sees something, and in the mind one recognizes that these objects of sensory perception are not what they appear to be. In this respect, verse 64 gives a two-stage cognition.

Where or how one attains this special ability of seeing is explained in the following verse. Here it is explained that the sun of knowledge rises in the sky of the heart and destroys darkness, i.e. the state of not knowing the truth.

---

[42] Sankaracarya (2002): 24.

[43] Ibid.: 25.

[44] Ibid.: 25.

[45] Sankaracarya (2002): 98f.

Let us now turn to the statements about the reality of the world, the universe and all worlds. Śaṅkara discusses this in the same teaching poem in verse 8:

> Worlds arise and submit in the cosmic primordial ground like bubbles in water.[46]

Here the comparison is made with a water bubble, which exists but is of a very transient nature. But in the very next verse, the quality of physical appearances is put into perspective again and it is explained that the water bubbles are not real either, but only appearances of the Self:

> The manifold appearances are all founded as conception in the eternal sustainer of the universe, Vishnu, in the all-pervading Self, in being and consciousness, just as bangles, clasps, etc., are potentially contained in gold.[47]

Accordingly, the world of appearances does exist, but they are themselves a conception of that power which sustains the universe, which is again called Being and Consciousness. This means that although the appearances appear, i.e. become visible, they have no existence of their own based in themselves, but are founded in the being and consciousness behind them. Here again a simile is used. It is about the material gold, which was familiar in ancient India. Many things can be made from this material, especially jewelry. But the material, the gold, always remains gold, regardless of what one forms from it, be it a bracelet or a hair clip. Just as gold is the material of the bangle, consciousness is the material of the visible world of appearances.

In the following verse 18 of the same poem, the faculties of the mind are addressed:

> The Supreme Self is to be distinguished from the body, senses, mind, intelligence and matter at all times, and to be recognized as the observer of their being, functions and action, just as the king distinguishes himself from the servants and observes them.[48]

Here, the senses, thinking and intelligence are explicitly addressed, and it is stated that there is additionally an observing consciousness that is distinct from the thinker.

We have already touched on several aspects of perception and cognition. In Vedānta, therefore, there is a three-stage cognition:

1. Sensory perception—perception of the world (example, snake in the dark).
2. Generation of an image by interpretation of the perceived as real (example, snake is a wrong image, it is a rope).
3. Cognition of the infinite (works via the eye of cognition as an inner method of cognition, independent of senses and mind).

Firstly, human cognition is possible through sense perception. There is talk of hearing and seeing. There is sense perception, note the example with the clay.

[46] Ibid.: 72.

[47] Ibid.: 72.

[48] Ibid.: 72.

Table 3 Three stages of cognition according to Advaita Vedānta

Stages of knowledge	What?	With what?	By what?	Result
1. Level	The world	Sensory perception	Senses	Incorrect
2. Level	The world (immanent)	An image is—probably cognitively—generated	Maya—power of deception works—probably in the brain or central nervous system	Incorrect
3. Level	The world beyond (transcendent)	Perception and knowledge of the transcendent truth	Inner eye, the "eye of knowledge"	True

Source: The author

Secondly, there is thinking, the intellect, as stated in verse 18 above, but here maya, the power of deception, is at work so that what is perceived turns out not to be finally true.

Thirdly, there is direct vision through the "eye of knowledge" as immediate knowledge, which functions neither through the sense organs or the human eye nor through the mind. As already mentioned, we are not concerned with this third possibility of cognition here.

Further relevant to the question of how cognition works in Śaṅkara's philosophy is the concept of "māyā," the cosmic **power of delusion**. Thus, Śaṅkara writes in verses 1 and 3 of the *"Māyā-panchakam,"* the five verse poem on Maya:

> Maya, extraordinarily skilled in creating the unreal, brings forth in Me—the Supreme Consciousness, free from all conceptions, the Incomparable, the Eternal, the Indivisible, the Undivided—the diversity of world, God and individual souls. (1)
>
> Maya, extraordinarily skilled in creating the unreal, binds the non-dual undivided consciousness from the essence of bliss to the ocean of rebirths, bounded by space, air, fire, water and earth, and confuses it beyond measure. (3)[49]

Māyā is a force that deceives the senses. It causes the senses to believe that objects and people are real. Thus, it is responsible for our senses perceiving the unreal as real.

The three stages of knowledge according to Advaita Vedānta in the sense of Śaṅkara are (see Table 3):

Vedantic epistemology was not explicitly specified as -ism in the Vedas over 3000 years ago. Because of its similarity to constructivism, the author therefore refers to it as "Vedantic constructivism."

As already mentioned, several times, in this work we do not deal with the third stage of cognition, which is reserved for those spiritually seeking and finding. Thus, on the part of the Vedic cognitive process, we consider only the first two stages, parallel to the two stages of Western epistemology.

[49] Sankaracarya (2003): 21f.

2.8 *Summary and Meaning of Advaita Vedānta*

While Vedānta asks how one can know the truth and primordial reality that lies beyond the visible world (transcendent), Western theories of knowledge ask how one can know the truth in the world (immanent)! Cognition in the sense of Advaita Vedānta only takes place after dealing with the question of what truth is in the first place. The statements of the four Upanishads such as "this is you," "this self is Brahman," "I am Brahman" complement each other alternately. According to Śaṅkara, consciousness is therefore the primary reality, nature secondary or an illusion in consciousness.

The core statement of Advaita Vedānta in the sense of Śaṅkara is that the essence of man and the truth of consciousness are not separate. There are not two things or beings here, there is only one, namely Brahman, the infinite being and consciousness. Every individual consciousness, and the human being is also directly addressed here by You, is Brahman. Every individual soul is the highest being. Here there is no difference, no division, no being placed separate in the world. Man, and God, individual and the Infinite are united and one in the ocean of infinite Being. The non-separateness of all being on the stage of the infinite is thus the core statement of Advaita Vedānta.

This non-separateness or existential connectedness of all that exists provides a world view in which people can have a new relationship with each other and with nature. Through this approximation of East and West in the field of knowledge, approaches to action can be questioned that disregard the totality of being, and such holistic approaches to action can be supported that take into account the non-separateness of everything.

3 Comparison of Western and Eastern Epistemological Approaches

Since Western epistemological approaches differ from that of Vedānta, the latter is compared with the various Western approaches and examined for commonalities.

First, rationalism is analyzed to see to what extent parallels to Vedānta are present here. René Descartes stated that we are primarily thinking beings. There is no parallel here. However, it is interesting that Descartes assumes that his senses could deceive him. Thus, Descartes assumes that there is an evil spirit that wants to deceive him and that everything is only played out for him. This approach is interesting, because if we compare this with Śaṅkara, we find a force here, although it is not called an evil spirit, that causes this pretense: Māyā, which is extraordinarily skilled in creating the unreal.

Locke's **empiricism** states that knowledge is always based first on the experiences of the senses. Our observing provides the mind with the material for thinking.

Locke does not deny that we need thinking, but it is downstream from sensory perception.

In Vedānta, there is also sensual perception, but this does not serve as a means of finding the truth. Therefore, there is no clear correspondence between empiricism and Vedānta.

In the field of tension between rationalism and empiricism, Kant's **transcendental philosophy** is to be viewed, which considers both a priori and a posteriori knowledge to be possible. The principles for pure cognition, in which sense perception plays no role, are postulated. In comparison with the Vedānta, there are no parallels here, because there is neither knowledge through the mind, and certainly not knowledge through special principles of the mind. In this respect, Kant's Critique of Pure Reason brings no further correspondences with Vedānta.

Realism, which is of minor importance in Western philosophy, also does not fit with Vedānta here either, because the Vedas say exactly the opposite, namely that this supposedly real world is not real.

However, **constructivism** is of interest. Constructivism assumes that the world is a construction in our brain. Since the approaches of Paul Watzlawick and Heinz von Foerster differ somewhat, they will be examined separately.

Watzlawick assumes a first-order reality, that is, that of sensory perception. In a second step, however, these impressions are processed in the brain and linked with meaning. Since this linking is an individual-subjective attribution of meaning and significance, the result is also subjective and cannot offer a picture of an objective reality. This is second-order reality. His core statement is thus that each person creates his or her personal reality and that this is always subjective and not objective. He leaves open the question of whether there is an objective reality, but we cannot recognize it. In comparison with Vedanta, both have sense perception as the first impression, but after that the two systems separate; while Watzlawick assumes subjective, personally colored images of reality, Vedanta assumes deception.

It is different when one looks at radical constructivism according to **Heinz von Förster.** Förster also sees cognition as two-stage, but much more precisely on the basis of scientific knowledge about sensory perception by means of receptors and about cognitive performance in our brain. Here, too, the formation of reality is individual and two-stage. In the first stage, it is not the sensory perception as such, but concretely the reaction of the receptor cells of the sensory organs. In the second stage, these impulses from the cells of the sense organs are transmitted to the central nervous system, in particular to the brain, where they are processed. He comes to the conclusion that the cognitive nervous system interprets and processes the impulses. Thus, the resulting image of reality is a purely constructed image and not reality itself. He does not go so far as to say that the world and all living things are also constructed. Of importance here is the conclusion that Heinz von Förster draws, in which he points to the relationship of different people with their individually formed realities, and that a common image is obtained through the community. Therefore, with his ethical imperative, he demands a behavior through which one opens up the greatest possible scope in the relationship between I and YOU.[50]

According to the Western and Eastern epistemological approaches described above, cognition thus has two or three stages. Firstly, perception takes place through the senses; secondly, processing takes place in the brain and an image of reality is created, which does not correspond to reality. And then there is a third stage, which exists in addition to these two in Vedanta, namely the realization of reality or the infinite, but this takes place neither through the senses nor through the mind. This third stage is excluded from this consideration in order to remain comparable.

Let us take a closer look at the meaning of these two constructivist theories: constructivism in the sense of Heinz von Förster and Vedic constructivism.

If one compares the two approaches, we find that there is a first common level of sensory perception in both, which conveys information about the world.

In a second step, according to Förster, the impulses are processed in the brain and constructed into an image, through the autonomously regulating neuronal circuits in the brain with the aim of creating the greatest possible inner order. In the case of Vedanta, an image of reality is constructed in the second step—the exact mechanisms (maya) through which this happens—must be left open at this point, but this is only a deception. Both assume that images are created that at most reflect a subjective reality, but in no way can an objective reality be derived from them. In this respect, Heinz von Förster's constructivism and the epistemology derived from Vedanta are most similar, if one disregards epistemological interpretations of quantum physics, which cannot be pursued further here.[51]

But here is the difference: Vedanta assumes that the images that arise in the mind are just images and ultimately not real, because we see the jewel made of gold, but in its essence, it is just gold. Only the consciousness behind it, the infinite, which is non-dual, has a claim to reality.

But what does this mean for human behavior? It means that all human beings and all living beings are not separate from each other, even if we construct images of the separate world. Nevertheless, the world exists, even if it is not real, and we have to get along with each other. The conclusion, therefore, is that we humans recognize the common consciousness behind the construction of separateness of beings.

While in most philosophical texts Indians deal with how to get out of this dual world again by means of enlightenment or reaching the state of nirvana and thus having true knowledge in the sense of the third stage, the consequence for people on the level of the second stage of knowledge is that we see the physical form like jewelry and not the material gold, of which it is made of. So according to this we should recognize that the material or the essence of ourselves and other beings is identical and take this into accout in our dealings with each other. In this context, it

---

[50] Cf. Förster (1998): 60.

[51] A similarity of the epistemological conclusions from quantum physics with the Vedanta teaching is pointed out above all by the Heisenberg student and friend Carl Friedrich v. Weizsäcker (cf. Schüz 1986: 195 f).

does not matter for the recommendation of action whether one assumes that this world is constructed (2nd level) or whether one says that it is only an illusion (3rd level). The point is to act with the knowledge from the 3rd stage in the constructed world (2nd stage). From this, from the combination of Heinz von Förster's constructivism and derived action guidance and Śaṅkara's constructivism and its indirect action guidance for human beings in a subjective world, a new non-duality ethical imperative can be formulated[52]:

Always act in recognition of non-separateness from others respectively interconnectedness with all beings!

4 Conclusions

In the spirit of Basarab Nicolescu, who respects the different cultures and religions as equal, this Eastern approach is compared with the Western approach. Like Parmenides and Plato, the philosophical concept of Advaita Vedanta according to Śaṅkara refers to the non-duality or non-separateness of everything that exists and offers an implicit epistemological approach. In this context, the Western epistemologies of the modern era also offer a multi-stage process of cognition, as does Vedantic constructivism.

Two of the constructivist theories mentioned are most similar: namely those according to Heinz von Förster and Śaṅkara. While Förster derives his ethical imperative from this as a recommendation for action, additionally a new, connecting ethical imperative can be derived: Always act being aware of interconnectedness with all other beings.

If one assumes that, on the one hand, the world as such does not exist, but only different images of it in our heads and, on the other hand, that everything and everyone are interconnected or, even more, are one in their essence, then this has direct implications for our living together on earth. For with this, everything that one does to other people and to nature, one does to oneself. From this state of mind of interconnectedness on a higher level, it can only follow that we should treat each other and nature with care, even with love.[53]

Looking at the non-separateness of all that exists gives us that inclusive worldview that can guide us to thoughts, words, and deeds in order to overcome a blind and even destructive materialism, on the one hand, and to jointly overcome the manifold crises of post-modernity on the other. Thus, this epistemological approach offers the basis for an ethical recommendation for action and thus hope for a transdisciplinary jointly implemented transformation toward a society characterized

[52] Cf. the "ethics of the unity of the real," as Schüz derives it from the findings of modern physics (1986: 251 ff).

[53] Cf. Schüz (1986): 263–266.

by well-founded peace and mutual appreciation and love for all human beings and all of nature.

References

Basarab Nicolescu, L. D. (1994). *Charta der Transdisziplinarität.* Abgerufen am 01. 04 2022 von Interdiscipliary enzyclopedia of religion & science. https://inters.org/Freitas-Morin-Nicolescu-Transdisciplinarity

Descartes, R. (1992). Meditationes de prima philosophia/Meditationen über die Grundlagen der Philosophie. (A. G. Gäbe, Hrsg.) Hamburg: Meiner.

Förster, H. V. (1998). Das Konstruieren einer Wirklichkeit. In P. Watzlawick (Ed.), *Die erfundene Wirklichkeit.* Piper.

Garbe, R. (1894). *Die Samkhya-Philosophie.* Verlag von H. Haessel.

Höffe, O. (2008a). Aristoteles. In O. Höffe (Hrsg.), Klassiker der Philosophie - Von den Vorsokratikern bis David Hume (Bd. 1). München: C.H. Beck.

Höffe, O. (2008b). *Kleine Geschichte der Philosophie.* C. H. Beck.

Immanuel Kant, I. H. (1978). *Kritik der reinen Vernunft.* Reclam.

Jorgen, R. (2013). *2052, eine globale Prognose für die nächsen 40 Jahre. Der neue Bericht an den Club of Rome, 2. Auflage.* München: oekom-Verlag.

King, R. (April 1998). Early Advaita and Madhyamaka Buddhism: The case of the Gaudapadiyakarika. International Journal of Hindu Studies 2,1, 67–83.

Larson, G. J. (1987). *The encyclopedia of Indian philosophies, volume 4—Samkhya, a dualist tradition in Indian philosophy.* Princeton University Press.

Locke, J. (1872). Ein Versuch über den menschlichen Verstand. In J. Kirchmann (Ed.), *Johne Locke's Versuch über den menschlichen Verstand.* Verlag von L. Heimann.

Luks, F. (2019). Chancen und Grenzen der Nachhaltigkeitstransformation. Ökonomische und soziologische Perspektiven. (F. Luks, Hrsg.) Wiesbaden: Springer Gabler.

Meadows Donella, J. R. (2020). Grenzen des Wachstums—das 30-Jahre Update. Signal zum Kurswechsel, *6. Auflage.* Stuttgart: Hirzel Verlag.

Mittelstraß, J. (Hrsg.). (2004a). *Enzyklopädie Philosophie und Wissenschaftstheorie* (Bd. 4). Stuttgard Weimar: J.B. Metzler.

Mittelstraß, J. (Hrsg.). (2004b). *Enzyklopädie Philosophie und Wissenschaftstheorie* (Bd. 3). Stuttgard: J.B. Metzler.

Pörksen, B., & Watzlawick (Heft 4. Jahrgang 30 1997). "Wir können ovn der Wirklichkeit nur wissen, was sie nicht ist" Bernhard Pörksen im Gespräch mit Paul Watzlawick. Communicatio Scialis - Internationale Zeitschrift für Kommunikation in Religion, Kirche und Gesellschaft.

Reynolds, E. T. (1975). *Masterthesis: On the relationship of advaita vedanta and madhyamakika buddhism.* University of British Columbia.

Sankaracarya, S. (2002). Sieben Kleinode geistiger Erkenntnis. (E. Meyer, Hrsg.) Argenbühl: Heinrich Schwab Verlag.

Sankaracarya, S. (2003). Das Herz des Vedanta. Zusammenfassung der Kerngedanken sämtlicher Upanishaden. (E. Meyer, Hrsg.) Argenbühl: Heinrich Schwab Verlag.

Scholz, R. W., & Steiner, G. (14 August 2015). The real type and ideal type of transdisciplinary processes: Part I—theoretical foundations. *Sustainability Science* (10).

Schüz, M. (1986). *Die Einheit des Wirklichen—Carl Friedrich v. Weizsäckers Denkweg.* Neske/ Klett.

Tola, F., & Dragonetti, C. (2005). Philosophy of mind in the Yogacara Buddhist idealistic school. *History of Psychiatry, 16*(4), 453–465.

Watzlawick. (2005). *Wie wirklich ist die Wirklichkeit? Wahn, Täuschung, Verstehen.* Piper.

Neuhold, Elfriede, Dr. iur., studied law at the University of Vienna, Austria, and graduated 1995, after completing an apprenticeship in agriculture. In her doctoral thesis, she dealt with environmental law and organic agriculture. She studied also European and Indian philosophy for several semesters with focus on Advaita Vedanta. During fifteen years in business, she worked in management in the fields of law, real estate, and purchasing. She was also responsible for environmental management and sustainability, in particular for the establishment and operation of an environmental management system according to ISO 14001 and for the preparation of sustainability reports. In 2013 she returned to university, concrete to the University of Continuing Education Krems. There she was course director in the Department of Construction and Environment for several years, dealing with environmentally and climate-friendly construction and operation of real estate, ethics and sustainability. She created the department's strategy and led the teaching team for several years. In this capacity, she created a conference, the Real Estate Future Day, which was held for several years and she also acted as editor of the corresponding conference proceedings. In addition, she was a member of the Senate as well as the Curricula Commission and the Task Force Sustainability. Now she deals with education for sustainable development in the Platform for Sustainable Development (SDGs) and with the significance of the natural world.

Transdisciplinarity for Mastering Poly-Crises by Mutual Learning Between Science and Practice

Gerald Steiner

1 Introduction

Neither the scientific community, practice, nor society as a whole are adequately equipped to address the **systemic risks** embedded within the interconnected **poly-crises** confronting the world today. Dealing with multiple—**either natural or man-made**—crises does not allow for a sequential approach given that **they tend to emerge intertwined and simultaneously.** Recent examples of systemic risks include climate change and the loss of biodiversity, or emergent viral threats such as the recent COVID-19 pandemic or the appearance of bird flu (H5N1) in Antarctica with potentially disastrous impact on the seal and penguin populations that have never been exposed to this virus, as well as the availability and accessibility of natural resources (biotic and abiotic). Other examples relate to mobility and migration movements between and within countries and continents, geopolitical conflicts and wars, and behavioral patterns of humans such as mistrust, or conspiracy thinking, leading to sociocultural and religious clashes of cultures and societies.

One might, therefore, ask whether it is the **simultaneous occurrence and interrelatedness of various crises** that puts limits to our scientific and practical understanding and the required problem-solving capabilities, and if we are fit enough to successfully deal with even a **single crisis** and their often inherent high degree of **systemic risks.** Major weaknesses in tackling such single crises have painfully become obvious for almost every single one of the recent past, be it the COVID-19 pandemic, recent geopolitical conflicts all across the world (e.g., Middle East, Ukraine), or natural disasters such as the major fires in Greece, Maui, and Canada, earthquakes in Morocco, floods in Libya and Europe, and so on.

G. Steiner (✉)
UWK Universität für Weiterbildung, Krems an der Donau,, Austria
e-mail: gerald.steiner@donau-uni.ac.at

© The Author(s), under exclusive license to Springer Nature Switzerland AG 2025

M. Schüz (ed.), *Transdisciplinary Thinking and Acting*, Transdisciplinary Management of Ecological and Social Crises,
https://doi.org/10.1007/978-3-031-70680-6_9

Further, various crises differ with respect to the potential **orchestrated (re)action time of human agents.** Climate change and biodiversity loss, for example, did not occur suddenly; rather, they have their roots decades ago. Accordingly, they were warned against early on in scientific circles. Viral threats, fires, floodings, and other events, in turn, often occur quite suddenly, although there are observable patterns in these cases as well (e.g., fires in California); moreover, the magnitude of such crises may put additional strain on the system concerned (i.e., when not only individuals or specific geographic areas or countries are affected, but when there are reciprocal (often exponential) effects between subsystems and even, between affected systems and their environment). As outlined below, **there are no easy and quick answers and solutions for dealing with systemic risks in the context of a single crisis, and certainly not for systemic risks in the context of multiple (poly-)crises**. In fact, the phases preceding an intervention, in which the complex phenomenon, its possible development paths, and the **implications of various potential interventions** and **how those should be orchestrated** between various concerned systems, must first be understood, require rather time. As learned from COVID-19, it was not only the country-specific interventions but the need for international coordination and synergistic interplay that showed shortcomings with partly devastating consequences (Zenk et al. 2020; Steiner et al. 2024). During that time, a call for a **coordinated governance structure** across different levels of society emerged, including stronger **science-diplomacy** ties.

2 Complex Societal Challenges

2.1 Characteristics

The **expectation that scientists, politicians, and decision-makers should know what to do** must therefore first take into account the preliminary work necessary to get a comprehensive but also an as accurate as possible picture of **complex societal challenges** before imagining or developing an action plan. A complex societal challenge, system, or problem is distinct due to its unique **systemic characteristics**:

Complex societal systems are (1) networks of large numbers of interacting elements and show (2) emergent patterns and behaviors, are (3) multilayered (from individuals, groups, organizations, and clusters to supra-national systems) and (4) coupled in nature (as interplay of ecosystems including humans, people, and plants within their man-made- and natural habitats), (5) interact with their system environments (feedback loops) and may be characterized as (6) ill-defined, ambiguous, uncertain, and open-ended. Because of its dynamically changing behavior over time neither the future nor the current system state of a complex system nor its transition path may be fully determined or predicted, but it may be understood regarding its general systems patterns and potential transition paths.[1]

2.2 Relevant Knowledge Carriers

Who are the **relevant knowledge carriers** who may contribute to an enhanced understanding of those general systems patterns? **Experts from science and practice, together with society at large,** need to be considered as essential elements, where a **mutual learning process** between those parties is a primary target for a societal good, in order to make use of each of these knowledge carriers' valuable inherent body of knowledge and experiences (expert is not synonymous with having a university degree, instead, for example, a smallholder farmer without any kind of degree but with vast experience regarding the interplay of soil and weather conditions of a given system must be considered an expert). Mutual learning between science and practice for socially robust orientations is the core of **transdisciplinarity** and makes it clear why **cross-boundary knowledge integration** is the logical prerequisite and consequence of crises management.

As evidenced by various previous events, **natural disasters** have cascading or/and networking effects on various parts of a system, across systems, regions, and scales. Natural disasters may encompass, among others, geophysical events (e.g., earthquakes, tsunamis, volcanic eruptions), meteorological events (e.g., tropical and local storms), hydrological events (e.g., floods, mass movement), climatological events (e.g., wildfires, droughts), and pandemic events (e.g., COVID-19). For example, taking COVID-19 shows that the risk of infection for **single individuals, groups, and cross-boundary communities** is related to supply and logistic risks (e.g., medical care and drugs, but also equipment and food), has impact on mobility patterns in general, and consequently on the climate, and affects sociocultural dimensions such as communication and collaboration patterns including interpersonal trust or the trust in governance structures. However, it seems that to ensure **trust** in times of crises, public discourse and the involvement of various parties beyond decision-makers are crucial.[2] **Science needs to rethink its role** by learning to better deal with, and communicate, uncertainties associated with global systemic risk, which also includes to enhance its own openness for relevant bodies of knowledge and experience outside the scientific system and by strengthening its social responsibility for the public good.[3]

[1] See also Steiner (2009, 2011, 2013): 18.

[2] Schernhammer et al. (2021), Schernhammer et al. (2023).

[3] Caniglia et al. (2021a, 2021b).

3 Systemic Risk Management

Traditional risk management, which is usually geared toward managing a specific type of risk, is **reaching its limits** when dealing with systemic risk because the latter exhibits highly complex system patterns. In general terms, according to Renn et al., **systemic risk** is characterized by "high complexity, multiple uncertainties, major ambiguities, and transgressive effects on other systems outside of the system of origin."[4] Or—as stated by the International Science Council, United Nations Office for Disaster and Reduction (UNDR), and Risk-KAN Working Groups,—"systemic risk is associated with cascading impacts that spread within and across systems and sectors (e.g., ecosystems, health, infrastructure and the food sector) via the movements of people, goods, capital and information within and across boundaries (e.g., regions, countries and continents)" and "all disaster risk is systemic, all disaster impact is systemic."[5]

In short, a single crisis or disaster (whether natural or man-made) can have far-reaching systemic implications that require **systems understanding and transdisciplinarity**. Taking this line of thought further, multiple and intertwined natural and/or man-made crises further increase the complexity of this system and challenge our understanding of underlying mechanisms, development patterns, and response to specific intervention portfolios.

From a **systems and complexity perspective**, systemic risk first requires an **understanding** of such complex systemic phenomena and associated **potential systemic risks and their drivers.** This includes an understanding of

1. The emergence of systemic risks
2. The related individual vulnerability spaces (i.e., which subsystem(s) might be vulnerable)
3. Potential coupling mechanisms between these vulnerability spaces
4. The often divergent or cascading effects on different actors and (sub)systems
5. A scenario-based understanding of potential development and transition paths and
6. An understanding of the implications of potential intervention scenarios such as imposed by the effects of innovations (from product-, technology-, process-, structural-, social-, business model-, to comprehensive system innovations)

Such a comprehensive understanding of a system cannot be acquired single-handedly: it requires (self-) orchestrated **collaboration across system levels**. In other words, the use of widely available knowledge in the form of comprehensive **knowledge integration** becomes crucial. Moreover, beyond merely responding to crises, being **proactive and forward-looking** in adapting as an integral part of a capacity building strategy is critical in turning **potential vulnerability spaces into innovation niches**.

---

[4]Renn et al. (2022).

[5]Renn et al. (2022); cf. Sillmann et al. (2022).

Effectively leveraging an understanding of inherent complexity and fostering creative problem-solving capabilities necessitates the establishment of effective multilevel governance structures. These structures should incorporate science-diplomacy approaches to address global grand challenges and facilitate appropriate forms of collaboration between science and practice, or science and society. This includes adopting functional-societal transdisciplinary methods or 'Mode 2' transdisciplinarity frameworks.

4 Challenges for Knowledge Integration in the Face of Complex Challenges Such as Poly-Crises

In a nutshell, complex systems are of interrelated, dynamic, fuzzy, coupled, open-ended, adaptive, and ambiguous nature. Therefore, deterministic problem-solving routines based on single disciplines and domains are not sufficient. Yet, while opening doors toward inter- and transdisciplinarity for extended knowledge integration sounds logical and promising, it is not easy going. There are challenges with respect to the (1) diverging perceptions of what is important for whom in general (including shock related distortions in perception), the (2) collaboration of scientists from different fields and backgrounds working together (characterizing interdisciplinarity), (3) differences in preferences and aims between scientists and practitioners (characterizing transdisciplinarity), and perhaps the biggest hurdle, (4) being willing to make the extra effort based on a needed sophisticated process design, competences, and resources to engage in the "adventure" of a cross-boundary knowledge integration process.

4.1 Perceiving Complex Challenges: There Is Not the One Single Picture

There is no single simple reality, not in terms of the current situation and particularly not in terms of the future (see Fig. 1). Even if stakeholder groups (e.g., households, businesses, and youth) and scientific experts have one common high-order strategic orientation in mind such as the sustainable development of a certain region, various studies show a tendency for the preference profiles of various stakeholder groups (which themselves are often positively correlated with each other) to deviate strongly from those of scientific experts. Although Fig. 1 focuses exemplarily on stakeholder and expert profiles of a specific Austrian region, its key message is the need to consider the potential divergence of preference profiles of stakeholders from that of other scientific and decision-making agents: these potential differences must be taken into account in any form of knowledge integration.

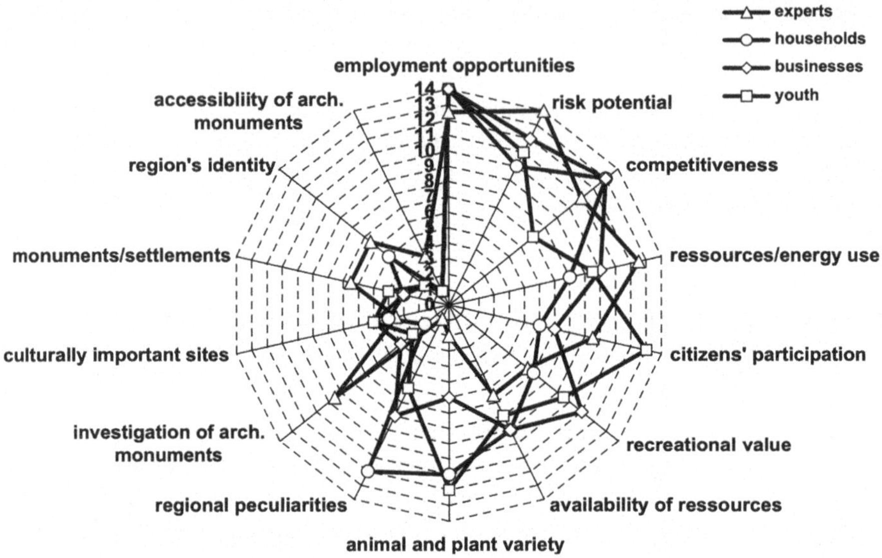

Fig. 1 Diverging preference profiles of stakeholders and scientific experts as an often occurring pattern Steiner and Posch (2006)

It also shows that societal complexity is very much context specific, goes beyond pure generalized theories and consequently, as already outlined by Roland W. Scholz,[6] needs to be analyzed at a case level. The so-called case or object, however, may not be misinterpreted as a kind of isolated or closed system, but needs to be understood in its relation to the relevant environment (see Fig. 2[7];). This is true with respect to different perceptions of agents (who take an active role for the system transition) or stakeholders (who are affected by actions taken within the system transition and react accordingly) regarding any complex system such as poly-crises (see Fig. 2). These differences may be better understood when having the perceptional patterns in mind: Based on **Brunswik's Theory of Probabilistic Functionalism (TPF),**[8] the potential poly-crises as a probabilistic entity are viewed as the object, the broader societal system as the relevant system environment, with the agents and stakeholders as the organisms, which exhibit reactive, proactive, or passive behavioral patterns.[9] In contrast to a deterministic view of perception and related decision-making processes, TPF emphasizes the relation between the perceiving individual (i.e., perceiving the poly-crises as the object, see Fig. 2) and its (multidimensional) environment (i.e., "ecological validity"), the probabilistic cues representing key aspects of the relevant environment, the adaptive nature of the

[6]Scholz (1995).

[7]Steiner (2018).

[8]See Brunswik (1943), Scholz (2017).

[9]Steiner (2018).

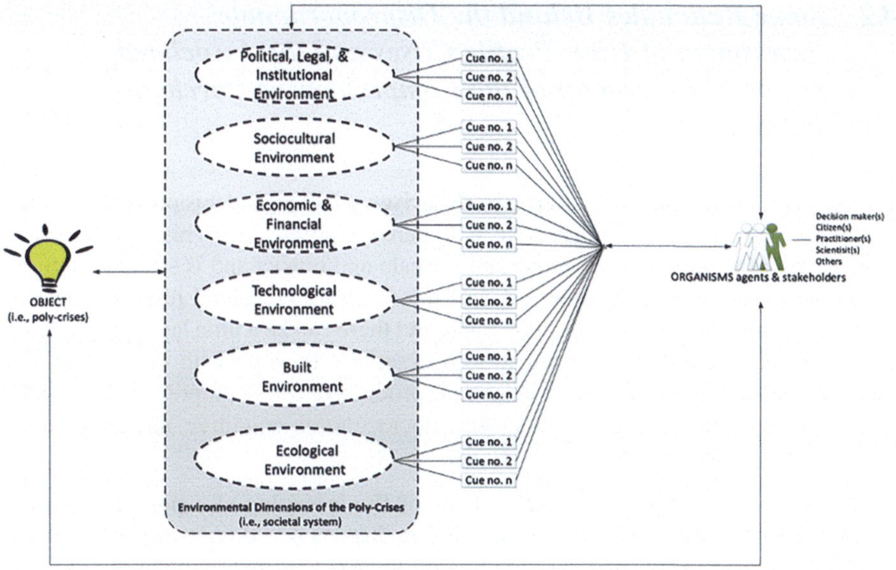

Fig. 2 Probabilistic functionalism—mental simulation of a crises: organism (i.e., decision-makers, citizens, practitioners, and scientists)–object (i.e., crises)–environment (i.e., societal system). Source: adapted from Steiner (2018)

agents (i.e., organism-representing decision-makers, citizens, practitioners, and scientists in Fig. 2), and in general, the probabilistic nature of the information available in dealing with a highly uncertain world.

Within this contextual picture, **"distance"** is a multidimensional concept related to "(1) space; (2) time; and (3) the state of the system in dependence of the specific agent or stakeholder position," consequently, gaining knowledge of both **overt and the covert areas of the object and the environmental system** is essential.[10] For example, knowledge about a potentially existing incongruence between self-reported attitudes and implicit attitudes with respect to a controversial intervention measured by, e.g., the Implicit Association Test (IAT), could be supportive in designing and implementing adequate interventions.[11]

---

[10] Steiner (2018).

[11] Steiner et al. (2018).

4.2 Some Rationales Behind the Disproportionate Assessment of High-Profile Crises and the Associated Distorted Perception of the Comprehensive Development Pathway

Poly-crises, and particularly **high-profile crises,** challenge decision-makers, citizens, practitioners, and scientists alike. Compared to poly-crises, a high-profile crisis goes beyond functional parameters such as scale and impact and is strongly defined by media coverage, public attention, and the resulting perceived relevance. However, not every crisis is a high-profile crisis, and there can be a time lag before a crisis is recognized and receives the attention it deserves. To cope with misleading and distorted perceptions, which may be counterproductive for a comprehensive system understanding, the interplay of underlying sociocultural, cognitive, and psychological factors is essential:

1. To understand the **general mechanisms of the interplay of a perceiving individual within its real-world environment**, Brunswik's **theory of probabilistic functionalism (TPF)** may provide a helpful basis (see Fig. 2 above[12]).
2. For **judgements under uncertainty**, the **availability heuristic** (i.e., availability bias) of Amos Tversky and Daniel Kahneman[13] shows that the magnitude of the consequences of an action strongly determines its **"mental availability"** in the future. **Insensitive and extensive media coverage** (focusing on sensational and shocking events in their reporting, as a means to increase viewer numbers) and increasingly, social media activities are correlated with greater mental availability. On the other hand and as exemplified by the **Israel Gaza/Hamas war of 2023**, previous potential peace developments such as the mutual endorsement of a "one-country /two-state solution" by the majority of both sides are more or less forgotten and no longer present: Accordingly, 70 percent of Israeli and 63 percent of Palestinian respondents accepted Israel as the state of the Jewish people and Palestine as the state of Palestinian people in a poll in 2004.[14] Now, perceptions are dominated by cruelties of this ongoing war. But even before this war, mistrust in the intentions of the other side prevented serious future-oriented negotiation processes. Or, in the words of the master mind of peace efforts Herbert C. Kelman, "we find ourselves in a situation in which majorities of both publics embrace a two-state solution and yet are not ready to give wholehearted support to negotiations because of profound **distrust** of the ultimate intentions of the other side." Against this backdrop and for future peace efforts, the role of **building up solid mutual learning and discourse arenas** prior to escalation turmoils such as based on the transdisciplinary processes discussed here should be given more attention for preventive war and conflict strategies.

---

[12] Brunswik (1943), Scholz (2017), Steiner (2018).

[13] Tversky and Kahneman (1974).

[14] See Kelman (2011).

3. For **decision-making as well as risk perception and risk evaluation** related to **potential future development scenarios** of complex systems and their environment (including the people affected), **affective forecasting** as proposed by Timothy Wilson and Daniel Gilbert helps to understand how people predict and anticipate their future emotional states including associated biases related to, for example, the **impact bias** (as individuals often overestimate the impact of both future positive and negative events on their emotion) or the **focusing illusion** (as the tendency of people to overemphasize the importance of one aspect of, e.g., risks without considering the broader contextual environment when perceiving and evaluating risks' impact on their individual well-being.[15]

4.3 Some Rationales Behind the Challenge of Interdisciplinary Knowledge Integration

Interdisciplinarity describes the integration of concepts and methods from different scientific disciplines.[16] As outlined by Scholz, even the most fundamental research questions, such as, e.g., how to model rain, require interdisciplinarity, in this case the interplay of physics and chemistry.[17] Interdisciplinary work is challenging, because the fusion of concepts and methods needs to be based on a joint basis and framework for communication and collaboration. In addition, there are sometimes very pragmatic reasons that prevent scientists from engaging in a disciplinary discourse, especially if the system boundaries for their own research are so narrow that interdisciplinarity is to be avoided.

Nonetheless, for research projects related to societally relevant challenges, a strengthening of the inter- and transdisciplinary field is slowly becoming apparent—perhaps testimony to the importance of these forms of extended knowledge in the light of the many crises the world is facing. Fig. 4 outlines one example, in the context of the overarching challenge of dealing with viral threats, which challenges not just a single but multiple disciplines. Intensive collaboration between different disciplines in understanding a complex problem, but also future-oriented developments, requires appropriate communication and joint learning based on suitable methods and processes. Here, systems thinking and -analysis may be of particular importance for achieving knowledge integration between and beyond scientific disciplines. Furthermore, bottlenecks with respect to real-world experiences by agencies and agents of practice (e.g., Red Cross, logistics firms) need to be integrated in order to avoid suboptimal decisions.

---

[15] Wilson and Gilbert (2003).

[16] Thompson Klein and Newell (1997), Scholz (2011).

[17] Scholz (2011): 21.

5 Which Form of Science-Practice Collaboration to Pick, and to What End?

A common feature of all forms of collaboration/cooperation between science and practice is the *attempt to go beyond a purely scientific perspective* based on the awareness that, particularly for real-world challenges, the relevant stakeholders and problem-solvers from practice should contribute their experience and knowledge as this could go beyond the understanding of knowledge of scientists alone.

5.1 Conditions for Choosing the Right Approach or the Right Combination of Approaches

There is no "one size fits all" approach. Instead, the selection of the science-practice collaboration approach that might be best suited for a particular situation depends on several dimensions:

1. *Degree of complexity* in the context of societal relevance.
2. *Evolutionary stage of the transition* or problem-solving process: this might be related to finding and discovering the real problem behind what might be more a symptom or, on the other side, the testing and experimenting of new technologies and innovations with respect to its application and use.
3. *Time resources* available for finding (societally) relevant and robust orientation within the transition or transformation process.
4. *Literacy to collaborate and co-create* as part of a mutual learning and problem-solving process (i.e., competences needed for working process design and -management, knowledge integration methods beyond Mode 2 Transdisciplinarity, etc.).
5. *Financial resources.*
6. *Working space and -environment.*

5.2 Overview of Approaches of Science-Practice Collaboration

For a general overview and systematization, Fig. 3 differentiates multiple approaches of science-practice collaboration by process ownership. In their extremes, they range from (1) *Applied science, partially including practitioners,*

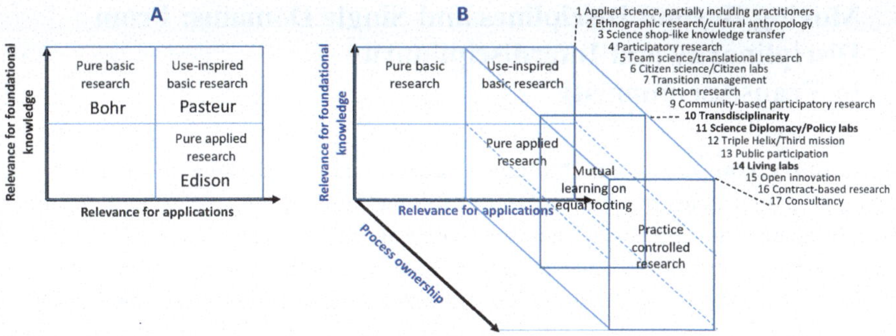

Fig. 3 Science-practice collaboration approaches: the role of transdisciplinarity and science diplomacy. Source: based on Scholz and Steiner (2023) (extended by 6 Citizen science in addition to Citizen labs; 11 Science Diplomacy; and 14 Living labs)

which is mainly under the control of the scientists, to (17) Consultancy, which is mainly under the control of the practitioners.[18]

To offer a framework for understanding the various forms of collaboration between science and practice, Scholz and Steiner expanded upon the conceptual foundation of Stokes' 'Pasteur's Quadrant' model, which aims to categorize scientific research based on the two dimensions relevance for (i) (generalized) fundamental knowledge and (ii) applications, by a third dimension "process ownership," which complements the two-dimensional categorization of Stokes.[19] To explain the two extreme positions of process ownership: **Applied research** implies that science takes control and **consultancy** implies that practice takes process ownership. **Transdisciplinary processes** stand exactly in the middle of process ownership based on the mutual learning philosophy to "include authentic co-definition, co-representation, co-design and co-leadership of science and practice" at eye-level. An in-depth analysis of science-practice collaboration approaches has been provided elsewhere.[20]

[18] For a detailed description and categorization of the single approaches, see Scholz and Steiner (2023).

[19] Scholz and Steiner (2023), Stokes (1997).

[20] Scholz and Steiner (2023).

6 Moving Beyond Disciplines and Single Domains: From Disciplinarity and Interdisciplinarity to Transdisciplinarity

In summary, complex social challenges cannot be tackled alone. From a scientific perspective this concerns the single disciplinary fields, but also agents within a discipline, neighboring disciplines with obvious interrelations as well as (at first glance) remote disciplines such as cell biology and anthropology. If we move on, this leads us to the involvement of the concerned domains and concerned fields of application, which opens the doors beyond science by relating it with the concerned real world and its stakeholders and decision-makers. But regardless how narrow or broad one considers the underlying form of **cross-boundary collaboration** to be, it always builds on the strengths of each single ingredient—akin to a chef who cannot prepare a **"great menu"** without having various exquisite, high-quality ingredients. What the cooking ingredients are to the chef, the individual disciplinary orientations in science and in-depth practical experience are to transdisciplinarity. As such, dealing with crises and risks also depends on the individual elements of science and practice but goes beyond the singularity by focusing on their orchestration as part of a strategic collaboration aimed at reactive and (if possible) proactive risk strategies.

6.1 Epistemological Differentiation

Generally speaking, a distinction between **disciplinary, interdisciplinary, and transdisciplinary** approaches can provide a basic orientation when distinguishing the specific roles of science, practice including decision-makers from public and private organizations, and society at large may provide for basic orientation.

By going beyond a singular disciplinary perspective, we are first moving toward interdisciplinarity and its limits. It seems that building bridges between different disciplines requires some common ground with respect to the underlying epistemology, disciplinary language, and/or methodology. Jean Piaget pointed out that interdisciplinarity is only possible between "neighboring disciplines" such as chemistry and physics. For a unity of knowledge including the various types of causation, a specific metastructure is needed.[21] **Mode 1 transdisciplinarity** aims to provide such a metastructure based on a pure inner science perspective,[22] which does not include stakeholders' perspectives. It was Jantsch who extended this view with *a* **functional**

[21] Piaget (1972). For a detailed reflection and analysis, see Scholz and Steiner (2015a, 2015b), Scholz and Steiner (2023).

[22] Gibbons et al. (1994).

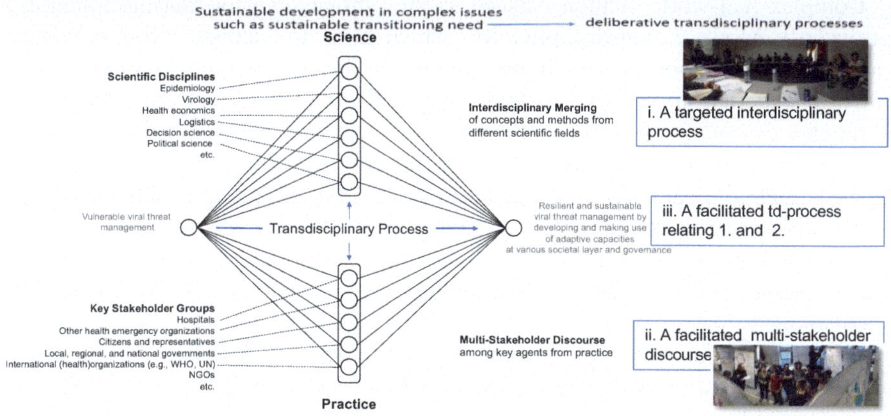

Fig. 4 From disciplines to interdisciplinarity and facilitated transdisciplinary processes. Source: Based on Scholz, R.W., Steiner, G. The real type and ideal type of transdisciplinary processes: part I—theoretical foundations. Sustain Sci 10, 527–544 (2015). https://doi.org/10.1007/s11625-015-0326-4

and societal perspective[23] in that **Mode 2 transdisciplinarity** is building a bridge between the varying epistemics of science and practice. For a deepened review of the evolution of transdisciplinarity.[24]

6.2 *Mode 2 Transdisciplinarity*

Mode 2 transdisciplinarity integrates diverse epistemic perspectives from both science and practice. It is defined as a facilitated process of mutual learning between science and society, combining targeted multidisciplinary or interdisciplinary research with multi-stakeholder discourse to develop socially robust orientations for specific real-world problems.[25] (see Fig. 4)

As mentioned before, it should be made clear that collaborative approaches cannot replace **individual capabilities and disciplines**; rather, they depend and build on them.

As outlined in the example of "viral threats" in Fig. 4, disciplinary and field-specific orientations are the core ingredients of any attempt of extended knowledge integration. The experiences from COVID-19 also show that an interdisciplinary perspective along a diplomatic discourse is not sufficient, particularly when dealing with societally relevant complex real-world challenges.

[23] Jantsch (1970, 1972).

[24] See Lang et al. (2012), Scholz and Steiner (2015a), Hernandez-Aguilar (2020), Lawrence et al. (2022).

[25] Scholz and Steiner (2015a).

Complex real-world challenges such as viral threats call for transdisciplinarity, particularly Mode 2 transdisciplinarity, which aims to facilitate cross-boundary knowledge integration by combining interdisciplinary scientific knowledge and practical expertise to address complex, societally relevant real-world challenges. Through a co-creation process, it fosters mutual learning among stakeholders from science—such as epidemiology, virology, health economics, logistics, decision sciences, and political sciences—and practice, including hospitals, health emergency organizations, the military, citizens and their representatives, governmental bodies, and international organizations. This collaborative approach between science and society serves as a foundation for developing socially robust solutions and orientations.

In its most advanced form (see Fig. 4 and approach no. 10 in Fig. 3), transdisciplinary collaboration between science and practice, or society at large, is characterized by:

- **Equitable partnership between science and practice/society** throughout the entire collaborative problem-solving process, from problem discovery and definition to the development of sustainable, solution-oriented innovations
- **Co-leadership**
- Comprehensive **mutual learning** process between science and practice/society.[26]

From problem-discovery/-definition to stakeholder-specific **system understanding** as a basis for

Future-oriented **scenario development** of the

- focused system as the **system scenario** including
- neighboring environments (beyond the given system focus and its system boundaries) as the **environment scenario** as well as
- the impact of **intervention scenarios** from the analyses of the underlying system (s) and scenario(s) and by means of collaborative creative problem-solving and innovation generation;
- both system understanding and scenario development are based on **cross-boundary knowledge integration** which is guided by a thorough design of the method-guided collaborative working process.

In addition to such a strong form of transdisciplinarity, settings are also possible that manage without this explicit form of co-leadership and still aim for comprehensive knowledge integration. Furthermore, and as outlined below in Section "Transdisciplinary Process Design," with the focus on the overall project, it may be that only the general framework (i.e., in terms of the general understanding of what is the underlying guiding question is or what the central challenge/s is/are to focus on) and specific phases may be characterized by a transdisciplinary setting, whereas others may require an interdisciplinary or even disciplinary setting.

[26] See also Scholz and Steiner (2023).

6.3 Aims of Transdisciplinarity

What are the **aims** of transdisciplinarity?. Mode 2 transdisciplinarity can contribute to **capacity building** for both science and practice. It facilitates **consensus-building** regarding the understanding of targeted systems and their underlying problems—rather than mere symptoms—including potential pathways for transition and transformation. Additionally, it can serve as a platform for **mediation, conflict resolution, and crisis management**, as well as for **preventive conflict and crisis mitigation**, where the nature of the crisis allows.

But how to achieve a shared perspective when facing a **"non-integratable" situation,** i.e. because of different cultures in science such as related to forms of reasoning, validation, language, communication, behavioral patterns (e.g., hermeneuts versus epidemiologists)? Here, a spiritual-philosophical meta-level as introduced by Nicolescu, which related modes of thoughts and cultures within his "spiritual-philosophical transdisciplinarity" approach and which might highly impact Mode 1 transdisciplinarity in the future,[27] may provide assistance. For an extended perspective on a spiritual-philosophical form of transdisciplinarity, see Schüz, Chap. 1 of this book.[28]

Consequently, in cases in which **Mode 2 transdisciplinarity reaches its limits**, such as when the common constitutional basis as related to democratic principles does not exist or is inadequate, a vertical integration such as by a spiritual-philosophical dimension might act as a needed shared basis. For example, in specific sociocultural and religious settings, people's values, belief systems, and identities are very much shaped by their cultural and religious backgrounds. However, these religious roots might be characterized by **historically grown religious conflicts** such as between Islam and Judaism, Buddhism and Hinduism, or between Judaism and Christianity as well as Islam. If we consider a spiritual dimension at a meta-level beyond religious orientation, spirituality might provide one possible means that allows for a joint basis for communication and collaboration despite the close relationship between the concepts of spirituality and religion.

6.4 Trust as a Prerequisite of Transdisciplinarity

Often, entering into an exchange with a party "on the other side of the bridge" requires moving beyond a certain boundary in order to establish some form of basic trust that there is a probability that it might be worth the effort. Consequently, trust is a prerequisite for mutual learning, facilitated through two key mechanisms: (1) trusting others, which encompasses interpersonal trust as well as trust in state

[27] Nicolescu (2000); and in this volume p. XXX.

[28] See Schüz chapter 1 in this volume.

structures, government systems, and scientific institutions; and (2) being perceived as trustworthy by others, reflecting one's trustworthiness.

Trust may be understood as a function of performance and reliability expectation regarding the other party, empathy for the other party, openness as a general personality characteristic, and can further be influenced by environmental triggers (such as supportive built environments and design measures)

trust = f (performance and reliability expectations; empathy; openness; environment)

As demonstrated by a most recent study in the D-A-CH region, interpersonal trust appears to have declined during the COVID-19 pandemic, with potential effects beyond interpersonal relations[29]: It indicates that the erosion of trust in Western Europe goes beyond the interpersonal dimension and also affects state structures, government systems, and scientific institutions. The rapid growth of anti-democratic movements based on or appealing to conspiracy theories observed during the COVID-19 pandemic, as well as distrust of scientific and government institutions, requires urgent action. The study further points to optimism, **complexity thinking** as well as the ability to take perspectives, extroversion, and agreeableness as important correlates of higher trust values. Complexity thinking is understood as the ability to recognize even non-evident structures and interrelationships in complex real systems, as well as the emergent properties and adaptive capacities associated with them. This thinking is then the basis for understanding, for example, the interactions between developments in biodiversity, water, climate change, and health in relation to specific economic and energy use scenarios.

To ensure mutual trust and to counter conspiracy thinking, the involvement of various parties such as practitioners and citizens in interaction with decision-makers appears essential.[30]

6.5 Knowledge Integration Strategies

As pointed out, today's crises and their intertwined character together with a strive to aim for a betterment of society such as based on a one-health paradigm which focuses on the well-being and health of humans, animals, and plants as well as soil and the overall environmental conditions, seem to strongly depend on our future success in joining our capabilities (i.e., knowledge, experiences, wisdom, etc.) to tackle these challenges. In other words, bringing together the individual stakeholders and their knowledge is one thing, but the synthesis of the collected knowledge also plays a key role. Hereby, several dimensions may appear as focus areas in attempting knowledge integration:

[29] Schernhammer et al. (2023).

[30] Schernhammer et al. (2021), Weitzer et al. (2021), Schernhammer et al. (2023), Han et al. (2024), Steiner et al. (2023a, 2023b).

1. Integrating knowledge from various disciplines (i.e., interdisciplinary science).
2. Integrating knowledge from interdisciplinary science and practice/society at large (i.e., transdisciplinary knowledge integration which is called Mode 2 transdisciplinarity); transdisciplinary multistage system modeling offers a practical guideline for establishing system-understanding among various agents and stakeholders and allowing for a comparison of the potentially diverging perceived system representations including, ultimately, their integration.[31]
3. Integrating functional-societal transdisciplinarity (Mode 2 transdisciplinarity) with spiritual-philosophical transdisciplinarity (i.e., Nicolescu's meta-level approach to transdisciplinarity).
4. Integrating various approaches from (2) and (3) as part of a comprehensive process design.
5. Integrating governance processes from various societal layers (i.e., from groups, families to states, countries, regions, continents, global perspective).
6. Integrating methods to be applied for socially robust orientations which originate from a specific approach (e.g., transdisciplinarity, science diplomacy) or a scientific field or discipline or a specific field of practical application. Often single domains, approaches, schools, disciplines, etc., show a closed-box mentality when choosing the appropriate method to be applied, a mentality which has its origin not only in discipline-oriented behavioral patterns.

In order to cope with today's and tomorrow's complex societal challenges, comprehensive knowledge integration is a precondition. This makes it possible to make use of the best that science and practice or society as a whole have to offer.

7 Transdisciplinary Process Design

By reflecting on 41 mid- and large-scale transdisciplinary studies at seven universities on sustainable transitions, and different societal systems from organizations to urban and regional systems including policy processes, six phases of transdisciplinary processes were identified: (1) **Initiation phase**, which includes successful networking, the finding of stewards and key partners from practice and science who (enthusiastically) share the basic idea. If a critical momentum has been attained, (2) the **preparation phase** can start. The **core phase** (3), then, is a directed work of producing the outcome that will be (4) disseminated, evaluated, etc., in the **post-processing phase**. The detailed steps of the template and what tasks, challenges, and barriers have to be addressed are subjects of Part II.[32] Eventually, a triggering phase may precede those four phases.[33]

[31] See Satalkina et al. (2022).

[32] Scholz and Steiner (2015b).

[33] See Scholz and Steiner (2023).

If you want to build a ship, don't drum up the men to gather wood, divide the work, and give orders. Instead, teach them to yearn for the vast and endless sea. — Antoine de Saint-Exupéry

The essential core within this quote is the awareness building process of the so-called big picture of a challenge which calls for non-standard, innovative approaches. This philosophy might be a **credo for a transdisciplinary processes design**: It seems to be crucial to get agents from science and practice "on board" in order to work together in a mutual learning processes for the betterment of society by providing guidance on complex societal challenges based on socially robust orientations. Jointly developing a clear **guiding question**, which is tackling a societally relevant complex challenge and which is appealing to both science and practice is crucial (see Fig. 5 below). This guiding question encapsulates the overarching key problem, which justifies undertaking the challenging and time-intensive transdisciplinary journey. Much like an adventure, this process lacks a single, predetermined path and does not guarantee precisely predictable outcomes (see Fig. 5).

In a certain way, this transdisciplinary adventure requires from its **co-leaders, respectively, organizers** to act similar to a Hollywood film producer when building up your team: you do not only take what is available "locally," respectively, within the own "working habitat" at the moment (more or less by accident), but you need to be aware about the competences, knowledge, and experiences necessary to—as best as possible—succeed along this journey which encompasses transdisciplinary multilevel system modeling and stakeholder analysis, -scenario-development, and ultimately (work in progress), a decision-theater like decision support system based on a solid dynamically adaptable Agent-Based-Modeling (ABM) simulation (see Fig. 5 above).

When tackling a complex challenge, the agent's and stakeholder's knowledge and experience will show different profiles with varying strengths and weaknesses. To have them sufficiently involved in dealing with the complexity, systemic and process-specific order is crucial. As an example, to deal with potential vulnerabilities regarding accessibility and availability of a certain mineral resource such as lithium, a **supply chain perspective with agents and stakeholders** from science, practice, and society at large for each single phase, from exploration, mining, processing, use, and recycling, will help oversee the knowledge- and experience map including potential white spots and shortcomings (see Structural Ordering in Fig. 5).

For a comprehensive **stakeholder/agent management** as the basis for the design of a **facilitated stakeholder discourse** from pre-check and consistency check, identification, analysis, classification, including a phase-specific action plan, see Steiner[34] and Scholz and Steiner.[35] Fig. 4 outlines how a stakeholder discourse may be orchestrated together with a targeted interdisciplinary process as part of a facilitated transdisciplinary process.

---

[34] Steiner (2008).

[35] Scholz and Steiner (2015b).

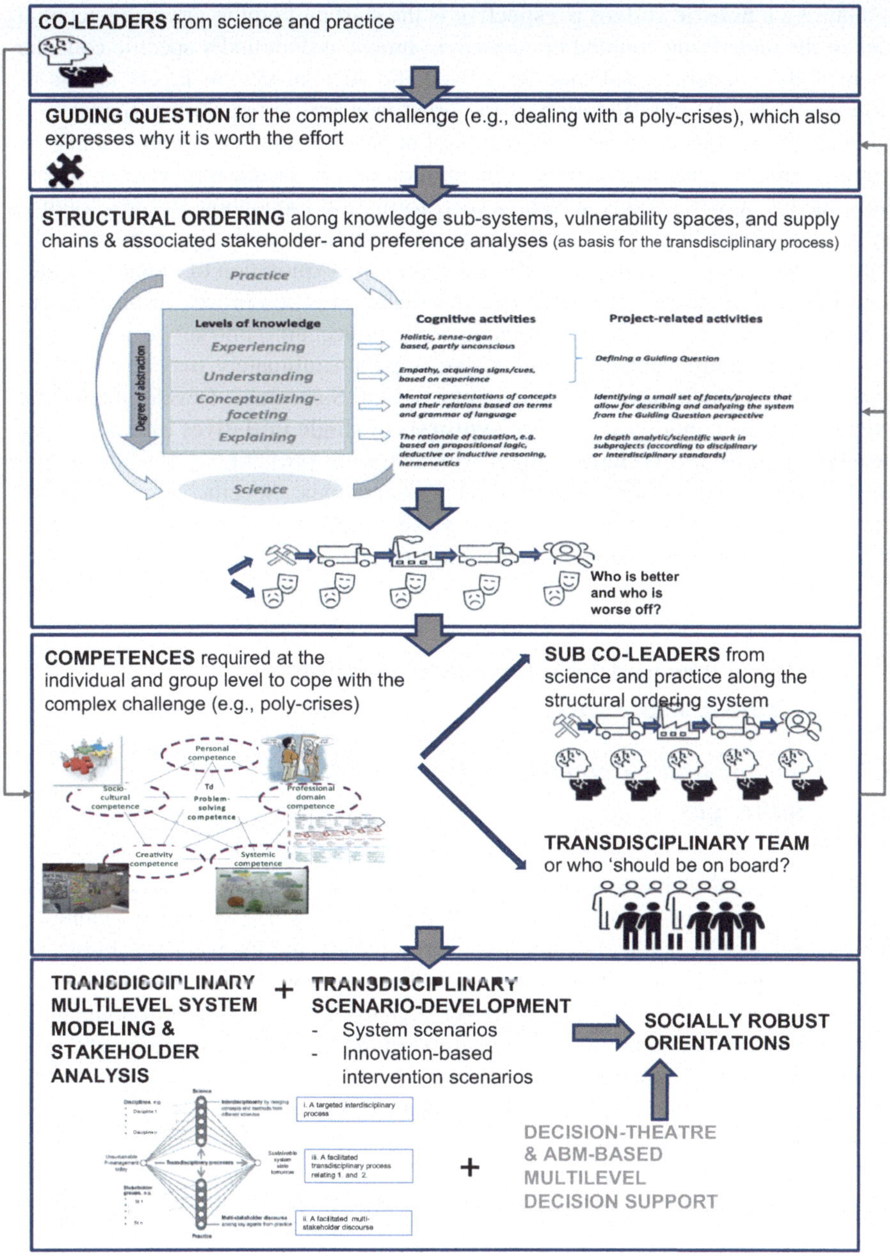

Fig. 5 Transdisciplinary processes design. Source: the "architecture of knowledge" in the upper part of the segment "structural ordering" and *the "transdisciplinary process" in the bottom segment of "socially robust orientations" are* based on Scholz and Steiner (2015a).

Gaining a **holistic system perspective** is the starting point to grasp the peculiarities of the underlying coupled human-environment system under specific consideration of the concerned stakeholders of specific societal system layers and is the prerequisite for the **identification of the most relevant vulnerability spaces** (e.g., flooding and its functional and sociocultural impact on elderly). A transdisciplinary learning process, characterized by collaboration and co-leadership between science and practice across various system levels, forms the foundation for developing a holistic understanding of the challenges, vulnerabilities, and opportunities within different societal (sub)systems. It also facilitates the exploration of potential system scenarios and interventions related to the societal transition process. Therefore, the **faceting** of the overall system (in relation to the project aim) into subprojects related to specific challenges and vulnerability spaces, which, ultimately, may be turned into opportunities [which may be seen under the light of Sustainable Development Goals (SDGs)] and innovation spaces. The **synthesis of single initiatives into a comprehensive system- and scenario model** is essential for providing a decision support needed which allows for socially robust orientations. Specific educational and training programs for the various case stakeholders and agents are a particular means of **empowerment and capacity building**.

8 Competences and Framework Conditions for Transdisciplinarity

8.1 Literacy and Competences for Complex Societal Challenges

Complex societal challenges such as the various single crises we had and have to face or the coupled poly-crises which challenge today's world will not and cannot be successfully met by neither experts from science alone, nor by experts from practice alone, nor by decision-makers alone (such as at international or national or local governmental levels but also at various organizational levels from industry to NGOs), nor by citizens alone. The acquisition of the necessary knowledge and skills for the design and implementation of transdisciplinary collaborative and mutual learning processes requires appropriate and in some cases new pedagogical tools, which call for an appropriate theoretical-conceptual perspective and clearly outlined learning cases.

The good news is, we can get it done, namely by collaboration across different boundaries such as of disciplinary, geographic, sociocultural, religious, interest and preference specific, and visionary kind. Transdisciplinarity may play an important role here. The realistic or skeptical news is, we are not yet well enough equipped yet to get it going. This means we are having a **shortage of necessary competences** for dealing with complex societal challenges at every societal level, from individuals, households, organizations, governments, and international organizations.

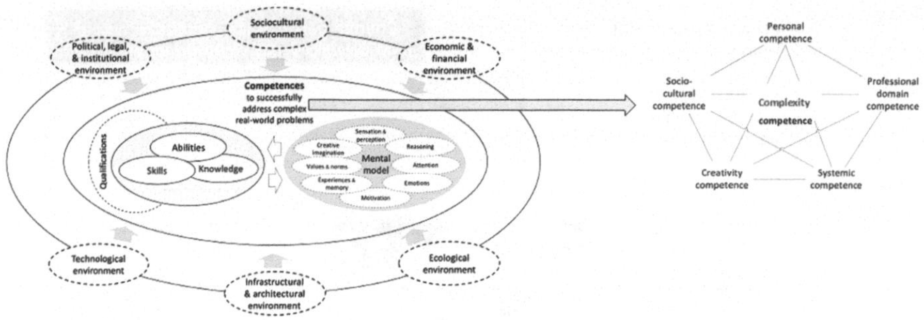

Fig. 6 Competences for dealing with complexity in a transdisciplinary manner—a dynamic perspective. Source: Steiner (2013), Steiner et al. (2013/2014), Risopoulos et al. (2020)

As outlined in Fig. 6, **complexity competence** as a multidimensional concept is considered as a person's "abilities, knowledge, and skills needed to adequately deal with a complex real-world problem," whereby competence is affected and simultaneously affects underlying mental models, which change over time and are depended on (1) political, legal, and institutional; (2) sociocultural; (3) economic and financial; (4) technological; (5) infrastructural and architectural; and (6) ecological environmental dimensions.[36]

It is essential to be aware that every single competence dimension depends on a specific repertoire of **abilities, knowledge, and skills**. Complexity competence is based on the synergetic relation between personal competence, professional domain competence, systemic competence, creativity competence, and sociocultural (collaborative) competence. In brief, **personal competence** is needed to "be aware of, and manage oneself," **professional domain competence** is a prerequisite to "comprehend the peculiarities of the specific domain in which the problem is embedded,", **systemic competence** allows to understand the entire system with subsystems and interdependencies in its environment, **creativity competence** is the precondition to enable us to step outside old habits and to "synthesize convergent and divergent thinking in order to create something new or find a new solution to complex real world problems such as crises," and **sociocultural (collaborative) competence** to allow for collaboration between various disciplines but also to enable mutual learning and, eventually, joint problem-solving in the interplay of science and practice, respectively, society.[37]

---

[36] Steiner (2013), Steiner et al. (2013), Steiner et al. (2014), Risopoulos et al. (2020).

[37] Steiner (2013). For a comprehensive review of frameworks for competences for complex problem-solving, see Bates et al. (2022).

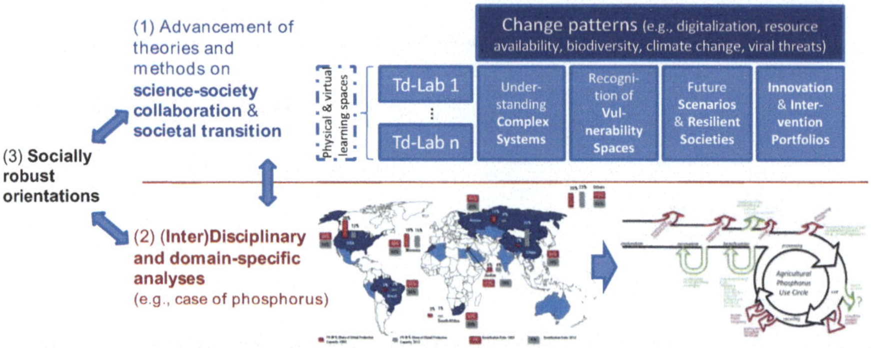

Fig. 7 Institutional settings for transdisciplinary research and—teaching

8.2 *Institutional and Organizational Setting for Transdisciplinarity*

Transdisciplinarity orientations, processes, and projects can be massively supported by appropriate framework conditions, the so-called **transdisciplinary ecosystems**. Let us take, for example, the case of universities. Transdisciplinarity research, transdisciplinarity education, and transdisciplinary educational research initiatives need a supportive habitat to ensure a transdisciplinary orientation that extends beyond single projects:

1. **Strategic orientation and leadership** such as expressed in the official **"University Development Plan"** and by the University's **leadership** constellations are critical for an effectful transdisciplinary program.[38] This seems of particular importance for being recognized internally but especially by external transdisciplinary partners of practice as **trustful partner** for a mutual learning process between science and practice (i.e., transdisciplinarity) aimed at providing socially robust orientations for specific complex real-world challenges. Current establishments of **transdisciplinary inter- and supra-institutional networks of institutions and joint activities** in research and education provide a meaningful basis of exchange and may play an important role for the enhancement of transdisciplinarity.
2. **Organizational settings** as **"physical and virtual learning space hot spots (Td Labs) for transdisciplinary research and education"** such as by the installation of theme-related transdisciplinary organizational units, for example, **Transdisciplinarity Laboratories** for various core themes such as "Sustainable Mineral Resources," "Sustainable Digitals Environments," or the

[38] In the case of UWK—Danube University Krems see: https://www.donau-uni.ac.at/dam/jcr:10d1678a-f3a9-49d7-bd01-fe7d98432a3c/Strategy_University%20for%20Continuing%20Education%20Krems_2021.pdf.

"GovLabAustria" as "Innovation Space for Management Solutions.".[39] These units (see Td-Labs 1-n in Fig. 7 below) act as interfaces between science and practice (i.e., between the university and partners of practice) and help to identify, collect, and organize thematic challenges of societal relevance and allow for organized communication between agents and stakeholders involved in transdisciplinary educational and/or research initiatives.

3. **Transdisciplinary education and training** needs explicit **curricula, study programs, and didactical strategies** such as the **transdisciplinary Joint Erasmus Mundus Excellence Master** in "Transition, Innovation and Sustainability Environments" (**TISE**),[40] which may be supplemented by a comprehensive **transdisciplinary case study didactics,**[41] including a **didactical case study strategy** which makes use of the paper-based **Harvard Case Studies** and the real-world-based **Transdisciplinary Case Studies** in order to enhance complexity literacy as outlined by Steiner and Laws[42] (it shows that the Harvard case study can be a good and protected preparation for doing real-world transdisciplinary case studies; for transdisciplinary education programs in the field of complexity literacy for sustainable digital transition, see Steiner et al.,[43] and for climate change related education and science implication, see Bohunovsky et al.[44]).

TISE is an international joint initiative of four European higher education institutions (UWK—Danube University Krems/Austria (coordinator), Poznan University of Economics and Business/Poland, Universidade Nova de Lisboa/Portugal, and the University College Dublin/Ireland) and various associated partners and approximately 30 students from 24 countries in each cohort with diverse professional and sociocultural backgrounds. The **rotational program** leads the whole student group (cohort) for each a single term to Portugal, Ireland, Poland, and Austria as part of the four terms joint master program. Here, the goal is to integrate not only different learning themes, but especially to integrate the different context situations and experiences in the **superordinate learning goal of being capable to orchestrate themselves transdisciplinary learning processes on complex societal challenges**. For this purpose, TISE aims to educate these future decision-makers in making use of transdisciplinary theory and methodology, systems sciences, social and cultural sciences, economics, technology studies, data analytics, and innovation and creativity studies (e.g., including design thinking as well) and other fields and disciplines as

[39] See: https://www.donau-uni.ac.at/en/university/faculties/business-globalization/research/transdisciplinary-research-platforms-and-projects.html.

[40] https://www.donau-uni.ac.at/en/studies/transition-innovation-and-sustainability-environments-tise.html); (see also Mühlmann et al., 2022; Steiner et al., 2023a, 2023b).

[41] Scholz and Tietje (2002).

[42] Steiner and Laws (2006).

[43] Steiner et al. (2023a, 2023b).

[44] Bohunovsky et al. (2022).

well as the applied knowledge and experiences of practitioners. These knowledge components are all relevant for dealing with complex societal challenges and vulnerabilities such as related to digital transformation, health, and urban-rural disparities at a **case-based level** (see Section "Transdisciplinary Process Design").

Beyond the key characteristics described in Section "Transdisciplinary Process Design," the innovativeness of TISE lies in[45]: A combination of **experimental physical and virtual learning spaces** (for transdisciplinary teams from university and practice, e.g. students, lecturers, mentors, practitioners, scientists) is used as the habitat for mutual learning processes. Particularly regarding the identification for (sub)project relevant stakeholders and organizations, the cross-link with the other **physical and virtual learning space hot spots (Td Labs)** (see the Section (2) Organizational Settings above) plays an important role for communication- and collaboration purposes. Within the experimental learning spaces also the **Transdisciplinary Field Research Training (TFRT)** takes place in parallel to the regular coursework. The integration of **digital and analog tools** (e.g., simulation-based learning, augmented and virtual reality, digital playful learning and edutainment, inverted classrooms, digitally extended makerspaces, digital planning tools, feedback digitization). Over all, the program aims to address the challenges of our increasingly complex world by dynamic knowledge integration between the various involved agents, from society as well as from science as part of a comprehensive mutual learning process.

9 Conclusion and Outlook

This chapter discussed the concept of a **functional-societal form of transdisciplinarity** (i.e., Mode 2 transdisciplinarity) and its potential for addressing complex challenges and poly-crises in today's world. The key argument is that neither science nor practice is adequately prepared to deal with complex challenges and multiple crises that we are facing, such as climate change, biodiversity loss, viral threats, and geopolitical conflicts. These crises are interconnected and require a simultaneous and collaborative approach for effective decision-making- and problem-solving processes.

Further, the limitations of traditional disciplinary but also of purely science-based interdisciplinary approaches for dealing with complex societally relevant real-world problems and, consequently, the need for **cross-boundary knowledge integration** are highlighted. This opens the door for transdisciplinarity, which involves the collaboration between experts from different disciplines, practitioners, and society at large. **Transdisciplinarity** aims to develop socially robust orientations by mutual learning of science and practice which strives for integrating scientific knowledge and practical knowledge to tackle complex societal challenges. After outlining

[45] See also Mühlmann et al. (2022).

cornerstones of a **transdisciplinary process design**, the concept of **complexity competences** is discussed, which refers to the abilities, knowledge, and skills needed to address complex real-world problems. It emphasizes the importance of personal competence, professional domain competence, systemic competence, creativity competence, and sociocultural (collaborative) competence for effectively dealing with complexity. The challenges faced by education systems in dealing with real-world complexity and how these can be overcome through innovative learning strategy and didactic measures as well as virtual and physical learning environments form the conclusion of this section.

The complexity of poly-crises challenges all agents and stakeholders of society at every societal level. For **future developments of comprehensive societal knowledge management systems, system- and complexity literacy** is not only of relevance for decision-makers and/or scientists, but **for society as a whole**, although in adapted forms. Only then our society will be able to establish or adapt the interrelated **multilevel governance systems** that are not only top-down oriented entities but allow for mutual learning processes not only among single governance levels but, most important, between the knowledgeable and experiences agents and stakeholders across whole society.

The **key message** of this contribution is the necessity for well-developed concepts of collaboration between science and practice, or society at large, that integrate theory- and method-driven systems and scenario thinking as the foundation for mutual learning processes. There is no one size fits all approach, however, transdisciplinarity (mode 2) has substantial potential when aiming for an integral perspective of epistemics form science and practice, which, based on **equal footing** of both sides, allow for method-guided mutual learning processes aimed to provide socially robust orientations for complex, societally relevant challenges.

References

Bates, R., Brenner, B., Schmid, E., Steiner, G., & Vogel, S. (2022). Towards meta-competences in higher education for tackling complex real-world problems—a cross disciplinary review. *International Journal of Sustainability in Higher Education, 23*(8), 290–308.

Bohunovsky, L., Keller, L., Zint, M., Michelsen, G., & Steiner, G. (2022). Bildung und Wissenschaft für ein klimafreundliches Leben. In: *APCC special report* 'Strukturen für ein klimafreundliches Leben.'

Brunswik, E. (1943). Organismic achievement and environmental probability. *Psychological Review, 50*, 255–272.

Caniglia, G., Jaeger, C., Schernhammer, E., Steiner, G., Russo, F., Renn, J., Schlosser, P., & Laubichler, M. D. (2021a). COVID-19 heralds a new epistemology of science for the public good. *History and Philosophy of the Life Sciences, 43*, 59. https://doi.org/10.1007/s40656-021-00413-7

Caniglia, G., Zenk, L., Schernhammer, E., Bertau, M., Steiner, G., Kainz, M., Jaeger, C., Schlosser, P., & Laubichler, M. D. (2021b). Scientists' responsibility for global futures. *Science & Diplomacy* (AAAS), https://www.sciencediplomacy.org/perspective/2021/scientists-responsibility-for-global-futures, 01/22/2021.

Gibbons, M., Limoges, C., Nowotny, H., Schwartzmann, S., Scott, P., & Trow, M. (1994). *The new production of knowledge*. Sage.

Han, E., Weitzer, J., Birmann, B. M., Bertau, M., Zenk, L., Caniglia, G., Laubichler, M. D., Schernhammer, E. S., & Steiner, G. (2024). Association of personality traits and socio-environmental factors with COVID-19 pandemicrelated conspiratorial thinking in the D-A-CH region. *SN Social Sciences 4*(41): https://doi.org/10.1007/s43545-023-00790-9

Hernandez-Aguilar, C., Dominguez-Pacheco, A., Martínez-Ortiz, E. J., Ivanov, R., López, J. L., Bonilla, et al. (2020). Evolution and characteristics of the transdisciplinary perspective in research: A literature review. *Transdisciplinary Journal of Engineering & Science, 11*. https://doi.org/10.22545/2020/00140.hal-03017605

Jantsch, E. (1970). Inter- and transdisciplinary university: A systems approach to education and innovation. *Policy Sciences, 1*, 403–428.

Jantsch, E. (1972). Towards interdisciplinarity and transdisciplinarity in education and innovation. In L. Apostel, G. Berger, A. Briggs, & G. Michaud (Eds.), *Interdisciplinarity: Problems of teaching and research in universities* (pp. 97–121). University of Nice.

Kelman, H. C. (2011). A one-country/two-state solution to the Israeli-Palestinian conflict. *Middle East Policy Journal, 18*(1), 27–41.

Lang, D. J., Wiek, A., Bergmann, M., Stauffacher, M., Martens, P., Moll, P., Swilling, M., & Thomas, C. J. (2012). Transdisciplinary research in sustainability science: Practice, principles, and challenges. *Sustainability Science, 7*, 25–43. https://doi.org/10.1007/s11625-011-0149-x

Lawrence, M. G., Williams, S., Nanz, P., & Renn, O. (2022). Characteristics, potentials, and challenges of transdisciplinary research. *One Earth, 5*(2022), 44–61. https://doi.org/10.1016/j.oneear.2021.12.010

Mühlmann, K., Satalkina, L., Zenk, L., & Steiner, G. (2022). Transition, innovation and sustainability environments, TISE: A showcase of education geared towards societal challenges and future orientation. In M. E. Auer, H. Hortsch, O. Michler, & T. Köhler (Eds.), *Mobility for smart cities and regional development—challenges for higher education* (ICL 2021. Lecture notes in networks and systems) (Vol. 390). Springer. https://doi.org/10.1007/978-3-030-93907-6_94

Nicolescu, B. (2000). Transdisciplinarity and complexity: Levels of reality as source of indeterminacy. In *Bulletin Interactif du Centre International de Recherches et E´tudes Transdisciplinaire*, 2014 (November 29).

Piaget, J. (1972). The epistemology of interdisciplinary relationships. In L. Apostel, G. Berger, A. Briggs, & G. Michaud (Eds.), *Interdisciplinarity: Problems of teaching and research in universities* (pp. 127–139). OECD.

Renn, O., Laubichler, M., Lucas, K., Kröger, W., Schanze, J., Scholz, R. W., & Schweizer, P. J. (2022). Systemic risks from different perspectives. *Risk Analysis, 42*(9), 1902–1920. https://doi.org/10.1111/risa.13657. Epub 2020 Dec 16.

Risopoulos, F., Daghofer, F., & Steiner, G. (2020). Competences for solving complex problems: A cross-sectional survey on higher education for transformative learning and transdisciplinarity. *Sustainability, 12*, 6016. https://doi.org/10.3390/su12156016

Satalkina, L., Zenk, L., & Steiner, G. (2022). Transdisciplinary multistage system modeling: Migrant entrepreneurship in the digital economy. *Kybernetes, 51*(13), 219–240. https://doi.org/10.1108/K-02-2022-0255

Schernhammer, E. S., Weitzer, J., Han, E., Birmann, B. M., Bertau, M., Zenk, L., Caniglia, G., Laubichler, M. D., Steiner, G. (2023). *Determinants of trust in times of crises: A cross-sectional study of 3,065 German-speaking Adults from the D-A-CH region* [where published?]

Schernhammer, E., Weitzer, J., Laubichler, M. D., Birmann, B. M., Bertau, M., Zenk, L., Caniglia, G., Jäger, C. C., & Steiner, G. (2021). Correlates of COVID-19 vaccine hesitancy in Austria: Trust and the government. *Journal of Public Health*. (Oxford University Press), fdab122. https://doi.org/10.1093/pubmed/fdab122

Scholz, R. W. (1995). Zur Theorie der Fallstudie. In R. W. Scholz, T. Koller, H. A. Mieg, & C. Schmidlin (Eds.), *Perspektive "Grosses Moos"—Wege zu einer nachhaltigen Landwirtschaft. UNS-Fallstudie 1994* (pp. 39–46). Vdf.

Scholz, R. W. (2011). *Environmental literacy in science and society. From knowledge to decisions*. Cambridge University Press.

Scholz, R. W. (2017). Managing complexity: From visual perception to sustainable transitions—contributions of Brunswik's theory of probabilistic functionalism. *Environment Systems and Decisions, 37*, 381–409. https://doi.org/10.1007/s10669-017-9655-4

Scholz, R. W., & Steiner, G. (2015a). Transdisciplinarity at the crossroads. *Sustainability Science, 10*(4), 521–526. https://doi.org/10.1007/s11625-015-0338-0

Scholz, R. W., & Steiner, G. (2015b). The real type and ideal type of transdisciplinary processes: Part I—theoretical foundations. *Sustainability Science, 10*(4), 527–544. https://doi.org/10.1007/s11625-015-0326-4

Scholz, R. W., & Steiner, G. (2023). Process ownership in science–practice collaborations: The special role of transdisciplinary processes in sustainable transitioning. *Sustainability Science*. https://doi.org/10.1007/s11625-023-01291-7

Scholz, R. W., & Tietje, O. (2002). Embedded case study methods: Integrating quantitative and qualitative knowledge.

Sillmann, J., Christensen, I., Hochrainer-Stigler, S., Huang-Lachmann, J., Juhola, S., Kornhuber, K., Mahecha, M., Mechler, R., Reichstein, M., Ruane, A. C., Schweizer, P., & Williams, S. (2022). ISC-UNDRR-RISK KAN briefing note on systemic risk. *International Science Council*. https://doi.org/10.24948/2022.01

Steiner, G. (2008). Supporting sustainable innovation through stakeholder management: A systems view. *International Journal of Innovation and Learning, 5*(6), 595–616.

Steiner, G. (2009). The concept of open creativity: Collaborative creative problem solving for innovation generation—a systems approach. *Journal of Business and Management, 15*(1), 5–34.

Steiner, G. (2011). *The planetary model of collaborative creativity: Systemic-creative problem solving for complex challenges (Das Planetenmodell der kollaborativen Kreativität: Systemisch-kreatives Problemlösen für komplexe Herausforderungen)*. Gabler Research (in German).

Steiner, G. (2013). Competences for complex real-world problems: Toward an integrative framework. Working paper 2013-0002, Weatherhead Center for International Affairs, Harvard University, April 2013.

Steiner, G. (2018). From probabilistic functionalism to a mental simulation of innovation: By collaboration from vulnerabilities to resilient societal systems. *Environment Systems and Decisions, 38*(1), 92–98. https://doi.org/10.1007/s10669-018-9674-9

Steiner, G., Geissler, B., Schreder, G., & Zenk, L. (2018). Living sustainability or pretending to? From explicit self-report measures to implicit cognition. *Sustainability Science, 13*(4), 1001–1015. https://doi.org/10.1007/s11625-018-0561-6

Steiner, G., & Laws, D. (2006). How appropriate are famous concepts from higher education for solving complex real-world problems? A comparison of the Harvard and the ETH case study approach. *International Journal of Sustainability in Higher Education, 7*(3), 322–340.

Steiner, G., & Posch, A. (2006). Higher education for sustainability by means of transdisciplinary case studies: An innovative approach for solving complex, real-world problems. *Journal of Cleaner Production, 14*(2006), 877–890.

Steiner, G., Risopoulos, F., & Mulej, M. (2013). Competences for citizen-driven innovation in crisis resolution. *Systemic Practice and Action Research, 26*(6), 571–577.

Steiner, G., Risopoulos, F., & Mulej, M. (2014). (2014): Social responsibility and citizen-driven innovation in sustainably mastering global socio-economic crises. *Systems Research and Behavioral Science*. https://doi.org/10.1002/sres.2255

Steiner, G., Steffelbauer, I., Laubichler, M. D., Zenk, L., Schernhammer, E. S., Birmann, B. M., Bertau, M., Caniglia, G., Mühlmann, K., Satalkina, L., & Weitzer, J. (2023a). Complexity literacy for a sustainable digital transition: Cases and arguments from transdisciplinary education programs. In L. Keller et al. (Eds.), *Handbook digitalization, new media, and education for sustainable development—global and National Perspectives and Actions*. IGI Global.

Steiner, G., Weitzer, J., Bertau, M., Zenk, L., Caniglia, G., Ziganshina, D., Laubichler, M. D., Schernhammer, E. S. (2023b). Transdisciplinary collaboration as a policy tool in times of crisis? A study of public opinion in the D-A-CH region. (forthcoming).

Steiner, G., Laubichler, M. D., Bertau, M., Zenk., L., Weitzer, J., Han, E., Birmann, B. M., Ziganshina Lienhard, D., Caniglia, G., & Schernhammer, E. S. (2024). Navigating through poly-crises towards One Health: Mirage or tangible prospect? Insights from the Transatlantic Research Lab on Complex Societal Challenges. *GAIA-Ecological Perspectives for Science and Society 33*(1):202–203.

Stokes, D. E. (1997). *Pasteur's quadrant. Basic science and technological innovation*. Brookings Institution Press.

Thompson Klein, J., & Newell, W. H. (1997). Advancing interdisciplinary studies. In J. Gaff & J. Radcliff & Associates (Eds.), *Handbook of undergraduate curriculum* (pp. 393–413). Jossey-Bass.

Tversky, A.; Kahneman, D. (1974). "Judgment under Uncertainty: Heuristics and Biases". *Science*. 185 (4157): 1124–131. Bibcode:1974Sci...185.1124T. https://doi.org/10.1126/science.185.4157.1124.

Weitzer, J., Laubichler, M. D., Birmann, B. M., Bertau, M., Zenk, L., Caniglia, G., Jäger, C. C., Steiner, G., & Schernhammer, E. (2021). Independent evaluations and transparent communication are crucial to reduce COVID-19 vaccine hesitancy. *International Journal of Environmental Research and Public Health, 17*(21), 7884. https://doi.org/10.3390/ijerph17217884

Wilson, T. D., & Gilbert, D. T. (2003). Affective forecasting. *Advances in Experimental Social Psychology, 35*, 345–411.

Zenk, L., Steiner, G., Pina e Cunha, M., Laubichler, M. D., Bertau, M., Kainz, M. J., Jäger, C., & Schernhammer, E. S. (2020). Fast response to superspreading: Uncertainty and complexity in the context of COVID-19. *International Journal of Environmental Research and Public Health, 17*, 7884. https://doi.org/10.3390/ijerph17217884

Steiner, Gerald, Prof. Dr. habil., As a systems and sustainability scientist, former Schumpeter Professor at Harvard, and member of the European Academy of Sciences, I have dedicated my career to addressing complex societal challenges by transforming vulnerabilities into innovation opportunities. Integrating diverse knowledge—connecting citizens, practitioners, and scientists—across supply chains and multilayered knowledge systems is critical, particularly in areas like food security and sustainable resource management. This work underscores my belief in transdisciplinarity as vital for managing global challenges. Currently, I serve as Professor of Innovation and Organizational Communication, Dean of the Faculty of Business and Globalization, and Chair of the Department for Knowledge and Communication Management at UWK - Danube University Krems and as a member of the Associate Faculty at the Complexity Science Hub in Vienna. I also co-initiated and co-lead several Transdisciplinarity Labs, which aim to foster collaborations that bridge scientific and practical knowledge. Most recently, I founded the Transatlantic Lab for Complex Societal Challenges. My goal is to develop forward-thinking educational programs and promote cross-disciplinary innovation to drive sustainable global change, empowering society as a whole.

Idea of University: Idea of Transdisciplinarity – An Encounter

Yusuf Örnek

1 Introduction

The idea of the university is a philosophical problem acquired and analyzed by classical German philosophy in the nineteenth century. This problem, which developed together with the political problems of the Prussian State and the ideals of the Romantic Period, occupied many thinkers and writers and continued to be a subject in the academic life of the twentieth century. This issue, which has developed together with politics since its emergence, has both philosophical (theoretical) and political (practical) aspects. In the following text, the philosophical foundations of the idea of the university will first be expressed, and then this problem will be compared with the idea of transdiscipline, which was added to it as a new dimension in the twentieth century.

2 The Idea of the University in the Age of Modern Enlightenment

2.1 I. Kant

In his 1798 work *The Conflict of Faculties,* **Immanuel Kant** distinguishes between positive science and the science of reason. Positive science is based on ready-made, given rules, while the science of reason is critical. The relationship and negotiation between the two must be purely scientific. Since the common aim of the sciences is

Y. Örnek (✉)
Bilim University, Antalya, Turkey
e-mail: yusuf.ornek@antalya.edu.tr

© The Author(s), under exclusive license to Springer Nature Switzerland AG 2025

M. Schüz (ed.), *Transdisciplinary Thinking and Acting*, Transdisciplinary Management of Ecological and Social Crises,
https://doi.org/10.1007/978-3-031-70680-6_10

to produce knowledge, they should be sought where they maintain their sovereignty. This place is the university; the joints of the university, which is a large organism, are the faculties. According to Kant, the division of the university into faculties is not based on the purpose of science itself, but on its relationship with the state, and the hierarchy of faculties is determined accordingly. There are sciences which, because of their significant overlap with the interests of the state, are bound to its power in a special way because they concern the practical side of public life. On the other hand, there are sciences whose content does not concern the state at all; they are completely independent and autonomous from the power of the state. This led the state to distinguish between "higher faculties" and "lower faculties."[1] Higher faculties include sciences that are dependent on the state, while lower faculties include sciences that are not dependent on the state. The reason why faculties are characterized as superior and inferior lies in the fact that they serve the aims and objectives of the state. The sciences used by the state and useful to it are represented in faculties with practical functions. The sciences that do not directly serve a purpose of the state are, in Kant's terms, non-practical or purely theoretical sciences. Thus, the positive sciences are practical and the critical sciences are theoretical.

According to Kant, the main purpose of the state is the happiness of its citizens. Happiness is of three kinds: The well-being of the body, the well-being of the citizen, and happiness after death, that is, happiness for eternity. The happiness of the body is health, that of the citizen is justice, and that of the believer is eternal peace. Since the state is obliged to ensure the happiness of the people living within its borders, it opens faculties of medicine, law, and theology at universities to train individuals to undertake this task. The faculty of medicine trains physicians, the faculty of law trains jurists, and the faculty of theology trains clergy. Thus, the citizens' need for happiness is met through the students trained in these faculties. These faculties, which meet the practical needs of public life, are characterized as "higher."

Kant is not against the placement of the faculty of theology as the highest faculty in universities, followed by the faculty of law and then the faculty of medicine. What he is against is the placement of the faculty of philosophy at the bottom. It is obvious that the faculty of philosophy does not contain branches of knowledge that can respond to the daily needs of the state. It has no practical utility, it plays no role in public life. However, the whole existence of philosophy is reason: Autonomous reason, capable of self-determination and self-duty, unrestricted by foreign ranks or positions. The faculty of philosophy is universal because it includes all the wisdom of reason, that is, all the sciences, and is capable of questioning them all. It reveals knowledge through reason and grounds it. It is not static anywhere, it is inquisitive and questioning everywhere. The faculty of philosophy is therefore not only universal, but also critical and independent of the practical aims of the state. Because it is universal, it encompasses all "higher faculties." Because it is critical, it tests their preconditions. Since the "higher faculties" approach an object positively, while the

[1] Immanuel Kant (1907), Der Streit der Fakultäten, pp. 18–20.

"lower faculty," philosophy, approaches the same object critically, a "Clash of Faculties" emerges. This is also the age-old conflict between praxis and theoria.

Kant brought the faculty of philosophy, which was in the lower category, to the top with its critical use of reason, that is, its ability to question and think independently. This is because the faculty of philosophy has the capacity to question all sciences without prejudice, to determine their limits, and in doing so, it has no expectation of profit. Its only expectation is to serve the truth. In doing so, Kant stated that the faculty of philosophy should be completely free and free from the orders of the rulers of the state.

The problem that Immanuel Kant examined in this work was about the nature of the relationship between the sciences and philosophy, one of the main problems of his time, and it was no coincidence that the philosopher focused on this issue in 1798. Kant wrote his thoughts by extending the problem he saw and experienced in his own university life to the entire institution called the university. He said that the hierarchy of sciences should be organized and handled as a whole, and this meant that the entire university should be subjected to a reform. Because universities were institutions that represented all sciences and served the state with these qualities.

2.2 F. W. J. Schelling

In his 1803 *Lectures on the Method of Academic Study,* **F.W.J. Schelling**, one of the great figures of idealist German Philosophy in the nineteenth century, referred to these ideas of Kant.[2] In his time, there were those who were skeptical of the increasing specialization of the sciences and their divergence and differentiation in detail. Schelling, like Kant, defended the unity of the sciences against this tendency. For this, the unity and roof of the university was necessary. Since the sciences under this roof carried their own purposes within themselves, Schelling demanded that they should be independent from the state, and that university professors who did science should have intellectual freedom.

Schelling says that traditional educational institutions, which have a mass of dead knowledge, are old and have completed their lives. According to him, education and training should be done with a brand new spirit. For this, the first demand is for professors and the second demand is for students: A university professor is someone who knows the relationship between his/her discipline and other disciplines, who is aware that all sciences are part of a whole. Universities have to protect such professors. Because the state demands this from universities. The state expects free scientific research and teaching, the generation and dissemination of new ideas, not the doing and teaching of conventional things.

According to Schelling, academic work should be free, that is, the individual should do science of his or her own free will, not as an official. This is only possible

[2] Schelling (1958): 305.

if the living unity of the sciences is understood. Just as the universe is a living and organized whole, so is science, which is its copy. The specializations of individual faculties are not cogs in the wheels of a factory machine, but parts of a living being. Sciences must act from the spirit of the whole. Their living unity must be properly understood by university professors and students, because this idea underpins any method of academic learning. Only in this way can the individual realize the spirit of the whole.

In these lectures given in 1803, Schelling, influenced by Kant, started from the idea of science and said that the university should function as a whole because it should pursue this idea.

> *We all together stipulate, and we must stipulate, that the state must see in its teaching institutions true scientific institutions, and only on this condition is its reputation in our eyes valid.*[3]

Schelling was in favor of the university being an intellectual aristocracy. He wanted the teaching staff to be free and science to be independent from the interests and goals of the state. According to the thinker, science carries its own purpose within itself and cannot be at the service of the state. Nothing other than science should be valid in universities.

Schelling, like Kant, gave philosophy a place above all sciences and called it "the knowledge of all sciences."[4] According to the thinker, the unity of sciences and arts can only be achieved through philosophy. This is because philosophy sees the sciences not as a pile of dead knowledge, but as a single tree of principles that has emerged from a living tree of knowledge and has spread and branched out into various branches of science. This single and fundamental principle is the system of identity in Schelling's work, absolute reason, eternal subject and object, the first knowledge that opens itself to the universe. It is this original knowledge that is absolute. The knowledge of the whole is but a copy of it. Nothing is as real as this knowledge in its eternity and development: This is the thought of the perfect and absolute unity of the ideal and the real. The sciences are nourished by this thought. General common sense means being aware of this identity.

The years 1803–1815 were the years of the so-called Napoleonic Wars in European history. Napoleon defeated Russia and Austria in 1805 and the alliance of Prussia, Saxony, and Sweden in 1806. The Treaty of Tilsit was signed between France and Prussia on July 9, 1807. According to this agreement, Halle, one of the most important cities of Prussia, and of course Halle University, founded in 1694, became France's. The loss of this university, one of the most important bastions of neo-humanism, is a heavy loss for Prussia, and the Prussian king decides to establish a new university in Berlin, true to his saying "what we have lost physically we must

[3] Ibid.: p. 251.

[4] Ibid: p. 277.

regain culturally."[5] Before that, Kant and Schelling's writings on the university inspired Fichte, Schleiermacher, Steffens, and Humboldt.

2.3 J. G. Fichte

In 1807, **J.G. Fichte** clarified the classical, neo-humanist German idea of the university in his essay *A Plan, Deduced from First Principals, for an Institution of Higher Learning to be Established in Berlin, Connected to and Subordinate to an Academy of Sciences*. Fichte argued that a modern university must break free from the scholastic tradition and defined it as follows: "Schools of the arts that use reason scientifically."[6] These schools should not only increase the student's ability to perceive, but also his or her power of judgment and decision-making. The purpose of this institution is not to transmit knowledge to the student, but to develop the training of judgment and the art of critical thinking for the use of knowledge. Fichte recommended the use of the Socratic dialog method in lessons. According to the thinker, students in these schools should lead a socially isolated life, studying only for science, not to pursue their own interests or fulfill their own needs.

2.4 F. Schleiermacher

Like Fichte, Friedrich Schleiermacher was another philosopher who wanted the Socratic dialogical method to be used in lectures. In his 1808 *Occasional Thoughts on German Universities in the German Sense,* Schleiermacher wrote that the main task of the university was to awaken in young people the spirit of science and the search for truth, and to encourage their special talents. The aims and objectives of this institution of higher learning are personal education organized around individual talents and broad academic freedom. Academic freedom will enable students to acquire a very different way of thinking from the traditional environment in which they were brought up.

Schleiermacher identifies the following as the work of the university: To awaken in the young the idea of science; to enable the student to master the branch of science to which he or she wishes to devote himself or herself; to be able to see detailed knowledge not in isolation, but in its closest scientific connections and to look at it as part of a larger whole; to make students aware of the uniqueness and unity of knowledge within the framework of the fundamental laws of science; to search for, find, present and finally assimilate within oneself the ability to realize all this.[7]

[5] Jaspers (2016): IX, Footnote 7.

[6] Fichte (1971): 102.

[7] Schleiermacher (1958): 238.

Schleiermacher's ideas on how the university-politics relationship should be conducted are still relevant today: According to the philosopher, if governments prioritize politics in the relationship between politics and scientific institutions and do not give science its rightful place, this is an erroneous understanding that can cause harm. Especially if the administration wants to be above the university, this is a sad sign that the state does not sufficiently recognize the highest institution of science and education and prefers some empty mechanisms to real life.

> *Yes, if the state destroys the center of all knowledge, that is, the nursery where it is nurtured, and virtually severs the living links between educational institutions, there is no doubt that the goal, or at least the unconscious effect, of such a process is the suppression of the entire spirit of science and the highest level of free education...*[8]

Schleiermacher believes that the natural direction in the relationship between the university and the state is for the state to retreat to its own limits, so that the unity and harmony in the scientific institution will come to the fore.

> *The university must be free to formulate its own law and change it according to the circumstances; in this the state cannot assign itself a management role... The state is obliged to supervise the facilities, advantages and real estate it gives to the university. It does this through experts chosen by the university. Anything beyond that would be tutelage.*[9]

2.5 *H. Steffens*

In 1809, the German philosopher **Heinrich Steffens**, who gave seven lectures entitled *On the Idea of Universities,* defined a university as follows: "a state institution that allows all the minds within it to move in their own direction without coercion."[10] Defining the state as "the common organization of the minds of all its citizens,"[11] Steffens states that the state has the obligation to take care of universities because "the entire supreme existence of the state depends on the free spirit of inquiry prevailing in the university without interference."[12] Calling universities "schools of wisdom,"[13] Steffens writes that wisdom is "the soul of the state."[14] He emphasizes that while the state has a duty to look after the universities, the state expects from the universities "important expressions of wisdom about its own future."[15] Through this reciprocal relationship, the nation can look to its future

[8] Ibid.: 249.

[9] Ibid.: 272.

[10] Steffens (1958): 353.

[11] Ibid.: 338.

[12] Ibid.: 352.

[13] Ibid.: 318.

[14] Ibid.: 346.

[15] Ibid.: 353.

with confidence. Without the support of the state, universities would collapse, which, according to Steffens, "would mean the collapse of the nation."[16]

Steffens expresses the vital importance of establishing a new university in Berlin in the following words:

> *The universities are the institutions of the state that want the young people of the state to be free by determining their own measure of emancipation. And I believe that it has never been more imperative than it is now to restore the dignity of the universities, which are the most important and primary institutions of the state.*[17]

The new university to be established in the capital of the defeated Prussian state must restore both its own honor and the honor of the state. For this, the state is obliged to give the university all the support it needs, otherwise the future of the state will be jeopardized.

> *Where the state does not recognize science, and where scholars are alienated from the life of the state, both become nothing; there is no fruit of science and life from a barren soil.*[18]

Thinking as Kant envisioned, Steffens gave the first place among the faculties of the university to the faculty of philosophy: "... because wisdom differs from knowledge of various things in that it is the spirit that organizes and animates that knowledge."[19] And it is philosophy that provides this spirit. The faculties of economics, medicine, law, and theology follow.

2.6 W. v. Humboldt

In 1809, **Wilhelm von Humboldt** wrote an unfinished 10-page paper titled *On the Internal and External Structure of the University in Berlin and its Relationship to other Organizations,* and as Prussian minister of education, he was personally charged with the establishment of the university. During his 18-month ministry, Humboldt founded the University of Berlin and emphasized the ideal of humanism. Wilhelm von Humboldt was the embodiment of the humanist ideal. For him, the perfect human being is one who is nourished by all aspects of cultural life, but universal education does not mean knowing a lot. To be educated is not just to be intellectual. For Humboldt, who was inspired by the humanist education of the ancient world, "general humanity" does not materialize anywhere and remains an idea, but each individual shares in it and is part of it. Thus, humanity is an ideal that includes both universality, individuality and wholeness. The only place where this ideal is realized as an institution is the university.

---

[16] Ibid.: 347.

[17] Ibid.: 346.

[18] Ibid.: 331.

[19] Ibid.: 358.

Humboldt's writing constituted the program of the neo-humanist ideal of the university. According to the thinker, the university is "the moral summit of the nation's culture."[20] The main task is "mental and moral education."[21] University life rests on two foundations: "solitude and freedom."[22] In all his/her research and educational tasks, the scientist is alone and free from all kinds of coercion and influence. The most important basis of scientificity is the absence of any coercion. Within the university, science "should not be a problem that has been solved completely,"[23] that is, the unceasing search for the truth and the criticism of all kinds of authority are the most basic elements of scientific research. For this reason, the university must be dissolved and independent from "any institutional structure of the state."[24] The state should not use the university for anything related to its own political interests. The state's duty is to have common sense and trust the university because its ultimate purpose is to serve the state.

Humboldt's text was written in 1809, but was not sufficiently recognized.[25] It was only when Humboldt, a philosopher and man of culture, became minister himself that the Prussian state got an institution of higher education worthy of the university ideal. The text was recognized nearly 100 years after it was written and laid the foundation for the modern German university in the twentieth century. This ground-breaking text was unknown to the nineteenth century.

2.7 *E. Spranger*

In the twentieth century, it was the philosopher **Eduard Spranger** who made the text known and contributed to the publication of the university ideal. A student of Wilhelm Dilthey, a pedagogue and educational politician, Spranger, as a representative of a new humanism and idealism in the cultural sciences, republished in 1910 (on the 100th anniversary of the founding of the University of Berlin) the most important philosophical texts of the university ideal and emphasized the importance of Humboldt.[26] According to Spranger, the University of Berlin had succeeded because it was a form of organization in which the ideal of science and modern statecraft went hand in hand.[27] Humboldt's great success was due to the organic unity of the interconnected sciences with the state. The founding of the University of Berlin was therefore a stroke of genius and a living force that had a great impact on

[20] Humboldt (1956): 377.

[21] Ibid.

[22] Ibid.

[23] Ibid.: 377, 379.

[24] Ibid.: 378.

[25] The earliest editions of this article are found in at Bruno Gebhardt (1903).

[26] Spranger (1910).

[27] Ibid.: XL.

the future.[28] This new concept of science had two foundations: (1) Unlimited scientific research that serves no purpose. (2) The organic unity of all scientific activities. The idea of the unity of sciences necessarily demands that all sciences be represented within the university. Philosophy as the highest discipline unites them.[29]

The main features of the German neo-humanist ideal of the university, which developed in the early 1800s, are the following[30]:

1. The unity of research and education: Humboldt was the first to articulate this clearly.
2. Freedom of research and education: This has been articulated since Kant.
3. The university is an institution that serves the ideal of pure science, independent of practical aims.
4. The university educates not only scientists, but also people of moral integrity.
5. The unity of all sciences is the most important subject of the university ideal, and this unity is held together by the disciplines of the faculty of philosophy.

To summarize the idea of the university in the nineteenth century according to Sylvia Paletschek[31]: **Fichte** wanted to create a Prussian university that was sharp and uniform against individualism. **Schleiermacher** aimed for a comprehensive German university. **Steffens**, on the other hand, was after a university with church-like universality. **Humboldt** rejected all these narrowing views and wanted to create a "general university of culture."[32] But Humboldt's concept of science was the same as that of Fichte, Schelling, Schleiermacher, and Steffens:

> *Science is an organic whole; the pursuer of true knowledge must not only seek specialized training, but must identify himself with the spirit of this whole... From this point of view the university is a university of science. Every individual thing is not a finished thing, but something in constant becoming, something that can never be found as a whole. It is created only and only by the internalized action of scientists, which presupposes the complete freedom of mental creativity.*[33]

Neo-humanism's ideal of a holistic education had to be relevant and open to the spirit of the modern times. The ideal of a university that transcends the age in which we live has greatly enhanced the value of the faculty of philosophy and the cultural sciences. This ideal was developed by philosophy and the cultural sciences, which did not have to serve a practical purpose. This university understands itself first of all as a research institution and then, through scientific methods, is oriented toward taking on other tasks.[34]

---

[28] Ibid.: XLI.

[29] Ibid.: XL.

[30] Palatschek (2002): 184.

[31] Ibid.: 188, Footnote 17.

[32] Ibid.: 188.

[33] Ibid.: 188.

[34] Ibid.: 194.

2.8 K. Jaspers

Karl Jaspers was the German philosopher who dealt most intensively with the idea of the university in the twentieth century. He published three books under this title in 1923, 1946, and 1961, as well as numerous writings on the development of higher education. In 1923, Jaspers wrote his first work on the university "in the name of the people who put their lives at the service of the ideal of the university"[35] and categorized this ideal under two concepts: "humanitas" and "universitas." The concept of humanitas seeks to ensure the humane development of human beings who have an unlimited desire to acquire knowledge. This requires unlimited communication, questioning, searching for answers, discussion, and self-criticism. The concept of Universitas, on the other hand, aims at the orientation of science toward universality and integrity. The only discipline that can do this is philosophy.

In his inheritance from the university ideal of the nineteenth century, Jaspers writes that the university has three important purposes: (1) To provide education to provide a profession. (2) To provide an education aimed at becoming a human being. (3) To conduct research to increase knowledge and to train scientists.[36] Jaspers writes that the university is in a constant state of change due to the needs of society and the directives of the state, and states that behind all these changes, the ideal of the university remains unchanged. Because this ideal is an expression of knowing, research, and creativity that carries its purpose within itself.

In 1946, Karl Jaspers, who published his book *The Idea of the University* for the second time, wrote that the future of humanity could be guaranteed if unity was achieved in the universe of sciences. According to him, universities should seek the unity of the sciences with a holistic view and should not be a collection of schools that train professional specialists.

2.9 H. G. Gadamer

Hans-Georg Gadamer, Karl Jaspers' successor at the University of Heidelberg, in his 1986 lecture *Idea of University—Yesterday, Today, Tomorrow*, said that in industrial society there is a gap between research for a purpose and research for science. The university is now constantly torn between practical utility and education. Because society and the state have expectations from the university. Gadamer complains that university teaching systems are becoming more and more bureaucratized. What is important is that students and professors find their own free space.[37]

[35] Jaspers (2016): 3.

[36] Ibid.: 39.

[37] Gadamer (1988): 22.

Gadamer refers to Humboldt's ideal of the university and states that his concepts of solitude and freedom are very important. Solitude is necessary for the researcher to work efficiently in calm and peaceful solitude. Freedom, on the other hand, is the freedom to choose one's studies and research. According to Gadamer, Humboldt's main task for the university is not to teach and train researchers, but to "learn to live with ideas."[38] Because Humboldt says that the research of scientists never ends. Scientists begin to question the results of their research again and the horizons that open before them are never exhausted. This shows that scientists live with the idea of science. The function of ideas, which are products of the human mind, is to guide the researcher, to shed light on the path he or she is traveling, to ensure that the goals remain alive and are constantly redefined. Since ideas are guiding, they are constantly pursued. This is what Gadamer means when he says "ideas and facts always belong together and are far from each other."[39] The human mind creates ideas based on facts and pursues the ideas it creates. Ideas are ultimate goals that can never be fully attained but nevertheless live on in reality, albeit partially.

3 The Transdisciplinarity Idea of the Twentieth Century

When Hans-Georg Gadamer said "learning to live with ideas," he meant that human beings live with the ideas that their own intellect is constantly producing, sometimes moving away from them, sometimes coming closer to them. Just as ideas such as justice, freedom and equality are products of human reason, and just as human beings share in and contribute to them, they sometimes move away from them. The same is true for the ideas of science and the university. These ideas are lived within the university as an institution; at certain times one approaches these ideas, and at other times one moves away from them.

3.1 B. Nicolescu

Toward the end of the twentieth century, a group of scholars, including Basarab Nicolescu and Edgar Morin, came up with a new idea that was of great interest to both scientific disciplines and the university: Transdisciplinarity. The work accelerated when **Nicolescu**, a theoretical physicist and one of the leading figures of this group, founded the *Centre International de Recherche et Etudes* Transdisciplinaires[40] in 1987 and organized *The First World Congress of*

[38] Ibid.: 12.

[39] Ibid.: 2.

[40] A short chronology of transdisciplinarity is outlined by Nicolescu (2008a, 2008b): 253–255.

Transdisciplinarity[41] in 1994, where the *Charter of Transdisciplinarity* was declared.

There are 200 years between Kant's *Conflict of Faculties* and the *Charter of Transdisciplinarity*, and we know that today's world is much more complex than the one Kant lived in. The thinkers of the nineteenth century also complained about the constant specialization and fragmentation of the sciences. Because in this way, the unity of scientific work and the holistic approach to the research topics of the disciplines were lost, which made it difficult to realize both the ideal of science and the ideal of the university due to the difficulties created by the organization of research and teaching. However, the effort to look at the sciences as a whole is a goal that has remained the same for 200 years. The idea of transdiscipline is a philosophical idea that tries to achieve this with a new approach.

It is interesting to note that the transdisciplinary literature hardly mentions the idea of the university as an idea of classical German philosophy. Neither Kant's "conflict of faculties," Schelling's "philosophy of identity," nor Humboldt's "solitude and freedom" are of interest to transdisciplinary theorists. However, the discussions of the university idée fixe of the nineteenth century addressed both the holistic approach to the sciences and the organization of the sciences in these institutions, as well as questioning the relationship between universities, which were administratively autonomous, and the state. The authors of the transdisciplinary approach do not see the organization of science in universities in historical perspective, but only point to the problems of the specialized fragmentation of sciences in modern life. These authors criticize the fact that each discipline tries to grasp reality with its own methodology, resulting in an abundance of fragmented knowledge.

It is true that today's world is much more complex than the world of the Age of Enlightenment. Descartes' principles of method and content are far from meeting the needs of modern man in this complex world. Today's pluralistic life cannot be understood by rational methods and approaches alone. Since Descartes, the methods of using reason have led to the specialization of the sciences, their increasing division and fragmentation, and the organization of universities.

It is clear that Rationalism and the Enlightenment, far from eliminating the world's problems, only deepen them. The crises in nature and societies cannot be overcome with the idol of science leading the way. The rapid disappearance of the diversity of species in nature, the widening gap between the rich and the poor in societies, and the increase in racism and discrimination are reducing confidence in the sciences that are supposed to find solutions to these problems. On the other hand, the faster growth of scientific and technical knowledge in some countries than in others has a negative impact on the distribution of prosperity in the world.

In the nineteenth century, the thinkers who thought and wrote about the idea of the university all focused on the relationship between science and the state. They wrote that science should not have to serve the aims of the state, and that the ideal of pure science should be pursued. However, in today's world, we are witnessing the

[41] Nicolescu (2008a, 2008b): 261–265, reprinted in this volume, pp. XI–XV.

state completely instrumentalizing the sciences and the university as an institution. It is evident that while the ideal of science remains pure and clean as an ideal, it can easily be instrumentalized when put into practice. As an institution, the university, which aims for the nineteenth century ideal of science, has fallen behind humanitarian goals. This makes it necessary to re-examine the idea of the university. Universities, which largely provide education to acquire vocational skills, are instruments of the neo-liberal economic system. This process, accelerated by globalization, is increasingly handing universities over to multinational power centers. Humboldt's ideal of neo-humanism seems to have given way to the ideology of neo-liberal economics at the end of the twentieth century. Whereas the nineteenth century ideal of the university was that scientists should be free to pursue their own research interests, today their research interests are determined by governments, scientific research institutions, or foundations of industrial corporations.

Article 12 of the *Charter of Transdisciplinarity* recognizes that: "Economy must serve the human being and not the reverse."[42] The future of the earth has been jeopardized by the fact that the technical sciences are entirely at the service of the economy, and therefore of production. This is because it has led to a naïve notion that technology can completely dominate nature and that everything is doable. Kant's critique was forgotten, and science and technology were left with an unrestricted free range of action. At the beginning of the twentieth century, Max Weber was the first to draw attention to this and warned scientists against "disenchanting the world"[43] and questioned the limits of the sciences. The prevailing view in the world of science and technology today is that technology knows no boundaries and that all goals can be achieved through technology. In the twentieth century, one of the thinkers who drew the sharpest attention to this optimism was Martin Heidegger. Referring to the belief that everything is possible thanks to technology[44], Heidegger writes that the only thing that is not possible is "the promise and design of the impossible."[45] Vittorio Hösle, one of the most recent thinkers, states that in order for the relationship with nature to improve, people need to change their approach to their own lives and draw limits to their desires and needs.[46]

Aware of all these developments, Nicolescu writes that man has a dual nature: "on one side physical, biological and psychical nature, and on the other side, transcendental nature."[47] In parallel, he states that man lives in a dual cultural sphere: "technoscientific culture and spiritual culture."[48] This second sphere has no place in technoscientific culture, which is based on rationality. Spiritual culture involves the human relationship with the "sacred." Nicolescu argues that it is essential to

---

[42] Ibid.: 264.

[43] Weber (1975): 17.

[44] Heidegger (1989):136.

[45] Ibid.: 442.

[46] Hösle (1994): 64.

[47] Nicolescu (2008a, 2008b): 14.

[48] Ibid.: 14.

establish a link between these two cultures and writes that the only way to do this is to question the disciplines with a new approach. Since this has not been done, technoscientific culture, together with globalization, has led the whole world to uniformity and uniformity.

Nicolescu writes that this new approach is the idea of transdisciplinarity, which he believes will bring a new vision to the sciences:

> *Transdisciplinary knowledge is able to bring a new vision, not only of academic disciplines but also of cultures, religions, and spiritual traditions.*[49]

Nicolescu sees transdisciplinary knowledge as "complementary"[50] to scientific knowledge. Transdiscipline itself is not a new discipline, but an attitude. This attitude is open to all disciplines. Its aim is: "understanding the present world of which one of the imperatives is the unity of knowledge."[51] This is why Nicolescu writes that the prefix "trans" includes "between," "across," and "beyond."[52] This is an approach that questions the basic paradigm of science rather than its specialization, differentiation, and elaboration. According to this approach, the researcher is not outside of the subject he/she is researching; he/she constitutes a unity with it and is a part of its results. Because the transdisciplinary approach sees the activity of knowing as a more comprehensive process than a subject's orientation toward an object. This process transcends all kinds of dualist approaches such as subject-object, material-cognitive, simple-complex, diversity-unity and seeks communication between these seemingly opposing elements.

Nicolescu's understanding of science leads him to rethink academic organization. The idea underlying this conception is the following:

> *Academic disciplines study fragments of level of Reality and a multitude of disciplines are associated with a single level of Reality.*[53]

The fact that reality is multilayered and that the objects of the sciences are perceived and investigated only within the layers in which they exist implies that disciplinary or, optimistically, multi-disciplinary work is being done. The individual sciences, with their specific and distinctive methods, turn to the object of their research, focus on it, pay attention to measuring it, and take into account the ability to express the results of their research in numerical terms. The information obtained in this way achieves the purpose of the research, but it does not always allow for the emergence of other knowledge possibilities. The digital revolution has accelerated this process. Each discipline objectifies an object of the world according to its own method and expresses it numerically. Since each of the individual disciplines claims the absoluteness of the truths it arrives at, each of them carries within itself the danger of moving away from the scientific attitude. According to Nicolescu, this

[49] Ibid.: 14.

[50] Ibid.: 4.

[51] Ibid.: 2.

[52] Ibid.: 2.

[53] Ibid.: 14.

claim denies the fact that the universe is in a unity at its origin. As a way to overcome this, Nicolescu and the proponents of this approach emphasize the need to work under the leadership of the transdisciplinary ideal. We can summarize what this means as follows:

The world of experience of the human species is vast and diverse. The interdisciplinary research of individual sciences is conducted in the immanent world. However, individual human beings and cultures also have an experience of transcendence. In order to address both immanent and transcendent experiences together, our perspective needs to change. This can be achieved through a *transdisciplinary approach*. It is not always possible to explain people's experiences through myths, religions, and works of art with scientific methods. Transdisciplinary knowledge can provide the transition between immanent and transcendent worlds and the communication between layers. This is what distinguishes the idea of transdisciplinarity from the enlightenment, rationalist idea of science. In addition to natural sciences, social sciences and cultural sciences, worldviews, religions, myths and rituals that approach the world spiritually and are excluded by enlightened rationalism should be considered as "complementary" to the knowledge provided by individual disciplines and their truth should be utilized. Kant was making room for faith by demarcating knowledge. Enlightened science has so far been unable to benefit from this complementary dimension due to its passion for rationalism.

Advocates of the transdisciplinary approach argue that because each of the individual sciences looks only from its own perspective, the depth of the crises that concern the whole of the earth and humanity is not sufficiently seen. Add to this the fact that economic interests are not wanted to be harmed, and a willful blindness is created. Interdisciplinary studies have tried to overcome this disadvantage, but without success. When it comes to producing knowledge, a transdisciplinary approach is the healthiest. This approach sees reality as an open system with no definite boundaries. "A total system without stable boundaries."[54]

A transdisciplinary approach does not minimize the value of individual disciplines and their research methods. On the contrary, it puts their theoretical frameworks, methods, and practical applications under the magnifying glass. Ultimately, the aim of this research is to develop thinking in a multidimensional way and thus to increase the ability to foresee. This approach is philosophical in that it goes to the roots of knowledge and the activity of knowing, and questions knowing. It questions our ways of thinking, research methods, and the process of knowing as a whole. Here, the process of knowing is considered together with the whole of human existence, that is, not only as a mental activity, but also as a whole, together with the human body, emotions, and all psychic activities.

The prerequisite for a transdisciplinary approach is that disciplines must be able to criticize themselves. Scientific research has to be part of this critique. In this way, the boundaries of each discipline become clearer. This contributes to both research and teaching.

---

[54] Ibid.: 11.

3.2 E. Morin

As a matter of fact, **Edgar Morin**, one of the founders of the transdisciplinary approach, claims in his article *The Reform of Thought, Transdisciplinarity, and the Reform of the University*[55] that a reform in the education system can be achieved with the transdisciplinary idea. Once the principles of this approach are determined, their realization should be ensured from the beginning of the entire education system to the university. In the examples he gives, Morin focuses more on pre-university education, questions the future of the university and writes that it should be turned "in the right direction."[56] The adaptation of modernity to the university is, according to Morin, "to rebalance the tendency toward professionalization."[57] She refers to Wilhelm von Humboldt and writes that the university is a transnational institution because it is the bearer of a common human heritage; therefore, it should be approached with a transdisciplinary approach.

> *In order to do this, it must introduce the principals and the operators of the reform of thought that it evoked. These are the principals and the operators that will permit reconnecting the disciplines across an organic, systemic relation, all the while leaving them free to develop themselves.*[58]

While Morin proposes reforming universities, he links this to a change of thinking and does not say how the concrete change will take place.

This issue was addressed at an international congress on transdisciplinarity held in Locarno/Switzerland in 1997 with the support of UNESCO and the *Centre International de Recherche et Etudes Transdisciplinaires,* which sought to answer the question *Which University for Tomorrow*?[59] The subtitle of the congress was *Towards a transdisciplinary Evolution of the University*. In addition to seeking "a new vision"[60] for the university, the congress aimed at "the creation of a new art of living."[61] It was stated that the university has lost an important "meaning"[62] due to its "disorientation."[63] For this reason, it was proposed to establish "Institutes of the Research for Meaning"[64] in universities. In order to reform universities in the field of transdisciplinarity, it was suggested to "train the trainers,"[65] to allocate 10% of the

---

[55] Morin (2008): 23–32. Ibid.: 30.

[56] Ibid.: 29.

[57] Ibid.

[58] Ibid.: 30.

[59] Declaration and Recommendations of the International Congress "Which University Tomorrow: Towards a Transdisciplinary Evolution of the University" in: Nicolescu (2008a, 2008b): 267–273.

[60] Ibid.: 267.

[61] Ibid.: 269.

[62] "Loss of meaning" (art.2), "absence of meaning" (art.4), "decline of meaning" (art.7), see: *Declaration* in: Nicolescu (2008a, 2008b): 267–273.

[63] Ibid.: 268.

[64] Nicolescu (2008a, 2008b): 269.

[65] Ibid.

teaching process of each discipline to transdisciplinary approach,[66] to create centers and workshops for this[67] and finally to organize transdisciplinary forums in fields such as "history, philosophy, sociology of science and contemporary art."

Indeed, Basarab Nicolescu defines the aim of transdisciplinary education as "to discover the Eros of learning ... in states of permanent questioning and permanent integration."[68] This statement shows that this education is actually based on a philosophical approach. Because philosophy is the key to all branches of science as it deals with the most central problem for human beings, the problem of knowing/knowledge. This is where philosophy's unifying characteristic stems from. Philosophical thought is inherent in every branch of science with its questioning characteristic. Its critical approach enables branches of science to become conscious about themselves. Through philosophy, disciplines become aware of their own limits and question beyond them. In the nineteenth century university, it was philosophical knowledge that questioned and criticized disciplines and showed their limits.

4 Where the Idea of the University and the Idea of Transdisciplinarity Meet and Diverge

In the late eighteenth and early nineteenth centuries, thinkers who pondered on the idea of the university had similar conceptions of science and philosophy. The unifying, binding role of philosophy stems from Kant's concept of reason. Reason is a questioning, critical, boundary-pointing, dialog-seeking, communication-observing faculty. This approach toward sciences and philosophy is also valid for the twentieth century. The sine qua non condition of the transdisciplinary ideal is that the scientist recognizes the limits of his/her own field of research and continues to question; thus, new areas of research and problems emerge. The fact that it is scientists who put forward this idea does not change this fact. Because when science questions its own limits, it has stepped into the domain of philosophy. The most important commonality of the university ideal and the transdisciplinary ideal is that they are both philosophical approaches. This philosophical attitude, which questions science as a whole, wants to know the limits of all disciplines. While the ideal of the university stays only within the boundaries of the ideal of science under the influence of rationalism, the ideal of transdiscipline deals with areas that cannot be explained rationally, such as religion, faith, myths, and rituals. The sciences are prioritized in both, but the transdisciplinary ideal has a more comprehensive perspective.

Although the idea of the university emerged out of a necessity arising from the war, its institutional background goes back to Kant's *Conflict of Faculties*. Kant's

[66] Ibid.: 271.

[67] Ibid.: 271; see also Voss (1998), who argues that these efforts would create a kind of "awakening" in students.

[68] Nicolescu (2008a, 2008b): 11.

ideas about the faculties of theology, law, medicine, and philosophy are echoed in outline by all the thinkers of German Idealism and reflected in their ideas about the organization of the university. Although similar ideas were put forward by the proponents of the transdisciplinary ideal, it is observed that this philosophical thought developed toward the end of the twentieth century did not pay enough attention to the university ideal and did not benefit from it. However, a detailed study of the works of those who put forward the idea of the university could have helped to develop the transdisciplinary idea in a universal sense. With its current ramifications, the idea of transdiscipline lacks historical depth.

From a multifaceted perspective, transdisciplinarity aims to establish a normative principle for the organization of the sciences within the university. This approach can give new meaning to scientific research. It is not clear what is meant by "respect for the Collective and individual Otherness united by common life"[69] in Article 13 of the Charter and "ethics of the universal"[70] in the *Locarno Declaration.* This normative principle may coincide with Humboldt's goal of training "highly moral"[71] scientists, but the ethical dimensions of both philosophical approaches need to be explored. The researcher who acts with a transdisciplinary approach approaches what is happening in the world not only as a researcher but also as an ethical subject. Because this approach is not only about the activity of knowing, but also about doing and realizing.

Both the university ideal and the transdisciplinary ideal have a holistic understanding of science. Representatives of both ideals complain about the specialization and excessive separation of disciplines and yearn for a holistic understanding of science.

One difference of the transdisciplinary approach from the university ideal of the nineteenth century is its idea of serving humanity. The transdisciplinary approach, which claims that the problems faced by humanity cannot be solved by the university, which is constructed and organized with the understanding of the university ideal of the nineteenth century, has a humanist aspect with this feature. This can also be understood from the expression "Ethics of the universal." In the years when the ideal of the university was developed, the primary motive was to serve the ideal of science, not the state or humanity.

At the beginning of the nineteenth century, the reason why so many thinkers wrote about the idea of the university was due to the necessity brought about by the Napoleonic Wars. The main issue was the organization of the new university to be established in Berlin to replace the University of Halle, which had been abandoned to the French. At the end of the twentieth century, the main motive for the emergence of the transdisciplinary ideal was the loss of hope that the current understanding of science, technology, and the university could solve the great problems facing humanity. The idea of the university is a product of the Age of Enlightenment, of

[69] Nicolescu (2008a, 2008b): 264.

[70] Ibid.: 272.

[71] Humboldt (1956): 377.

enlightened Rationalism. It has an optimism that people will have a more livable world with the advancement of sciences. The idea of transdiscipline emerged as a criticism of this. With the awareness that the dominance of reason cannot solve the world's problems, it has a richer design of human and nature than in the nineteenth century. Nature is seen not only as a field of research and a tool to be used for consumption, but as a sphere of existence that has the right to live on its own. The idea of transdiscipline seeks to overcome the problem of loss of meaning in the idea of science in this way.

As can be easily understood from this brief comparison, the "encounter" between the idea of the university and the idea of the transdiscipline can help to better understand the problems of our age and to offer more productive solutions. However, more detailed research is needed on the philosophical foundations of these two ideas and the conditions for their realization.

References

Fichte, J. G. (1971). Deduzierter Plan einer in Berlin zu errichtenden höheren Anstalt (1807). In I. H. Fichte (Ed.), *Fichtes Werke* (Vol. VIII, pp. 97–204). Walter de Gruyter.

Gadamer, H.-G. (1988). Die Idee der Universität - gestern, heute, morgen [1986]. In *Die Idee der Universität. Versuch einer Standortbestimmung* (pp. 1–22). Springer.

Gebhardt, B. (1903–1936). Wilhelm von Humboldt als Staatsmann [1896], in: Albert Leitzmann [ed.] Wilhelm von Humboldt, Werke, Vol. I; Akademie Ausgabe.

Heidegger, M. (1989). In F. W. von Herrmann (Ed.), *Beiträge zur Philosophie (Vom Ereignis)*. Vittorio Klostermann.

Hösle, V. (1994). *Philosophie der ökologischen Krise. Moskauer Vorträge* (2nd ed.). C. H. Beck.

Humboldt, W. v. (1956). *Über die innere und äußere Organisation der höheren wissenschaftlichen Anstalten in Berlin [1809], see: Anrich (1956)* (pp. 375–386).

Jaspers, K. (2016). Die Idee der Universität [1923]. In O. Immel (Ed.), *Karl Jaspers. Schriften zur Universitätsidee* (pp. 1–68). Schwabe.

Kant, I. (1907). *Der Streit der Fakultäten [1798], Akademie Ausgabe* (Vol. VII, pp. 1–116). Georg Reime.

Morin, E. (2008). The reform of thought, transdisciplinarity, and the reform of the university. In B. Nicolescu (Ed.), *Transdisciplinarity—theory and practice* (pp. 23–32). Hampton Press.

Nicolescu, B. (Ed.). (2008a). *Transdisciplinarity—theory and practice* (pp. 23–32). Hampton Press.

Nicolescu, B. (2008b). In vitro and in vivo knowledge—methodology of transdisciplinarity, in: see Basarab Nicolescu (ed.), Transdisciplinarity—theory and practice, Hampton Press, pp.1–22.

Paletschek, S. (2002). Die Erfindung der Humboldtschen Universität. Die Konstruktion der deutschen Universitätsidee in der ersten Hälfte des 20th Jahrhunderts. In R. Johler & M. Mitterauer (Eds.), *Historische Anthropologie. Kultur—Gesellschaft - Alltag* (Vol. X, pp. 183–205). Köln/Weimar/Wien, Böhlau Verlag.

Schelling, F. W. J. (1958). Vorlesungen über die Methode des akademischen Studiums [1802]. In M. Schröter (Ed.), *Schellings Werke* (Münchener Jubiläumsdruck) (Vol. III, pp. 229–374). Beck.

Schleiermacher, F. (1958). Gelegentliche Gedanken über Universitäten im deutschen Sinn *[1808], in: see* Anrich, E. *(Ed.). (1958). Die Idee der deutschen Universität. Die fünf Grundschriften aus der Zeit ihrer Neubegründung durch klassischen Idealismus und romantischen Idealismus.* Wissenschaftliche Buchgesellschaft. (pp. 220–308).

Spranger, E. (1910). *Fichte, Schleiermacher, Steffens über das Wesen der Universität*. Verlag der Dürr'schen Buchhandlung.

Steffens, H. (1958). Über die Idee der Universitäten [1809], *in: see* Anrich, E. *(Ed.). (1958). Die Idee der deutschen Universität. Die fünf Grundschriften aus der Zeit ihrer Neubegründung durch klassischen Idealismus und romantischen Idealismus.* Wissenschaftliche Buchgesellschaft.(pp. 309–337).

Voss, K.-C. (1998). The university as a space of possibility. In *Rencontres transdisciplinaires*. Bulletin Interactif du Centre International de Recherches et Etudes transdisciplinaires. 12 Février. Available: https://ciret-transdisciplinarity.org/bulletin/b12c10.php, retrieved 15 Sept 2023

Weber, M. (1975). *Wissenschaft als Beruf* (6th ed.). Duncker & Humblot.

Yusuf M. Örnek, Prof. Dr. phil. habil., was born in Izmir in 1955. After high school, he studied sociology and philosophy at Hacettepe University in Ankara. He graduated from the university in 1977 and traveled to Germany to pursue a doctorate in philosophy. In 1983, he graduated from Johannes Gutenberg University in Mainz with a doctoral thesis entitled *Karl Jaspers' Concept of Existential Freedom* and returned to Turkey. Between 1984 and 1987, he worked as an assistant professor at Hacettepe University, Department of Philosophy. In 1987, he became an associate professor and left the university to start a business life. During the years he worked in the tourism industry, he gave lectures at several universities in Istanbul and Antalya. Between 2009 and 2013, he worked as a consultant to the Antalya Chamber of Commerce and Industry. Since 2014, Örnek has been lecturing in the Department of Philosophy of Law at Antalya Bilim University, Faculty of Law and is currently working as a full-time professor at this university. Örnek has translated I. Kant, M. Heidegger and K. Jaspers into Turkish and has published books and articles on the history of philosophy.

Part II
Applications

Culture Comparative Research as a Possible Approach to Transdisciplinary Thinking

Petra Y. Barthelmess

1 Introduction

In January 2000, the famous physicist Stephen Hawking (1942–2018) was asked in a "millennium" interview what he expected for the next century. He replied, "I think the next century will be the century of complexity."[1]

Natural disasters, economic crises, pandemics, biodiversity loss, wars, the climate crisis, social inequality, and political polarization, ... these things cannot be considered as isolated phenomena, they are taking place in different domains (society, environment, economy, etc.), at different scales (local to global and back to local again), and with even greater dynamism as the pace of change increases. These crises are not only inherently global and multi-layered, but often interconnected. A good example for this is the Covid-19 pandemic. We cannot understand let alone manage or even solve this crisis by virology or epidemiology alone. What about mobility contact networks, human social behavior, political and economic dynamics, and chain reactions, . . . What different local responses can be expected, and how does this unfold on a global scale? All these developments are intertwined, in a dynamic interrelationship that influences and reinforces each other. In short, all this is a rather complex matter. This level of complexity has led to a general insight that complex problems and their impacts cannot be understood, analyzed, let alone tackled by each discipline alone.[2]

[1] Cit. in: Brockmann (2021).

[2] For literature on complex systems and the need to for a transdisciplinary approach see: Gibbons et al. (1994), Klein (2006), Jaeger et al. (2011), Scholz et al. (2006), Sterman (1994).

P. Y. Barthelmess (✉)
Zurich University of Applied Sciences (ZHAW), Winterthur, Switzerland
e-mail: Petra.barthelmess@zhaw.ch

© The Author(s), under exclusive license to Springer Nature Switzerland AG 2025

M. Schüz (ed.), *Transdisciplinary Thinking and Acting*, Transdisciplinary Management of Ecological and Social Crises,
https://doi.org/10.1007/978-3-031-70680-6_11

Fig. 1 Itcho Hanabusa, Blind Monks Examining an Elephant. Source: Library of Congress Prints and Photographs Division Washington, D.C. 20540 USA

There is a very appropriate Zen parable for this limitation: The parable of the blind monks examining an elephant (see Fig. 1). It is a parable about how each blind monk comes to a different conclusion depending on which part of the elephant he has examined. To the one who touches the leg, the elephant has the shape of a column; to the one who touches the belly, the elephant looks like a wall; to the one who feels the trunk, the elephant seems to have the shape of a snake, and so on. All of them are right in their observation, but only the sum of all impressions gives the real shape of the elephant.

Knowledge builds on each other in a vertical axis. We have implemented this in academic practice by forming disciplines and delineating ever narrower subject areas to be as precise and accurate as possible. But knowledge also builds on the horizontal axis, as we connect different disciplines to create a multi-perspective view, to achieve a comprehensive understanding, to include "hidden" aspects of other fields, and thus avoid tunnel vision and blind spots, and to design comprehensive solutions. As obvious as it sounds, complex problems can only be addressed with complex approaches. Working at multiple levels (from the local to the global and back again), in multiple and coupled domains (society, environment, economy, technology, etc.), and in different social and cultural systems (values, needs, worldviews, preferences, etc.) has become a mandatory prerequisite for first understanding what is going on and second finding ways to address or at best solve complex problems.[3]

But do we even understand each other across the boundaries of individual academic disciplines? Do we understand the language and way of thinking of the other professionals? Are we even able and willing to listen to others? Are we willing and able to make the journey into other academic disciplines and explore their language, perspectives, and ways of thinking? The field of cross-cultural studies has made this boundary crossing and perspective shifting its very object of inquiry. Let me introduce the practice of cross-cultural studies as both a possible and an

[3]Naess (2010).

enjoyable approach to transdisciplinary thinking. Before diving deeper into the topic of cultural comparative studies, I would like to familiarize you with the basics of the function of culture.

2 The Function of Culture

What is culture? The term "culture" comes from Latin (colere) and means "to cultivate" in an agrarian social context. Meanwhile, the term has undergone a quiet evolution. In their review on concepts and definitions of culture, Alfred Kroeber and Clyde Kluckhohn compiled a list of more than 150 different definitions; a multitude stemming from the different emphases of the many disciplines that provide us with a definition of culture.[4] The common denominator, according to Kroeber and Kluckhohn, is that culture refers to the social world[5] and determines how groups of people structure, organize, and give meaning to their lives.[6]

According to Albert Blumenthal (1940), the anthropological concept of culture is a "non-genetically generated form of adaptation."[7] Anthropologist Bronislaw Malinowski (1944) defined culture as "an instrumental apparatus enabling man to cope better with the particular concrete problems he faces in his environment and in the satisfaction of his needs."[8] We humans have an urge to find explanations for what lies behind the inner workings of our environment. We want to make sense of what happens to us by constructing causal mechanisms to deepen our understanding of the environment. To this end, we seek to identify patterns and thus develop solutions for dealing with our external world to achieve satisfactory outcomes.[9] Culture in the anthropological interpretation means, first and foremost, an active modification of the threatening nature, both internal and external, of human beings.[10] In short, while animals and plants survive a changing environment primarily through physical adaptation (somatic), humans rely less on their anatomy than on their culture, which is an extra-somatic means of adaptation and survival.[11] As humans, we share...

[4] Kroeber and Kluckhohn (1952).

[5] Also see Bailey and Peoples (1998), p. 23.

[6] Kroeber and Kluckhohn (1952), p. 291; or from an organizational and occupational culture perspective, see for example: Schein (1990), p. 73 seqq.

[7] Wiese, L. of 1939. Review of six essays on culture by Albert Blumenthal. American Sociological Review, vol. 4, pp. 592–94.

[8] Malinowski (1944), p. 74 seqq.

[9] For the concept of human as "the model-making organism par excellence," see Hall (1989), p. 9 seqq.

[10] Pedersen (1995), p. 258.

[11] Jochim (2013), p. 14; White (1949).

1. A universal desire for control over the environment and control through **observation** of the environment.
2. A need to **understand** the environment and its inner workings through the recognition of correlations and causal relationships.
3. The ability to **solve** problems by introducing new measures of adaptation and control.[12]

Taking our nutrition habit as an example:

1. Observation: Eating certain fruits can cause stomach upset and even death.
2. Understanding: There must be something poisonous in these fruits that is not good for our bodies.
3. Solution: Knowledge is power! The older people have more experience when it comes to which fruits are good and which are not. They show us the way. They are the keepers of the accumulated experience of the group and all the ancestors of the group. We must respect and honor our elders. We must revere and worship our ancestors because we live by the accumulated memory of them all.

Once we agree as a group on our observations, our understanding of what we observe, and the actions we should take based on our understanding, the shared culture of the group becomes a stable foundation for group learning and group collaboration. No one eats the poisonous fruit anymore. Any new, additional knowledge about the poisonous fruits is shared among the group and incorporated into the group's accumulated memory—a memory whose keepers are revered and valued.

Cultures, however, can also be created upon wrong observations, misconceptions, false causal constructs, and consequently nonfunctioning notions of control.

Example:

1. Observation: Strangers come to our island. These strangers cut down trees and build leveled strips of earth. From time to time, they stand on these strips. They wear strange hats and wave sticks colored white and red, on which suddenly loud vessels fly down from the sky—loud vessels filled with delicious food.
2. Understanding: Cutting down trees, creating earth strips, wearing special hats, and waving colored sticks causes loud machines filled with delicious food to come down.
3. Adaptation and control: The so-called cargo cult as an indigenous millenarian belief system practiced on some Melanesian islands after the end of World War II. The locals proceeded to the airstrip, originally built by U.S. troops fighting the Japanese in the Pacific War. They waved batons. They carved headphones out of wood. They mimicked the movements of U.S. military airfield operators. All this to bring back the US planes with their magnificent, delicious cargo.[13]

[12] For a good summary of culture's function, see: Brumann (1999).

[13] See, for example, Encyclopedia Britannica: https://www.britannica.com/topic/cargo-cult.

Like the Zen parable of the blind Buddhist monks examining an elephant the above example can serve as an anecdote for the limits of knowledge that WE ALL face. Melanesians did not and could not know about the entire US military machine that was behind the Pacific War, dropping planes full of food and weapons. Their observation was limited to what was going on in their immediate environment.

And the same thing is happening to us. We all live in our little worlds. Embedded in the place where we live, in our upbringing, our professional activities, our professional socialization.... Our knowledge is limited. And as the Chinese Taoist philosopher Chuang-zi once said, "to pursue the limitless with the limited is dangerous."[14] In our age of complexity, with our limited knowledge, how can we observe, understand, let alone tackle the problems we face? Starting from the idea that culture is a means of human adaptation to the environment and thus the most important tool of mankind for adaptation and survival,[15] it would be interesting and most advisable to embark on the adventure of learning about and being inspired by other cultures and their perspectives and observations, other ways of understanding and making meaning.

As an example of such a cross-cultural investigation, I would like to share some of the results of our research on the multiple views on one's relationship with nature, the sense of urgency to create a more sustainable global economy, and the different ideas and plans on how to bring about the necessary change across two comparable and yet distinct cultures. We conducted the survey among 1700 Swiss and South Korean university students. Both cohorts live in economically advanced countries, both belong to the educated class, but they live in civilizations with very different cultural traditions regarding humanity's relationship with nature.[16]

---

[14] Our lives are limited/But knowledge is limitless/To pursue the limitless/With the limited/Is dangerous." 吾生也有涯，而知也无涯。以有涯随无涯，殆已。Original text from the chapter "Essentials for Nurturing Life."

[15] For the response of academic communities to the relationships between culture and ecology, see the rise in publications in this field described in: Bergmann (2017).

[16] The survey included 829 South Korean students at the following universities: Chungang, Hanyang, Sungkyungkwan, Korea, Soongsil, Kongju, Ajou, Sogang University as well as SolBridge International School of Business. The comparative part of this study included the opinions of 673 Swiss students at the Zurich University of Applied Sciences. Gender: South Korean survey participants 52% male and 48% female. / Swiss survey participants 42.6% male and 57.4% female. The age group of 20–26 included 93.4% of all South Korean and 75.4% of all Swiss survey participants. Mean age of all participants: 23.59 years (South Korea: 22.31 / Switzerland: 25.16). Study subject: South Korean survey participants: Social Science 30.6%, Engineering 27.3%, Humanities 12.6%, Natural Science 7%, etc. Swiss survey participants: Social Science 31.8%, Engineering 20.4%, Natural Science 15.5%, Humanities 4.0%, etc. (Departments of Architecture, Design/Construction Engineering, Applied Psychology, Business Administration, Health, Life Sciences and Engineering). The survey was conducted from February to April 2011. For the survey, see our publication Barthelmess et al. (2013).

3 Our Relationship with Nature

A comprehensive understanding of one's ecological orientation and behavior thus starts already with the very basic question: How connected do I feel with nature? Looking at the evolutionary history of mankind, humans have an innate need to connect with nature—a need based on the understanding that nature has ensured the continued survival of mankind as a species.[17] If we are aware of our dependence on nature, it would be reasonable to behave in an environmentally conscious manner. To quote Chuang Tzu, it would be "foolish" to engage in ecological depletion, pollution, and mass destruction of the natural habitat, as this would mean the destruction of our own livelihood. We would be sawing the branch we are sitting on, so to speak. But that is exactly what we are doing. We still are running on ecological deficit. We all know that can't be sustainable. But this cognitive insight has not necessarily translated into the necessary environmentally responsible action. How can we explain our complacency in this regard?

Various researches have been conducted to examine and understand the discrepancy between cognitive insight about nature's value and environmentally responsible behavior.[18] One of many foci of this research revolves around the question of human connectedness to nature, or the loss of that connectedness. The hypothesis is that environmental concerns are directly related to the extent to which individuals feel and sense their connection to the natural world.[19] For most of human civilization, cultural adaptation to the environment meant adaptation to the natural world in which man was immersed. This world has disappeared before our eyes. The creation of the modern world, with its division of labor and urbanization[20] led to an alienation from the natural world. At school we learn about the laws of nature, how tsunamis are created, what leads to climate change, what is the purpose of biodiversity ... but these insights remain abstract facts. Our modern, urbanized lives have disconnected us from the original need-based connection of humans to their natural habitat. We receive our monthly salary through work that does not involve nature; we organize our food not by hunting and gathering or growing crops, but by walking/driving to the nearby grocery store; the "natural" world that surrounds us is office and residential buildings, roads, cars, stores, etc. We use electric lights, central heating/cooling systems to optimize our environment for our specific needs. Our civilizational adaptation efforts nowadays mostly take place outside the natural world—in a man-made urban environment. In this sense, humanity's exodus to cities has detached us from nature; it has weakened our sense of dependence on nature and our sense of connectedness. We know that nature is still there. Theoretically, we

[17] Kellert and Wilson (1993).

[18] Dunlap and Mertig (1995), Kaiser et al. (1999), Nordlund and Garvill (2002).

[19] Dunlap and VanLiere (1978), Howard 1997, Dunlap et al. (2000), Clayton (2003).

[20] 2008, first time in the history of mankind the urban population has surpassed rural population in numbers. See: The World Bank Data on Urban Population (% of total population). https://data.worldbank.org/indicator/SP.URB.TOTL.IN.ZS?view=chart.

know that we ultimately depend on it. We are aware of the unsustainability of our lifestyle. But the sense of urgency to act is dampened by this disconnectedness.[21] The starting point of our study was therefore the very basic question: How connected do I feel with nature? And how or what should be measured to make a statement about my connectedness with nature?

In our cross-culture study, we used the widely used Nature Relatedness Scale (NRS) developed by Elizabeth K. Nisbet et al. (2009) and tested based on North American data and culture.[22] The NRS scale was calculated using a self-report measure designed to assess respondents' sense of connectedness to nature by integrating the following three aspects:

1. **Behavior:** One's experience and behavior which reflects a physical familiarity with the natural world, the level of comfort with and desire to be in nature.
2. **Cognition:** One's nature relatedness perspective, reflecting an external, nature-related worldview, a sense of agency with respect to individual human actions and their effects on all living things.
3. **Identity:** One's internalized identification, feelings, and thoughts about the personal connection to nature.

The purpose of the NRS scale was to assess the individual's physical, cognitive, and affective relationship to the natural world and to examine the correlative relationship between these three scales. The study found out that all three scales were interrelated in the sense that positive experience coincides with positive worldview as well as with positive feelings and strong identification.[23]

In our study we asked ourselves, would a North American scale work in a different cultural setting? What other insights would it yield? In our study we did the following: We used the NRS to measure our Swiss and South Korean cohorts' connection to nature (OBSERVATION), we analyzed the figures and conducted qualitative interviews to understand the inner working of their sense of connectedness (UNDERSTANDING). In a final step, we asked our survey participants to recommend what actions should be taken and who should be the agent to initiate this change (SOLUTION).

[21] See Soga and Gaston (2016) on their findings on "extinction of [nature] experience" and its impact on one's affection toward nature.

[22] Nisbet et al. (2009).

[23] Nisbet et al. (2009), Nisbet and Zelenski (2014).

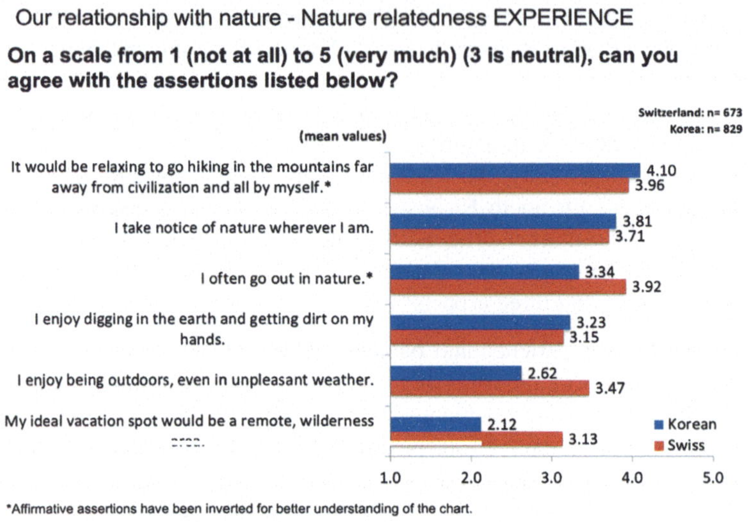

Fig. 2 Nature relatedness EXPERIENCE. Source: The author

4 Observation

4.1 Nature-Related Experience/Behavior

Do you take notice of nature wherever you are? Are you often out and about in nature? Do you enjoy getting dirt on your hands? Do you enjoy vacationing in a remote wilderness area? We asked app. 1500 students from both countries to reflect on their physical familiarity with and attraction to nature by answering the below listed questions (see Fig. 2).

The Swiss have more exposure to nature than the South Koreans. They also enjoy being outdoors more, even in inclement weather. Their appreciation to submerge into the remote and wild side of nature (as opposed to the familiar and cultivated experience of nature) is much higher than the South Korean students'.

5 Nature-Related Perspective/Cognition

Do you believe that the protection of nature is unnecessary because nature is strong enough to recover from any human influence? Do you believe that the condition of non-human species can serve as an indicator for the future of humans? Do you believe that animals and plants have the same rights as humans? The same students were asked to reflect on their relationship with the environment.

Here the South Korean students show on average a stronger sense of agency concerning individual human actions and their effects on all living beings (see

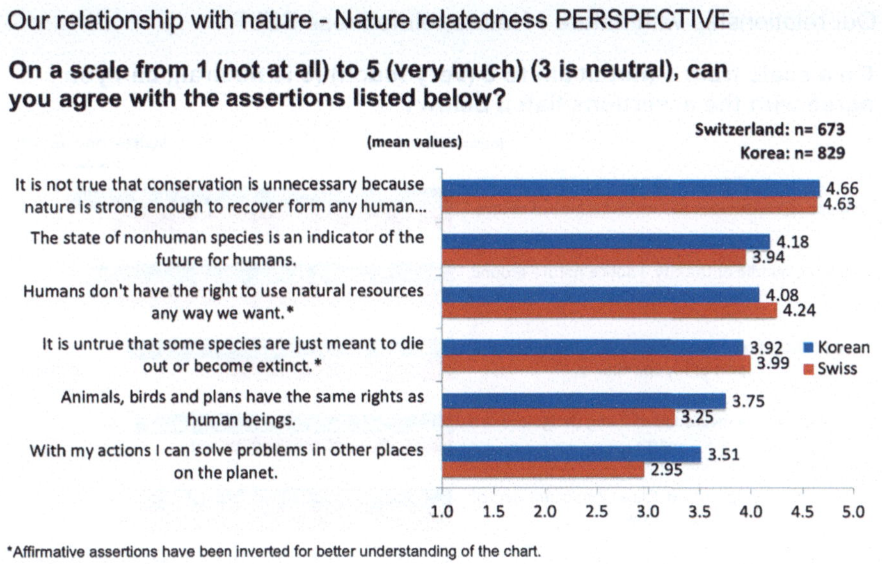

Fig. 3 Nature relatedness PERSPECTIVE. Source: The author

Fig. 3). Highly interesting is the similarity of Swiss and South Korean mean values in most questions except for the last two statements: (1) the equality of humans, animals, and plants in their rights to exist; and (2) their confidence in the effects of one's own actions.

6 Nature-Related Self/Identity

Are you a part of nature? Do you feel connected to all living things and the earth? Do you think your relationship with nature is an important part of who you are? Students in both countries were asked how strongly they identify with the natural environment.

In all statements, South Korean students show a stronger internalized identification with nature (see Fig. 4). Significantly different are the following two opinions: (1) the connection to nature as part of one's spirituality; and (2) the feeling of connectedness with all living beings and the earth.

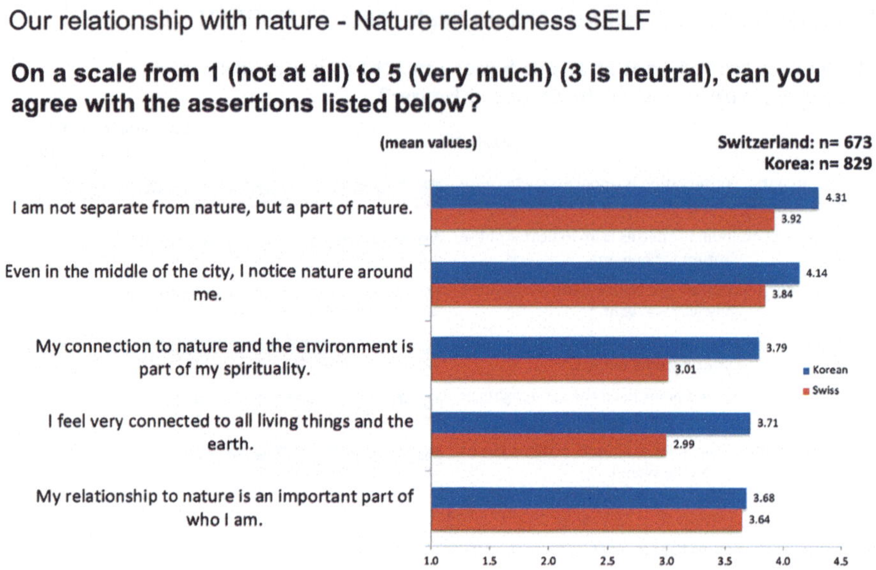

Fig. 4 Nature relatedness SELF. Source: The author

7 Overall Nature Relatedness Score

The original study of Nisbet, Zelensky, and Murphy with Canadian undergraduate psychology students provided evidence of a strong correlation between all three scales: (1) One's physical experience and joy of being outside in plain nature. (2) One's external, nature-related worldviews, convictions, and sense of responsibility vis-a-vis nature; and (3) One's feeling of connectedness with nature as being an integral part of one's self-identity.[24] Our internal consistency test shows that the NRS is also applicable to the Swiss and South Korean cohorts.[25] The table below offers an overview of the each scale (the higher the figure, the higher the scale) as well as the overall NR-scale (see Table 1).

What we can see is that both cohorts share a strong sense of connection with nature across all three scales. However, they do so with different emphases: The highest contribution to the Swiss cohort's nature relatedness scales is their physical experience and enjoyment of being in nature. A result that is consistent with the study of Nisbet, Zelensky, and Murphy that found the strongest correlation between the overall nature relatedness scale and one one's nature behavior (= one's physical

[24] Nisbet et al. (2009), pp. 715–740.

[25] The overall scale shows an acceptable Cronbach's α (value >0.7) for the South Korean data and a good Cronbach's α for Switzerland (value >0.8). The nominal data for all questions has been analyzed by Chi-square test and metrically scaled data by ANOVA. All results outlined in this article proved to be highly significant ($p \geq 0.01$) .From a statistical perspective, this means that the nature relatedness scores can be calculated and used for both respondent groups.

Table 1 Nature relatedness OVERALL SCORE

Scale	Korea	Switzerland
Behavior	0.586	0.719
Cognition	0.558	0.585
Identity	0.823	0.716
Nature relatedness total	**0.767**	**0.808**

Source: The author

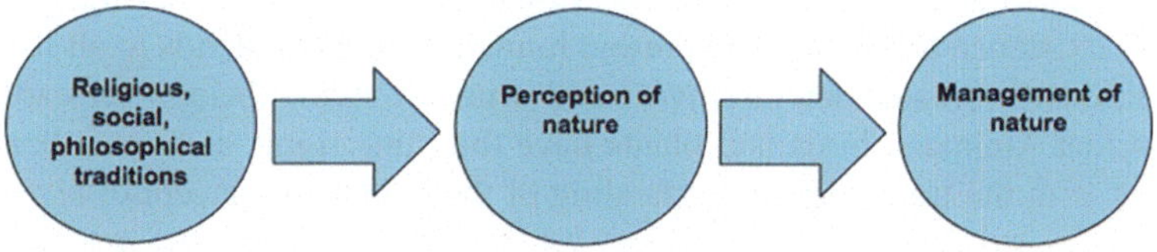

Fig. 5 Linkage society's nature relatedness traditions, perceptions, and management of nature. Source: The author

familiarity with the natural world, the level of comfort with and desire to be in nature).[26] The South Korean cohort falls significantly behind on the behavioral scale. This is not surprising, as nearly 90% grew up in the megacity of Seoul, with a population of over 10 million.[27] The cohort's even higher level of overall nature relatedness is primarily due to the South Korean students' high identity-scale.

Where does this high self-identification with nature come from? The assumption is that there is a fundamental connection between a society's perception of nature and its approach to nature, which in turn is closely linked to the society's inherent religious and philosophical tradition (see Fig. 5).[28]

For a better understanding of the South Koreans' nature identity score, see below an introduction to the Far Eastern culture tradition in a nutshell: The traditional East Asian ecological paradigm supports the concept of mankind's connectedness with nature. The Confucian "eco-ethics" (Yao, 2000) views humans and their natural environment as a cooperative relationship and emphasizes the state of harmony between humans and nature.[29] Similarly, traditional East Asian cosmology links humans with nature through the principle of *Gi* (in Chinese: *Qi* 氣), which can literally be translated as air, steam, and breath. Figuratively, however, *Gi* means energy or life force. *Gi* permeates and accompanies all things in the universe. All things in heaven and on earth are united in this one great stream of energy. The same *Gi* also flows through every existing thing, supporting the existence of all other things by passing on its *Gi*. (Thus, the *Gi* of the sun makes the plants grow, the *Gi* of

[26] Nisbet et al. (2009), p. 733.

[27] South Korean survey participants who spent most of their life in an urban environment: 89.5%; Swiss survey participants who spent most of their life in an urban environment: 34.5%.

[28] Colon (2008).

[29] Yao (2000), p. 175.

the earth supports the house, the *Gi* of the liver supports the blood circulation, the *Gi* of the mother nourishes the child, and so on).

The Buddhist concept of *karma* and *karmic* consequences also emphasizes human's embeddedness in nature. Buddhism sees one's life as a human being as only transitory. Being human, this individual person at this moment, constitutes just one stage on one's long journey through various biological structures. Birth–life–death–rebirth/reincarnation—after myriads of years and myriads of life forms, the cyclical process ends in the attainment of Nirvana, the end of all suffering. The belief that one's existence transcends the present human form and extends to all life forms has traditionally created an intuitive bond with all living beings on earth. The assertion that "animals, birds and plants have the same rights as human beings" is consistent with the transitory understanding of the Buddhist concept of *kharma*.

> I am not separate from nature, but a part of it.
>
> My connection to nature and the environment is part of my spirituality.
>
> I feel connected to all living beings and the earth.

These questions might sound a little strange to Swiss students. Not so for South Korean students who grew up within the "religious environmentalist paradigm"[30] described above. The high internal consistency of the answers to these questions supports the assumption that the deficits of urbanized South Korean students in experiencing nature, which are due to the high degree of urbanization of this cohort, are compensated by their cultural and especially religious and ethical upbringing.[31]

8 Ecological Behavior Across Cultures

How does this closeness to nature then inspire ecological behavior? In a next step, we wanted to find out if the students' nature relatedness score influences the participants' everyday behavior. In addition, we wanted to find out how they think about who should advocate for a more environmentally sustainable economy and whether there is a cultural gap between the two groups. Our research focused on the following four themes: behavior change due to ecological considerations, the main agent of change, and the relevance of personal efforts to create a better, more sustainable world.

- Do ecological considerations influence your behavior?
- How do ecological considerations influence your behavior?

---

[30] Pedersen (1995), pp. 258–276.

[31] The importance of ethics in Korean education system becomes clear when, for example, looking at the primary curriculum, which consists of nine principal subjects: Ethical and moral education (도덕), Korean language, social studies, mathematics, science, physical education, music, fine arts, and practical arts—in this order.

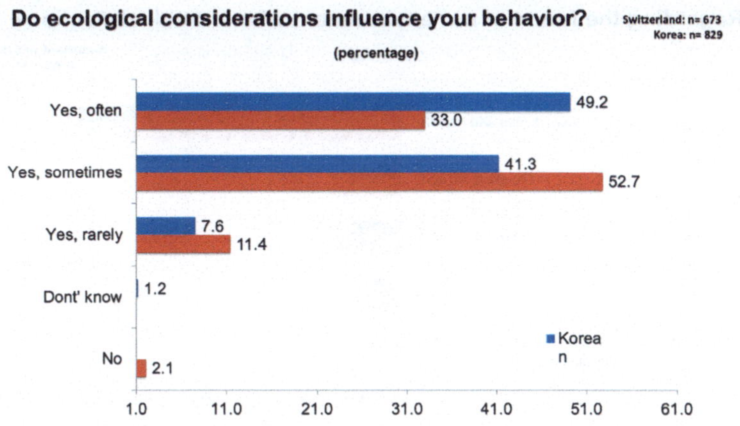

Fig. 6 Ecological considerations' influence on behavior. Source: The author

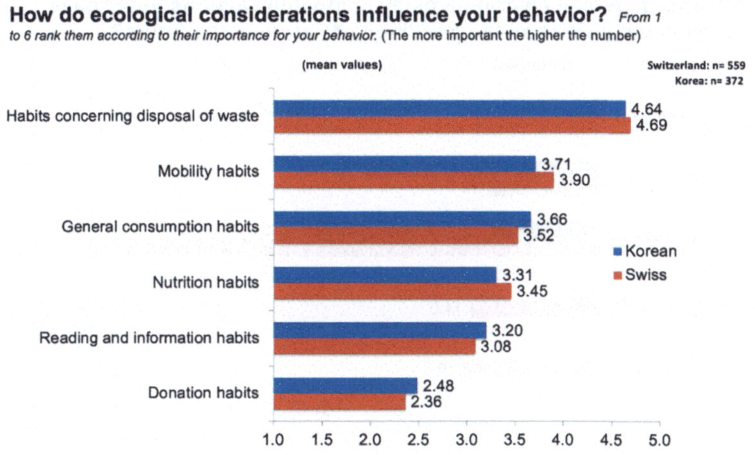

Fig. 7 The way ecological considerations influence behavior. Source: The author

- Who should take action to create a better, more sustainable world?
- Will your personal environmentally responsible actions today have a significant impact on the state of nature in the future?

Do environmental considerations influence your behavior? A large majority of Swiss and South Korean students answered this question in the affirmative (see Fig. 6).

How do ecological considerations influence your behavior?

Interestingly, South Korean and Swiss students set more or less the same priorities when it comes to certain measures (see Fig. 7): waste disposal in the sense of collecting one's own trash without external pressure (I don't leave my trash in the mountains/on the beach/in the park) is at the top of the list.

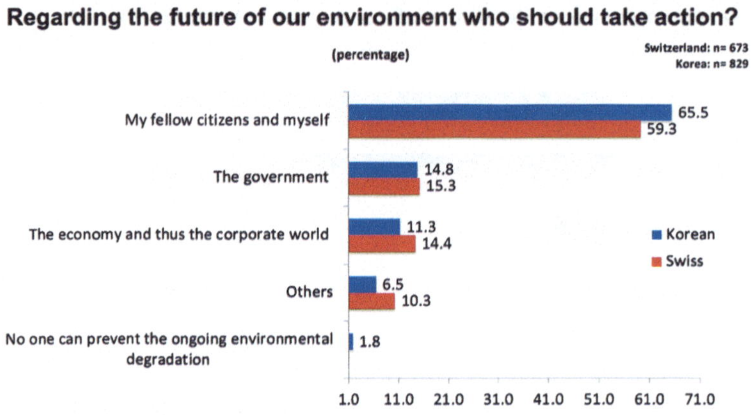

Fig. 8 Who should act? Source: The author

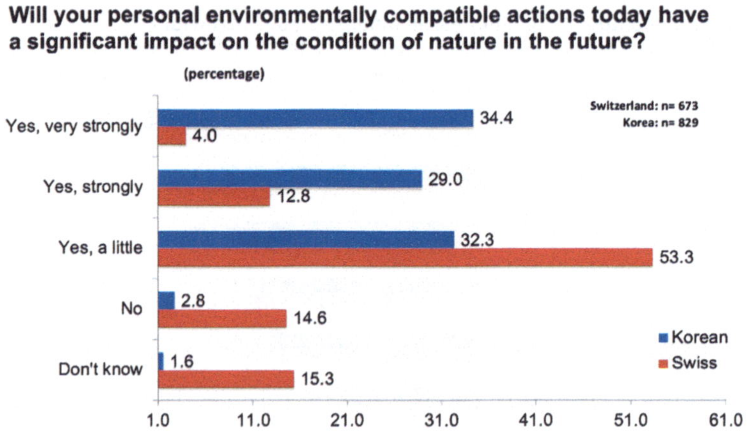

Fig. 9 One's personal actions' impact on nature's future condition. Source: The author

Who is in charge of creating a better, more sustainable world?

The majority of both South Korean (65.5%) and Swiss (59.3%) students believe that they themselves and their fellow citizens should take the lead (see Fig. 8). And? Can they make a difference? Based on the insight, that everyone should act, we asked them how they assess the influence of their action on the state of nature in the future.

Will your personal environmentally compatible actions today have a significant influence on the state of nature in the future? And here we were able to identify one further difference between the response patterns of the two groups is striking (see Fig. 9). While 63.4% of all South Korean students assess the influence of their actions as "strong" to "very strong," only 16.8% of all Swiss students share this conviction. The majority of Swiss participants estimate their influence as unknown (15.3%), non-existent (14.6%), or low (53.3%).

To summarize the last two questions: The South Korean majority believes that each individual should act (65.5%), (BECAUSE) each individual (or I) can also make a difference (63.4%). While the Swiss majority believes that each individual should act (59.3%), ALTHOUGH each individual (or I) [alone] cannot make much difference (83.2%). Again, to understand both positions, we looked at religious, social, and philosophical tradition—in short, we looked at the cultural underpinnings of both cohorts to explain their positions.

The religion or philosophy that is most important to the social organization of Korean civilization is Confucianism. Clearly, one must include this underlying value system in the explanation of the relationship of the individual to society. While in the West social change is ultimately brought about through normative intervention and institutional reform, Confucian societies value the concept of self-cultivation and role modeling as main contributor to all social change. Rather than relying on laws, Confucius relied on people's inherent goodness and benevolent attitudes conveyed through proper guidance. "Guide them with edicts, keep them in check with punishments, and people may keep out of trouble, but they will have no sense of shame. But if you guide them by your own virtue and keep them in check with decent actions, they will develop a sense of shame and control themselves." (Confucius, Analects, 29/30). "Guide them by your own virtue..."—The traditional East Asian approach is based on the inspirational power of example. By acting as a role model, the good, benevolent, and virtuous person inspires his/her fellow human beings to follow his/her example.[32] "If your desire is for good, people will be good. (...) When the wind blows, the grass bends." (Confucius, Analects, 12/19).

Like a drop of water that creates a wave, by acting righteously one serves as an example and can inspire others to behave in a more ecologically sensitive manner.

Guided by their own inspiration, they will do the right thing, which in turn inspires other people to do the right thing, which in turn inspires other people, and so on. The inspirational power of example knows no bounds. And so, the starting point of all social change lies in the powerful inspiration and unlimited influence of the role model, and in this sense social change can be triggered by anyone and everyone.[33] The last two survey questions can be understood as follows: If you act to save nature, you can make a difference through the power of your example. It's up to you! It is in your hands!

And the position of the Swiss majority: everyone should act, but each individual alone cannot achieve much. How can this be understood? In Switzerland, public space problems are traditionally solved by introducing new institutional frameworks and enforcing new regulations. By making the political process participatory, people can get directly involved to initiate new laws and create new institutions. In a direct

[32] For more on the importance of self-cultivation as a mean to establish authority, see Hall and Ames (1987), pp. 110–125; and Yao (2000), 155–160.

[33] "Virtue is not only a quality, but also an ability to transform oneself and a power to transform others." Cit. in: Yao (2000), p. 156. For a more detailed explanation of the importance of self-cultivation as a starting point for societal change, see: Yao, chapter "Morality as Transcendence" in Yao (2000), 155–160; also see Helgesen (2007), pp. 161–180.

democracy, every individual has a voice, every individual can initiate regulatory and institutional changes—therefore, every individual should act and participate in the democratic process to initiate institutional and regulatory changes that, once in place, bind every citizen. One person alone cannot make much difference. You have to change the overall framework to make a difference. In contrast to the Confucian approach of self-cultivation and role modeling, the Swiss students rely more on the extrinsic effect of institutionalized processes and services as well as regulatory enforcement as appropriate means of protecting the environment. People need to be given access; people also need to be controlled; a rather passive understanding of one's role. However, people can participate in the discussion about what and how they want to be controlled and thus as a collective self-impose the rules and processes they can agree to.[34]

9 Conclusion

The scientific revolution that marked the emergence of modern science has enabled us humans to make great strides in the way we observe and understand nature. The creation of a new way of gaining knowledge through experimentation and objective observation, logic and reasoning has changed our views of nature and our interaction with it from adaptation to control to subjugation. Given the catastrophic state of our planet in the twenty-first century, many of us have concluded that something must change. But what exactly? And how can this change be brought about? By whom?

According to anthropological understanding, culture determines how groups of people develop a shared understanding of what is happening, why it is happening, and how to deal with the challenges. Anthropologist Bronislaw Malinowski (1944) defined culture as an instrumental apparatus that enables man to cope better with the particular concrete problems he faces in his environment and with the satisfaction of his needs.[35] Cross-cultural studies attempt to identify the universal constants across cultures and the cultural characteristics that distinguish one group of people from another.

Climate change, loss of biodiversity, and pollution are not just a matter of science and technology. They have been brought about through social and cultural processes that stand behind human action and values. In reverse, the impact of nature's crisis can lead to change in our culture—our actions, rituals, and values.[36] In our study of nature connectedness and ecological beliefs and behaviors, we found many similarities between Swiss and South Korean students: they are all aware of the desolate state of our planet. They are all concerned about climate change and share a sense of

[34] For a deeper insight into the Swiss "bottom-up" political system, see: Linder and Mueller (2021), pp. 74–79.

[35] Malinowski (1944), p. 74.

[36] Bergmann (2017), p. 394.

urgency to create a more sustainable global economy. This sense of urgency stems from their experience—the experience of polluted water and air (South Korea) or the ongoing loss or destruction of natural habitat through the process of urban sprawl (Switzerland). In this sense, we can conclude that a cultural constant is emerging on a global scale. To return to the Zen parable of the blind monks and the elephant mentioned above: In our study, the blind monks, who judged the elephant's form differently depending on which part of the elephant they touched, seemed to have agreed on a consensus in their OBSERVATION of the elephant's form.

In line with other previous studies, our survey showed that the younger generation of both countries no longer follows the paradigm of unlimited economic growth at any price.[37] Both Swiss and South Korean students share the conviction that a linear continuation of our economic system threatens the existence of human civilization. Both cohorts are convinced of the need for fundamental and immediate change. Behind this are not only motives of their own survival. It goes beyond a transactional understanding of humanity's dependence on nature. Their concern is also strongly supported by emotional experiences (the physical experience and the joy of experiencing nature on the Swiss side) and religious traditions (the Taoist and Buddhist-influenced faith on the South Korean side). Transferred to the Zen parable mentioned above, we can state that all blind monks also agree in their UNDERSTANDING about the nature of this elephant: That it is a very large and wild animal.

What is to be done? Both cohorts of students agree that each individual is responsible for change, but for different reasons; the majority of South Korean students believe that each individual can become a role model through righteous behavior and thus inspire others to behave more ecologically, starting a whole movement that will continue to spread (The Power of Inspiration); the majority of Swiss students believe that every individual counts because everyone can participate in the legislative process through their political voice, which ultimately creates the legal framework for a greener economy (The Power of Political Initiative and Participation). Both cultural cohorts recommend different SOLUTIONS when it comes to how to deal with this very large and wild elephant.

Biodiversity forms the basis for a functioning ecosystem by increasing the adaptability of ecosystems to changing environmental conditions and thus stabilizing the ecosystem services provided by nature. Similarly, we can assume that a diversity of cultures increases the adaptive capacity of humans to their local particularities. The major challenges of the twenty-first century do not occur in isolation; they affect different domains simultaneously (society, environment, economy, etc.); they take place at different levels (from local to regional to global), and in different local environments. These challenges require multiple keys because multiple locks need to be opened. One of the sets of keys needed is the "key of local culture." Even the most perfect plans can fail if they are not embedded in the local culture. If culture

[37] For example see: OECD. Program for International Student Assessment (2009). Green at Fifteen? How 15-Year-Olds Perform in Environmental Science and Geoscience in Pisa 2006. Paris: OECD.

is man's primary adaptation tool, and adaptation occurs ultimately in the local context, knowledge of local culture is essential to bring about the desired change.

References

Bailey, G., & Peoples, J. (1998). *Introduction to cultural anthropology*. Wadsworth Publishing.

Barthelmess, P. Y., Schüz, M., Fuchs, R., Kucera, D., & Prandini, M. (2013). Different shades of green: A comparative study on nature relatedness and ecological consciousness among south Korean, Swiss, and Czech students. *Central European Business Review, 2*(2), 7–18.

Bergmann, S. (2017). Sustainable development, climate change and religion. In E. Tomalin (Ed.), *The Routledge handbook of religions and global development* (pp. 389–404). Routledge.

Blumenthal, A. (1940). A new definition of culture. *American Anthropologist, 42*(4), 571–586. http://www.jstor.org/stable/663653

Brockmann, D. (2021). *Im Wald vor lauter Bäumen. Unsere komplexe Welt besser verstehen*. dtv.

Brumann, C. (1999). Writing for culture: Why a successful concept should not be discarded. *Current Anthropology, 40*, 1–27.

Clayton, S. (2003). Environmental identity: A conceptual and operational definition. In S. Clayton & C. Opotow (Eds.), *Identity and the natural environment: The psychological significance of nature* (pp. 45–65). MIT Press.

Colon, R. L. (2008). *Religious groups and climate change. Drivers, barriers and religious engagement on climate change*. IIIEE Publications, Lund University.

Dunlap, R. E., & VanLiere, K. D. (1978). The new environmental paradigm: A proposed measuring instrument and preliminary results. *The Journal of Environmental Education, 9*, 10–19.

Dunlap, R. E., & Mertig, A. G. (1995). Global concern for the environment: Is affluence a prerequisite? *Journal of Social Issues, 51*, 121–137.

Dunlap, R. E., Van Liere, K. D., Mertig, A. G., & Jones, R. E. (2000). Measuring endorsement of the new ecological paradigm: A revised NEP scale. *Journal of Social Issues, 56*, 425–442.

Gibbons, M., Limoges, C., Nowotny, H., Schwartzman, S., Scott, P., & Trow, M. (1994). *The new production of knowledge: The dynamics of science and research in contemporary societies*. Sage Publications.

Hall, D., & Ames, R. (1987). *The Confucian creation of heaven: Philosophy and the defenses of ritual mastery*. State University of New York Press.

Hall, E. T. (1989). *Beyond culture*. Random House.

Helgesen, G. (2007). The case of moral education. In D. A. Bell & C. Hahm (Eds.), *Confucianism for the modern world* (pp. 161–180).

Howard, G. S. (1997). *Ecological psychology: Creating a more earth-friendly human nature*. University of Notre Dame Press.

Jaeger, C. C., Tabara, J. D., & Jaeger, J. (2011). *European research on sustainable development. 1: Transformative science approaches for sustainability*. Springer.

Jochim, M. A. (2013). *Strategies for survival: Cultural behavior in an ecological context*. Elsevier.

Kaiser, F. G., Wölfing, S., & Fuhrer, U. (1999). Environmental attitude and ecological behavior. *Journal of Environmental Psychology, 19*, 1–19.

Kellert, S. R., & Wilson, E. O. (1993). *The biophilia hypothesis*. Island Press.

Klein, J. T. (2006). Afterword: The emergent literature on interdisciplinary and transdisciplinary research evaluation. *Research Evaluation, 15*(1), 75–80.

Kroeber, A., & Kluckhohn, C. (1952). Culture: A critical review of concepts and definitions. *Papers, 47*(1), viii–223.

Linder, W., & Mueller, S. (2021). *Swiss democracy. Possible solutions to conflict in multicultural societies*. Palgrave Macmillan.

Malinowski, B. (1944). A Scientific Theory of Culture. New York: Chapel Hill.

Naess, P. (2010). The dangerous climate of disciplinary tunnel vision. In R. Bhaskar, C. Frank, K. G. Hoyer, P. Naess, & J. Parker (Eds.), *Interdisciplinarity and climate change: Transforming knowledge and practice for our global future* (pp. 54–84). Routledge.

Nisbet, E. K., Zelenski, J. M., & Murphy, S. A. (2009). The nature relatedness scale: Linking individuals‘ connection with nature to environmental concern and behavior. *Environment and Behavior, 41*, 715–740.

Nisbet, E. K., & Zelenski, J. M. (2014). Happiness and feeling connected: The distinct role of nature relatedness. *Environment and Behavior, 46*(1), 3–23.

Nordlund, A. M., & Garvill, J. (2002). Value structures behind pro-environmental behavior. *Environment and Behavior, 34*, 740–756.

OECD. (2009). *Programme for international student assessment. Green at fifteen? How 15-year-olds perform in environmental science and geoscience in Pisa 2006*. OECD.

Pedersen, P. (1995). Nature, religion, and cultural identity: The religious environmentalist paradigm. In O. Bruun & A. Kalland (Eds.), *Asian perceptions of nature: A critical approach* (pp. 258–276). Curzon Press.

Schein, E. H. (1990). Organizational culture: What it is and how to change it. In *Human resource management in international firms: Change, globalization, innovation* (pp. 56-82). Palgrave Macmillan UK.

Scholz, R. W., Lang, D. J., Wiek, A., Walter, A. I., & Stauffacher, M. (2006). Transdisciplinary case studies as a means of sustainability learning: Historical framework and theory. *International Journal of Sustainability in Higher Education, 7*, 226–251.

Soga, M., & Gaston, K. J. (2016). Extinction of experience: The loss of human-nature interactions. *Frontiers in Ecology and the Environment, 14*(2), 94–101.

Sterman, J. D. (1994). Learning in and about complex systems. *System Dynamics Review, 10*(2–3), 291–330.

White, L. A. (1949). *The science of culture. A study of man and civilization*. Farrar, Straus and Giroux.

Wiese, L. (1939). Review of six essays on culture by Albert Blumenthal. *American Sociological Review, 4*, 592–594.

Yao, X. (2000). *An introduction to Confucianism*. Cambridge University Press.

Barthelmess, Petra Youngzie, Prof. Dr., is a Professor of Culture and International Business at the School of Management and Law, Zurich University of Applied Sciences. She is the Program Director of the Master's Program in International Business and a member of the School's Committee for Academic Study Degrees with a focus on international education. As a lecturer, she teaches in the areas of International Relations, Political Economy, and Intercultural Cooperation and Communication. In research her focus lies on comparative and organizational cultural studies. She is also an independent consultant and coach for businesses and government institutions in matters of international and intercultural collaboration, communication, and marketing. Petra Youngzie Barthelmess has a multinational biographical background (Germany, South Korea, Switzerland) and an academic degree as Doctor of Economic History and Sinology from the University of Zurich. Publication examples: Barthelmess, Petra; Björck, Albena; Gysin, Emanuel; Dela Cruz, Jeremy, 2021, *Preparing agile transformation: a framework for assessment of the organizational culture readiness*. Journal of Applied Business and Economics. 23(6), pp. 232–246. Available from: https://doi.org/10.33423/jabe.v23i6.4689; Luescher, Raphael; Barthelmess Röthlisberger, Petra; Kim, Su-Yeong; Richter, Ulf Henning; Mittag, Michael, 2016. *Conceptualizing creativity: general and cultural biases in Gough's creative personality scale*. The Journal of Creative Behavior. 53(1), pp. 30–43. Available from: https://doi.org/10.1002/jocb.160. Contact: petra.barthelmess@zhaw.ch.

Crisis and Transformation: A Prerequisite for a New Paradigm in the Economy?

René Schmidpeter

Preliminary remark: We are on the brink of an epochal change in society and science. In the interplay between human creativity and artificial intelligence, we are just beginning to create things that defy the logic and evaluation of the "old" world. Who is the originator of the content created in this interplay? Artificial intelligence accesses the genuine thoughts of the author, while at the same time drawing on the inexhaustible cosmos of already published thoughts and contexts. It is not clear to either the AI or the human where exactly the added value is created. In the human mind of the author, in the algorithms of the AI or already in the multitude of thoughts in the global network of often nameless individuals? This question should not and will not be addressed in this hybrid essay—but it should be noted that this is exactly how it was created. Namely in the interplay between man and machine, in a symbiosis that transcends the previous rules of science. The author's thoughts have been supplemented, modified, and expanded by the answers and explanations of AI. The boundaries are fluid and incomprehensible. Therefore, the author of this text is the universe with the vicarious agent Rene Schmidpeter and the AI called ChatGPT as co-creator. This co-creation process is an experiment and in the new world (still) completly without guarantee. These thoughts will form the basis for the author's future considerations—the process is not complete, on the contrary, it has only just begun...

R. Schmidpeter (✉)
Bern University of Applied Sciences, Bern, Switzerland
e-mail: rene.schmidpeter@bfh.ch

© The Author(s), under exclusive license to Springer Nature Switzerland AG 2025
M. Schüz (ed.), *Transdisciplinary Thinking and Acting*, Transdisciplinary Management of Ecological and Social Crises,
https://doi.org/10.1007/978-3-031-70680-6_12

1 The Importance of the Paradigm Shift for the Development of Society and Science

A paradigm shift is a rare turning point in the history of thought and knowledge. It is therefore undoubtedly one of the most fascinating phenomena in the development of humanity and science. This term describes a rare and profound change in the fundamental assumptions, systems, and insights that guide a society, a scientific discipline, or any other group of people. A paradigm shift is not simply a superficial change, but marks a significant break with previous thinking and established structures. This essay will explain the key characteristics of a paradigm shift and emphasize the importance of this phenomenon for society, science, and the mindset in business.

Paradigm shifts lead to extraordinary changes not only in the sciences, but also in society. They occur at irregular intervals, usually when an old paradigm can no longer solve the problems at hand. They create remarkable turning points in the history or development of a particular field. An example of this is the scientific revolution of the sixteenth and seventeenth centuries, which was initiated by the groundbreaking work of Galileo Galilei and Isaac Newton. This revolution changed the way the world was understood and led to a paradigm shift in physics and astronomy. Another key characteristic of a paradigm shift is its fundamental nature. It is not just a superficial change, but affects the core aspects of a field. A paradigm shift is an overthrow of previous thinking and an introduction to completely new thought structures. Not only are existing views and methods revised, but they are replaced by completely new ones. The third characteristic of a paradigm shift is comprehensive transformation. During a paradigm shift, a fundamental change takes place that affects not only the scientific or social community, but often the whole of humanity. A good example of this is the industrial revolution, which changed the way the world produces, works, and lives forever.

In addition, a paradigm shift leads to the emergence of something new. This can include new scientific theories, social structures, technologies, or cultural paradigms. A prime example of this is the digital revolution that has taken place in recent decades, ushering in a new era of information processing and communication. Another key characteristic of a paradigm shift is its ability to influence people's behavior and thinking. A paradigm shift goes beyond superficial changes and has a profound impact on the perception of the world. This influence extends not only to people's behavior and actions, but also to their mindset and consciousness. A good example of this is the change in the way society thinks about environmental issues that has been triggered by the findings of environmental science.

Paradigm shifts are usually triggered by new findings or challenging events that question previous assumptions and ways of thinking. These can come from a variety of sources, including scientific discoveries, technological breakthroughs, social changes or crises such as the COVID-19 pandemic. The ability of society or a group to respond to such changes and accept new paradigms plays a crucial role in shaping the future. It is often difficult to recognize or describe a paradigm shift when it is just beginning, as the new views and ideas are not yet fully developed or accepted. It can take years or even decades for the effects of a paradigm shift to become clearly visible. Therefore, it is of great importance to foster the ability to adapt and remain open to new ideas and insights, as this is crucial to keep up with the changes in the world.

To summarize, paradigm shifts are rare and profound changes in the fundamental assumptions and systems of a society or science. They are of fundamental importance as they trigger a comprehensive transformation and create something new that did not exist before. Paradigm shifts influence not only people's behavior and actions, but also their way of thinking and their consciousness. The ability to respond to such changes and accept new paradigms is crucial for shaping the future of our society and science.

Assessment of the Section by Artificial Intelligence "The above description of a paradigm shift is apt and comprehensive. A paradigm shift is indeed a profound change that alters the foundations of thought and action in a particular domain. Here are some additional considerations and examples that might deepen your understanding:

I. **Thomas Kuhn and the structure of scientific revolutions**: The term "paradigm shift" was coined significantly by Thomas Kuhn in his book "The Structure of Scientific Revolutions." Kuhn argues that scientific progress is not continuous and incremental, but that it occurs in phases of normal science, crisis and revolution.

II. Examples of paradigm shifts:

- **Copernican revolution**: Copernicus' transition from the geocentric to the heliocentric view of the world was a paradigm shift in astronomy.
- **Einstein's theories of relativity**: Einstein's special and general theories of relativity changed the understanding of space, time and gravity and led to a paradigm shift in physics.
- **Digital revolution**: The introduction of computers and digital technology led to a paradigm shift in the way we communicate, work and live.
- **Genomics and genetic engineering**: The decoding of the human genome and the development of CRISPR-Cas9 led to a paradigm shift in genetics and biomedicine.

III. **Resistance to paradigm shifts**: Paradigm shifts are often met with resistance from established players and institutions that cling to the old views and methods. This resistance can delay the transition to new paradigms.

IV. **Long-term effects**: As you have already mentioned, the effects of a paradigm shift can often only be fully recognized in retrospect. They can last for decades or even centuries and shape the entire society, science and culture.

V. **Interdisciplinary nature**: Paradigm shifts often transcend the boundaries of disciplines and affect different areas of knowledge. For example, digitalization has not only influenced computer science, but also economics, medicine, sociology and many other fields.

VI. **Ethics and values**: Paradigm shifts can also raise ethical and societal questions, as new technologies and ideas are often associated with moral dilemmas. It is important to discuss and resolve these issues in order to manage the impact of change." (ChatGPT)

Overall, *"paradigm shifts are decisive moments in the history of knowledge and society, as they pave the way for innovation, progress and new perspectives. The table above provides a good overview of their characteristics and significance—as it was judged by Artificial Intelligence" as described by artificial intelligence.*

2 The Past Shows that We Are Reaching the Limits of Our Current Paradigms

The past decades have made it clear to us that our previous practices and ways of thinking are in many ways no longer sufficient to deal with the complex challenges of our time. Two key events, the financial crisis around 15 years ago and the recent COVID-19 pandemic, have served as wake-up calls and challenged us to develop new paradigms for dealing with crises and progress in various areas.

One of the most fundamental lessons learned from these crises is the need for an intersectoral and intersystemic approach. We have seen that we can no longer think and act in isolated silos to solve crises. Different sectors and systems are interlinked and changes in one area will inevitably have an impact on others. For example, the introduction of electromobility is only likely to be successful if alternative energy generation systems are developed at the same time to meet the increased demand for electricity. These and other findings underline the need for holistic thinking and action.

In such an interconnected world, the responsibility for a paradigm shift lies not only with institutions and governments, but also with each individual. Each of us must consciously participate in the transformation and be willing to work with others to change entire systems and their basic assumptions. This also requires a willingness to rethink and adapt economic organizations, scientific institutions, and political systems to meet changing demands.

Crises such as the financial crisis and the COVID-19 pandemic have a wide range of effects on society. They affect not only the material level, but also the psychological, political, and health levels. Everyone experiences crises in their own way, which leads to a diversity of perspectives and opinions. This can pose a challenge for society, as different opinions and views clash. Nevertheless, crises also offer opportunities for change and further development. The way in which a society copes with a crisis can lead to positive developments and help it to be better prepared for future crises.

In terms of digitalization, we are also experiencing a paradigm shift in various areas. Digitalization has fundamentally changed the way we communicate, work, and share information. It has enabled new business models, revolutionized education, and improved healthcare. But these changes also come with new challenges. A key aspect of this paradigm shift brought about by digitalization concerns the way we view data and privacy. Data protection and data security have become key concerns as more and more personal information is shared online. Digitalization has also raised the question of how we deal with artificial intelligence and automation and what impact they will have on the world of work. These questions require not only technological solutions, but also ethical considerations and social discussions.

Overall, the paradigm shift that is imminent or has already begun as a result of the current crises and digitalization is unavoidable. Current events require us to be ever more flexible and adaptable, to recognize the complexity of our world and to work together to find solutions that not only solve short-term problems, but also ensure long-term stability and sustainability. It is time to overcome old ways of thinking and work together to create a better future in which humanity can successfully overcome the challenges of its time.

Assessment of the Section by Artificial Intelligence "The reflections on the paradigm shifts and the challenges revealed by crises such as the financial crisis and the COVID-19 pandemic are very apt. Indeed, these events have underlined the urgency of rethinking in various areas of our society. Here are some key points highlighted in the statements:

I. **Intersectoral and intersystemic approach:** The idea that different sectors and systems are interlinked and that changes in one area have an impact on others is crucial. This is especially true for complex challenges such as climate change or the promotion of sustainable development. To tackle such problems successfully, governments, businesses and civil societies need to work together.
II. **Thinking and acting holistically:** The call for holistic thinking and action is indeed important, as many of our problems and solutions are intertwined. A silo mentality can compromise the effectiveness of interventions. Holistic approaches make it possible to develop solutions that are more stable in the long term.
III. **Individual responsibility:** Every individual has a role to play in the transformation of society and systems. This relates not only to personal behavior, but also to active participation in political processes and the ability to question and improve political systems.
IV. **Diversity of perspectives:** Crises often lead to different opinions and views in society. This diversity can be a strength if it is used constructively to find solutions that take into account the needs and concerns of different population groups.
V. **Opportunities for change:** Crises can indeed offer opportunities to drive comprehensive change. This is particularly true of digitalization, which brings with it new opportunities and challenges. However, shaping these changes in a positive way requires careful consideration and foresight.
VI. **Data protection and ethics in digitalization:** With digitalization, issues of data protection, data security and ethical concerns have come to the fore.

Society must find ways to take advantage of digitalization without neglecting privacy and ethical principles.

VII. **Artificial intelligence and automation:** The advancement of artificial intelligence and automation is having a significant impact on the world of work. This requires retraining and adaptability of the workforce as well as considerations regarding social security and labor legislation." (ChatGPT)

Overall, the thoughts highlight the importance of being flexible and adaptable in order to successfully tackle today's complex challenges and move society to a more sustainable and future-oriented level as described by the artificial inteligence.

3 Changing Values in a Dynamically Changing World

In a time characterized by rapid change and increasing uncertainty, questions about the opportunities and risks arising from this are of crucial importance. We are increasingly confronted with the realization that conventional, rational approaches and traditional strategic management are not always sufficient, as the world around us is changing at an astonishing speed. One of the most prominent aspects of this change concerns the future shift in values in our society. How will our world develop in terms of values? In view of the rapid developments, it is almost impossible to answer this question. But one thing is certain: we have the opportunity to actively shape the future. Each and every one of us is responsible for what this new world will look like.

A remarkable characteristic of our society is its openness and plurality. These characteristics enable different individuals to come up with innovative ideas and inventions that can change the world. It is often the "lateral thinkers" who develop pioneering products, services, and markets that eventually become so influential that they sweep others along with them. This influence can affect scientific discoveries as well as political movements. And it can be both positive and negative. It is therefore crucial to ensure that we preserve fundamental values, such as emphasizing human dignity and respect for one another. These values should not be abandoned, even if the world around us is changing. The question that arises, however, is what specifically should be preserved and what needs to be renewed. This evaluation is a collective process that requires communication, mutual appreciation, and empowerment to ensure that people can develop their full potential. Ethics play a crucial role here, as they guide us and ensure that we do not abandon our moral principles in times of change.

An interesting example of the change in values can currently also be found in companies. Values are often implicit and difficult to control directly. They develop gradually in a constant feedback process. Companies cannot simply change values from the top down or reorient employees. Instead, values often change when the corporate strategy and business approach are changed. Take, for example, a car manufacturer that moves from producing cars to sustainable mobility services. This strategic shift can cause the company's values to shift as it now focuses more on positive impact for society. This can lead to employees reflecting on their own attitudes and values and realizing that they better identify with the company's new direction.

Changing values in companies is a gradual process that is driven by changes in corporate strategy and the business model. Leadership plays a decisive role in promoting and implementing these changes. Managers should exemplify values and demonstrate behavior that has a positive influence on employees. However, this does not happen overnight, but requires wisdom and experience to make the right decisions in times of crisis and keep the team on track. In such times, it is important to minimize fears and bring employees along by communicating a clear vision and purpose. It is about bringing a team together and developing solutions together rather than dictating everything from the top down. This requires patience, foresight, and open communication to ensure that everyone is on board and moving in the desired direction.

In a changing world, change in values is inevitable, but it must be based on ethical foundations and accompanied by responsible leadership. Only in this way can we ensure that change leads us to a future in which our fundamental values and humanity are preserved.

Assessment of the Section by Artificial Intelligence "The thoughts on increasing uncertainty and the associated opportunities and risks are very apt. The rapidly changing world really does require us to adapt our way of thinking and our approach to solutions and management. The emphasis on human dignity and respect for each other are fundamental values that should be preserved in every society and in every company, regardless of the changes around us. These values serve as a guide for ethical behavior and can serve as a basis for dealing with the challenges of change.

Mentioning lateral thinkers who develop innovative products and services is also important. These individual and creative ways of thinking are often driving forces for progress and positive change. Companies and organizations should encourage this diversity and innovation and create space for new ideas. Emphasizing the role of leadership in promoting value change in companies is also crucial. Managers have a great responsibility to shape the corporate culture and guide employees towards shared values and goals. The ability to minimize anxiety and communicate a clear vision are particularly important in times of change to maintain trust and teamwork." (ChatGPT)

Overall, this paragraph emphasizes very well the importance of values, ethics and leadership in an ever-changing world. These principles are critical to the long-term success of organizations and to creating a positive and sustainable future according to the artificial inteligence.

4 The Integrative Sustainability Mindset: A Bridge to the Future?

Sustainability is a term that has become increasingly important in recent decades. But for me, sustainability goes far beyond the search for a conclusive definition. It is a way of thinking that encourages us to look at the present from the perspective of the future. This section will discuss the integrative sustainability mindset, which systemically links the environmental, social, economic, and knowledge-based dimensions.

Sustainability is more than just environmental protection. It means developing business models that have a positive impact in ecological, social, economic, and knowledge-based terms. These four dimensions should not be weighed against each other, but should be integratively intertwined. This means that long-term economic success is not possible if ecological and social aspects are neglected. Similarly, it is not possible to act in an ecologically and socially responsible manner without achieving economic benefits. Sustainability requires holistic thinking at a societal and corporate level that overcomes the analytical separation and ideologization of different areas.

A sustainable company is characterized by sustainable management, sustainable structures and leaders who integrate sustainability into their management style and decision-making. This extends to the sustainability of the business model, products, services, marketing, personnel development, and controlling processes. It is about integrating all these aspects into the corporate strategy and thinking. For example, the pandemic offers the tourism industry the opportunity to develop new concepts based on soft tourism and sustainability. These new offers should take into account the diversity of travelers' needs and leave traditional mass tourism behind.

Politics also plays a decisive role in promoting change toward greater sustainability. Subsidies should be used to promote innovative and sustainable ideas in the economy instead of maintaining outdated structures. The foresight of entrepreneurs and political incentives will determine whether this crisis becomes an accelerator or a boomerang for change in the economy.

Responsible leadership that is geared toward sustainability requires a change in thinking and action. Efficiency is important, but effectiveness, i.e. doing the right thing, is paramount. It requires an adjustment in the mindset of managers and a willingness to re-evaluate basic assumptions. Leadership should be integrative and holistic, moving away from purely analytical thinking. Personal qualities such as integrity and honesty are also crucial. Sustainability and social responsibility are becoming increasingly important as they can ultimately influence the success or failure of a company.

The integrative sustainability mindset offers a bridge to the future in which ecological, social, economic, and knowledge-based aspects are interlinked. Businesses and policymakers have a responsibility to put this mindset into practice to ensure long-term success and consider the needs of future generations. The COVID-19 pandemic has given us the opportunity to accelerate the shift toward more sustainable thinking and action, and it is up to us to seize this opportunity.

Sustainability is not only an ethical obligation, but also the key to a better future for all.

In a constantly changing world in which economic success often comes at the expense of the environment and society, responsible leadership and sustainability are becoming increasingly important in organizations. These two concepts are not just buzzwords, but an essential component of successful and ethically oriented corporate management.

Organizations should not be seen as isolated entities, but as integral parts of society. This requires integrity, because only organizations with integrity can gain and maintain the trust and recognition of society. Responsible leadership emphasizes integrity by ensuring that the values and principles on which an organization is built are reflected in every decision and action. Another crucial aspect of responsible leadership is the promotion of diversity and inclusion. Organizations that embrace different perspectives and viewpoints are able to make more informed decisions and create a more equitable and inclusive work environment. This diversity strengthens the ability to innovate and helps solve complex problems.

Assessment of the Section by Artificial Intelligence "In summary, the statements emphasize the need to embed sustainability and responsible leadership in organizations to ensure long-term success and social responsibility. This requires holistic thinking and the active involvement of managers and employees at all levels of the organization. The views presented on sustainability and the role of business and politics in promoting it are very detailed and reflective. It is clear that you view sustainability as a holistic way of thinking that links different dimensions together rather than looking at them in isolation. Here are some key points from your text:

I. **Sustainability as a way of thinking**: They do not see sustainability as a firmly defined concept, but as a way of thinking in which the present is viewed from the perspective of the future. This implies that sustainability should be anchored in many aspects of life and the economy.
II. **Integration of the sustainability dimensions**: They emphasize the importance of interweaving environmental, social, economic and knowledge based aspects of sustainability rather than weighing them against each other. This promotes a holistic approach to sustainable thinking and action.
III. **Responsible leadership**: They emphasize the role of managers and entrepreneurs in integrating sustainability into their decision-making and leadership style. Effective leadership requires a rethink and an adjustment of the mindset.
IV. **Political role**: They point to the important role of politics in promoting innovative and sustainable ideas in the economy instead of maintaining outdated structures. Political incentives can accelerate change.
V. **Organizational integration**: They emphasize the need to integrate sustainability into various aspects of companies, including the business model, products, services, marketing, HR development and controlling processes. This promotes the overall strategy of a sustainable company.

VI. **Honesty and self-reflection**: They emphasize the importance of honesty and self-reflection in organizations in order to address problems and not ignore them. An integrative understanding of organizations that reconciles individual goals with overall goals is emphasized." (ChatGPT)

The thoughts reflect a broad view of sustainability and its importance in the modern world. They also show how sustainability is not only an ethical goal, but can also have a positive impact on the long-term success of businesses and society as a whole according to artificial intelligence.

5 Responsible Leadership and Sustainability

A fundamental prerequisite for responsible leadership is the ability to lead oneself. Before managers can lead others, they must be able to lead themselves. This means making personal decisions and taking responsibility for their own behavior. Responsible leaders set themselves high ethical standards and act as role models for their team. In a rapidly changing world, setbacks and challenges are inevitable. Responsible leadership therefore also requires resilience. The ability to deal with setbacks and adapt is crucial to long-term success. Responsible leaders do not just rely on existing systems or bonuses, but actively take responsibility for change.

Sustainability and responsible leadership are closely linked. Sustainability is more than just ecological responsibility; it also encompasses social and economic aspects. Responsible leadership requires ethical reflection and the ability to make long-term, sustainable decisions. Integrating sustainability into the corporate strategy is essential. In order to successfully implement sustainability, transdisciplinary thinking is required. It means taking into account not only economic aspects, but also sociological, philosophical, and technological dimensions. This approach promotes innovation, resilience, and the development of practical solutions to complex challenges.

In summary, responsible leadership can be seen as a holistic approach that encompasses integrity, diversity, self-leadership, resilience, sustainability, and transdisciplinary thinking. This approach goes beyond short-term profit and aims for long-term success and social responsibility. Responsible leaders are ready to drive the necessary changes and ensure that their organizations are not only successful in the present, but also play a sustainable and responsible role in society in the future.

In a world where expectations of companies and leaders are constantly growing, responsible leadership and sustainability is not only an ethical obligation, but also a strategic necessity for the long-term success of organizations. It is the responsibility of leaders to live these principles and integrate them into the DNA of their organizations to create a better future for all.

Artificial Intelligence Assessment of the Section "The statements emphasize the importance of responsible leadership and sustainability in organizations. Here are some key points and thoughts that can be derived from your statements:

I. **Integration and integrity**: Organizations should be an integral part of society and integrate themselves into society. Integrity plays a decisive role in this, as it strengthens the trust and credibility of an organization.
II. **Diversity and inclusion**: Diversity and inclusion are important aspects of responsible leadership, as they bring different perspectives and points of view into decision-making processes and promote an equitable and inclusive working environment.
III. **Self-leadership**: Self-leadership is a cornerstone of responsible leadership. Individuals must be able to lead themselves before they can lead others. This means making personal decisions and taking responsibility for one's own behavior.
IV. **Resilience**: Responsible leadership requires resilience to deal with challenges and setbacks. The ability to adapt and stay on course despite difficulties is crucial.
V. **Sustainable leadership**: Responsible leadership can be seen as part of the sustainable leadership concept. Both approaches are closely linked and require ethical reflection and the ability to make long-term, sustainable decisions.
VI. **Transdisciplinary thinking**: In order to promote sustainability in organizations, transdisciplinary thinking is required that goes beyond economic aspects and takes sociological, philosophical and technological dimensions into account. This makes it possible to develop practical solutions to complex challenges.
VII. **Integrated sustainability strategy**: It is important not to treat sustainability as an isolated strategy, but to integrate it as an integral part of the corporate strategy. This requires the active involvement of all areas and the continuous adaptation of the strategy to reality.
VIII. **Ownership and engagement**: Employees at all levels should understand their contribution to the sustainable business strategy in order to encourage ownership and engagement.
IX. **Dynamic and interdisciplinary thinking**: In the face of a constantly changing environment, dynamic and interdisciplinary thinking is required to promote innovation, resilience and sustainability.
X. **Long-term perspective**: The aim is to create organizations that not only pursue short-term profit, but also act sustainably and responsibly in the long term." (ChatGPT)

6 Impact of the New Sustainability Paradigm in the Economy

In recent years, the way in which companies can operate and be successful has changed considerably. One of the most significant changes concerns the integration of sustainability into corporate strategy. The need for sustainable corporate management is underlined by various factors. Firstly, there is a growing expectation on the part of consumers and investors that companies should act ethically and sustainably. Public pressure on companies to live up to their social and environmental responsibilities is constantly increasing. Secondly, the legal framework has become stricter, particularly in the European Union, with laws such as the EU Taxonomy and the Supply Chain Act, which place sustainability aspects at the forefront. Companies are now legally obliged to act sustainably.

In addition, the understanding of corporate success has changed. Success is no longer measured solely in terms of turnover and profit, but also in terms of social and environmental impact. Companies that act sustainably can increase customer loyalty in the long term, boost their innovative strength, and better withstand unexpected changes in the business world. Despite the increasing importance of sustainability in the business world, sustainable management is not yet established across the board in German companies. There are differences between companies and sectors. Some companies have already invested in sustainable management at an early stage and developed a sustainable corporate culture. These companies do not see sustainability as a separate issue, but as an integral part of their corporate strategy.

However, there are also companies that view sustainability in isolation and treat it as a separate sustainability strategy that has little to do with their overall strategy. This can lead to inefficiencies and affect the company's long-term competitiveness. However, it is encouraging to see that more and more companies are moving toward sustainable leadership. Sustainable leaders are those who successfully integrate sustainability into their corporate strategy, paving the way for other companies to follow. These sustainable leaders can be found in a variety of industries, from technology companies to traditional industries.

Why should companies adopt a sustainable management approach? The answer lies in the numerous opportunities that arise. Firstly, sustainable leadership strengthens a company's competitiveness. It enables the creation of a positive reputation and increases customer satisfaction. These are all factors that contribute to economic success in the long term. Another argument in favor of sustainable leadership is that it makes economic sense. It should not just be seen as a cost factor, but as an investment. Sustainable measures offer a clear return on investment and reduce entrepreneurial risk. These arguments should not only convince ethicists, but also financial experts and CFOs.

Companies that practice sustainable leadership implement a variety of measures. These include the involvement of employees in decision-making processes, a fair distribution of profits, and the measurement of social and environmental impact. The incentive system and personnel development are being redesigned and managers are being selected according to different criteria than in traditional business management models. These measures show that sustainable leadership goes beyond what is taught in business management textbooks. It requires a holistic view that takes into account the role of the company in society and the connection between business and society.

Nevertheless, there are challenges on the path to sustainable leadership. One of the biggest is to change the thinking and thought patterns in the organization. Many people resist change and hold on to old paradigms. This is where education comes into play. Education is the key to promoting sustainable leadership. Companies need to invest in the education and training of their employees. There is a need to create a new awareness and sensitize people to sustainable thinking. This is an educational issue that needs to be addressed in order to accelerate the transformation toward sustainable leadership.

In recent years, the discussion about sustainability and responsible corporate governance has become a key topic in the business world. Companies are faced with the challenge of not only achieving economic success, but also assuming social and ecological responsibility. This essay is dedicated to the concept of sustainable leadership and its significance for the transformation of corporate management. Sustainable leadership, also known as responsible leadership, is more than just a buzzword. It is a paradigm shift that fundamentally changes the way managers lead companies. The approach emphasizes the integration of sustainability principles into leadership behavior and decision-making. The aim is not just to maximize short-term profits, but to create long-term value for society and the environment.

A key aspect of sustainable leadership is the mindset. Managers must undergo a fundamental change in their way of thinking. They need to understand that sustainability and profitability are not contradictory, but can go hand in hand. This awareness is the first step on the path to sustainable corporate management. However, it is not enough just to have the right mindset. Managers must also have the necessary skills to put sustainability into practice. This includes professional as well as personal and social skills. They must be able to develop sustainable strategies and implement them in their organizations. This transformation requires time and continuous training.

The question of how companies can successfully bring their managers into this sustainable leadership role is of crucial importance. One way is to offer targeted training programs. These programs can prepare young talent to be sustainable leaders from the start. They can also help existing leaders to develop and strengthen their sustainability skills. Mentoring and sparring are other important tools to accompany managers on their path to sustainable leadership. Experienced managers can pass on their knowledge and experience to younger generations. This creates an important link between the different age groups and promotes the transfer of knowledge.

Integrating sustainability into business education is another crucial step on the road to sustainable business management. Business education programs should integrate the principles of sustainability into the curriculum from the ground up. Students should learn how to develop business models that take into account not only economic success but also social and environmental responsibility.

It is important to emphasize that sustainable leadership is not just an optional supplement to conventional leadership. It is an integral part of sustainable leadership. Business management without taking sustainability into account is incomplete, as it ignores the real challenges and opportunities facing companies. In today's world, characterized by environmental problems, social injustices, and economic uncertainties, sustainable leadership is no longer a mere option, but a necessity. Companies that focus on sustainability can not only be successful in the long term, but also make a positive contribution to society.

It is time to think about the future from the present and actively drive forward the transformation of corporate management toward greater sustainability. This change requires a rethink of management culture, targeted training measures, and the integration of sustainability into business management. This is the only way for companies to fulfill their responsibility toward society and the environment and at the same time create long-term value. In 50 years' time, historians may look back on this period and ask how companies and managers managed to shape change. The answer lies in sustainable leadership that preserves and shapes the future.

This time, it is up to the reader to make the assessment. After all, it is all of us who think about this present from the perspective of the future and bring about this future in our decisions—or not.

7 Conclusion: Next Steps for Sustainable Corporate Governance

The paradigm shift toward sustainable business management is not a temporary phenomenon, but an essential component of future corporate management. It offers opportunities to increase competitiveness and make a positive contribution to society. The challenge is to change the way companies think and establish a new mindset. Education plays a crucial role in promoting this change in thinking. Companies that recognize and implement the importance of sustainable leadership are not only on the right track ethically, but also economically.

Not only top management, but also middle management and employees should be involved. If you believe in the shared intelligence of man and machine, here are some concrete measures and approaches that companies can use to promote a sustainable approach:

1. **Clear sustainability strategy**: Companies should develop a clear sustainability strategy that is integrated into the corporate strategy. This strategy should define the company's purpose and its long-term goals for environmental, social, and economic sustainability.
2. **Stakeholder engagement**: Companies should involve their internal and external stakeholders in the sustainability process to ensure that different perspectives and concerns are taken into account.
3. **Measurement and reporting of sustainability targets**: Companies should develop clear KPIs and measurement methods to track progress toward sustainability goals. This makes it possible to quantify and communicate the impact on an ecological, social, and economic level.
4. **Employee participation**: Involving employees in decision-making processes and encouraging ownership of sustainability goals is crucial. Employees should be encouraged to make suggestions and implement sustainable practices in their respective areas.

5. **Leadership and management**: Managers at all levels should actively promote sustainability and exemplify these values. This means taking responsibility for sustainable decisions and promoting the implementation of these decisions in their respective areas.
6. **Innovation and product development**: Companies should promote sustainable product and service development and ensure that products and services meet environmental and social requirements. This can lead to a competitive advantage.
7. **Incentive systems**: Companies should rethink their incentive systems and ensure that they reward sustainable performance. This can include the integration of sustainability targets into bonuses and performance appraisals.
8. **Cooperation and networking**: Companies can collaborate with other companies and organizations to promote sustainable initiatives and share best practices. This can enable the development of joint solutions and the scaling of sustainability efforts.
9. **Ethics and integrity**: Companies should promote ethical principles such as integrity and honesty and ensure that all business practices are in line with these principles.
10. **Cultural change**: Finally, a sustainable approach often requires a cultural change within the company. This can take time and effort, but in the long term it can lead to a chat.

It is now a matter of seeing sustainability across the entire breadth of the economy not just as a separate initiative or obligation, but as an integral part of corporate strategy and culture.

... and therefore the findings presented are probably not surprising to insiders of the new economic thinking, as it is precisely these pioneers who have established a new economic understanding and management mindset in recent years. However, this article is intended to serve as a basis and inspiration for further transdisciplinary considerations and impulses for the transdisciplinary transformation of economics for the author himself and the dear readers—thanks are therefore due first and foremost to the universe, including all the unknown thinkers who have contributed with the help of ChatGPT. It should be said again—this time the statements are all (still) without guarantee!

References

ChatGPT 3.5 - https://chat.openai.com; https://translaite.com/de; Wikipedia https://de.wikipedia.org.

Management series Corporate Social Responsibility (Series editor: Prof. Dr. Rene Schmidpeter) https://www.springer.com/series/11764

Podcast from July 12, 2022 "The big paradigm shift" Rene Schmidpeter https://gewinnezukunft.podigee.io/3-neue-episode

Podcast from November 23, 2020 "Stakeholder Podcast" Rene Schmidpeter https://www.listennotes.com/de/podcasts/the-stakeholder/rene-schmidpeter-N8l5CDyZzi_/
Series CSR, Sustainability, Ethics and Governance (Sam Idowu, Rene Schmidpeter) https://www.springer.com/series/11565
Encyclopedia of Sustainable Management (Sam Idowu, Rene Schmidpeter et al.) https://link.springer.com/referencework/10.1007/978-3-030-02006-4

Schmidpeter, René, Prof. Dr., is a Professor at BFH Bern (Switzerland) and research scientist at the interdisciplinary academic faculty of the *Parmenides Foundation* in Munich. He is guest professor/lecturer in Germany, Austria, China, USA, India, Australia, and England as well as series editor for Springer's *CSR, Sustainability, Ethics & Governance* book series, editor of the *Encyclopedia of Sustainable Management* (ESM), and editor of the Dictionary of *Corporate Social Responsibility* (DCSR). His research and teaching activities focus on sustainable business transformation, international perspectives on sustainability, social innovation and sustainable entrepreneurship with special focus on the relationship between business and society. René serves as an expert and keynote speaker in many academic, economic and political steering committees as well as sustainability advisory boards. He is Vice President of the *Global Corporate Governance Institute* (GCGI) in London, UK, and Founder of *M3TRIX/TETRANOMICS* in Cologne and Munich, Germany. Furthermore, he is co-founder of the Swiss Association for "Fostering Transdisciplinary Knowlede" as well as co-editor of the book series at Springer "Transdisciplinary Management of Social and Ecological Crises."

The Transdisciplinary Power of Informality: On Transformative Governance

Karl-Rudolf Korte

1 Introduction

Politics in democracies is characterized by the fact that it takes the diversity of social groups into account in its decisions. As Max Weber already pointed out in his famous speech on "Politics as a Profession," the politician should, in addition to "passion" and "a sense of proportion," above all have a "sense of responsibility."[1] It considers the consequences of his actions for all those affected and justifies them to his voters. The diversity of interests at stake is a challenge for democratic politics. With its legislation, it determines how people and diverse groups in a society live together. To ensure that this diversity is actually represented, politicians have not only formally predetermined paths at their disposal, but also informal ones. By taking these into account, the formation of political will always has a transdisciplinary component (cf. the horizontal approach to transdisciplinarity in the contributions by Schüz and Steiner in this volume).

This article aims to shed more light on the informal structures of politics, especially using the example of the current traffic light coalition (consisting of red—social democratic—yellow—liberal—and green—ecological parties).

The first step is to reflect on the mode of change. Because every political decision changes. On the other hand, every decision polarizes, which raises questions of power. Finally, the political management of the Berlin traffic light can be analyzed using the concepts of transformation and resilience. Further considerations are based on the concept of political rationality, which refers to decision calculi and heuristics. Thus, the context of the informal is spread out.

[1] Weber (2010): 49.

K.-R. Korte (✉)
NRW School of Governance, University of Duisburg-Essen, Duisburg, Germany
e-mail: krkorte@uni-due.de

© The Author(s), under exclusive license to Springer Nature Switzerland AG 2025

M. Schüz (ed.), *Transdisciplinary Thinking and Acting*, Transdisciplinary Management of Ecological and Social Crises,
https://doi.org/10.1007/978-3-031-70680-6_13

2 Change as an Opportunity

Informality is very powerful in politics. It is contained in the textures, the practices of policy management, which are handed down unwritten. But they are also being overwritten, new scripts developed—especially in times of transformative upheaval. "You haven't changed at all!"[2]—Who hasn't heard a phrase like that? If it's aimed at the outside and meant honestly, it comes across as a compliment—perpetual freshness and ageless dynamism? But if the sentence is aimed at the inside, it offends. One hears stagnation. And who wouldn't have to admit to being able to change constantly in order to flaunt complexity competence? Only modern bourgeoisie as a mark of distinction flirts with the neo-Biedermeier ever-same, the consciously conservative constant. But in order to supposedly preserve this, too, in dynamic times, it takes some effort, which in turn feels like change.[3]

In transformations in the sense of profound changes, we as citizens often take a reactive role, but sometimes also an active one. Policy research works with the concepts of measured change and apparent stability[4] . Other scientific perspectives can be found, for example, in the work of sociologist Hartmut Rosa. He has referred to the motivation inherent in any change in the concept of unavailability:

> The cultural driving force of the form of life we call modern is the idea, the desire and the desire to make the world available. But liveliness, touch and real experience arise from the encounter with the unavailable. A world that would be completely known, planned and controlled would be a dead world.[5]

The motive of change obviously leads to paradoxical effects that are destructive in terms of climate change, loss of biodiversity and some other planetary boundaries if we do not avert it. The irresolvable contradiction is that the aliveness of the drive for unavailability systematically destroys every other aliveness in the process of consummation.

Moreover, changes do not only have anthropological-cultural backgrounds. In terms of systems theory, the system as a modern society is only maintained in the mode of constant change.[6] Systems usually change more slowly than their environment. The reasons are mostly to stabilize themselves as a system in order to survive in volatile environments. Routines in turn stabilize us in social roles. We usually stop pressure to change only with such retarding and slowing down techniques. The inertia of the formation of orders and options for action is therefore to be expected in

[2]This formulation is also found in Bertold Brecht's "Geschichten vom Herrn Keuner. In context, it says there: "A man who had not seen Mr. K. for a long time greeted him with the words, "You haven't changed at all." ""Oh!" said Mr. K. and blanched" (Brecht 2020: 29).

[3]For further argumentation, see Korte, Karl-Rudolf (2022a). Transformatives Regieren in Zeiten der Krisenpermanenz, in: dms. Der Moderne Staat, H. 2, pp. 413–429.

[4]Cairney (2011).

[5]Rosa (2021): 8.

[6]Nassehi (2021).

principle. One can even exaggerate it: Even in disruptive moments of change through shocks, systems and we as citizens try to organize the continuation of routines rather than to deal productively with change as an offer. The consequence is sobering. For if we can only dynamically stabilize the systems—acting in the mode of increase (more, higher, further, faster)—constant transformations are needed. *So we have to change in order to at least maintain what seems important to us.*

Diversity research also provides arguments with similar force. The evolutionary researcher Charles Darwin already knew: "It is not the strongest species that survives, nor the most intelligent, but the one that reacts best to change." At least this quote is attributed to Darwin, without him ever having formulated it so precisely himself. Diversity instead of monocultures—in politics as well as in society—ensures lasting survival. There are two arguments in this. The ability to resist is greater when one has diverse, heterogeneous knowledge at one's disposal. And in addition to knowledge, the ability to adapt through change is a guarantee that new influences can be productively generated for the system.

We recognize how deceptive the greeting "You haven't changed at all" can be. Modern reality is obviously characterized by change as transformation. There is a history to this. "The Great Transformation"[7]—with this title Karl Polanyi described a paradigmatic change in society in 1944. What was meant was the differentiation of two subsystems, bourgeois society and the economy. Change existed, change occurred, transformation happened. The drivers of this process were described in a differentiated way by Polanyi—and the political concept of transformation has been associated with this title ever since, although transformation has always taken on new meanings.[8] For example, the term was applied to the process of German unification and the accompanying upheavals in 1989/90.[9] In comparative politics, the concept of transformation organizes standards of democracy between political systems.[10] The term has recently experienced a renaissance in the context of climate change.[11] In its main report "World in Transition: Social Contract for a Great Transformation," the German Advisory Council on Global Change (Wissenschaftlicher Beirat der Bundesregierung Globale Umweltveränderungen) redefined the term in 2011, and it continues to be used today. According to it, transformations of societies, politics, and economy are necessary to contain the planetary boundaries of climate change. The focus is on the transformative impact that paradigmatically leads to a digitalized sustainability society—globally and actively shaped. This understanding of transformation in a complex change process goes far beyond the term and the more policy-field-related understanding of reforms.

---

[7] Polanyi (1944, 2021).

[8] Kristof (2020).

[9] Kollmorgen et al. (2015).

[10] Ibid.

[11] Kristof (2020).

The German government's "traffic light" coalition also ties in with this with its coalition agreement "Dare more progress."[12] The agreement could also read "Dare more transformation" in reference to Willy Brandt ("Democracy dare," 1969). This is because the three traffic light parties, SPD, Greens and FDP, describe in many chapters how Germany should transform itself: Socially integrative, sustainable, and innovation-driven. For example, Germany is to be transformed as an industrial location into a social-ecological market economy, which leaves out no area of social, economic, and political life. That is a very high ambition. After all, how do you manage change as transformation in Germany? With the exception of specific phases since 1949, the Federal Republic of Germany is known to be rather resistant to change[13], which is due to institutional, politically entangled barriers and political-cultural sensitivities of the security Germans[14] . Here, changes take place rather slowly, which at the same time ensures a high standard of social peace.

However, disruptions and external shocks can lead to change. It is therefore surely no coincidence that the traffic light coalition was first constituted at the federal level under the sign of "coronacracy."[15] The virus had been testing all areas of our lives since 2020, both forcing change and inspiring a longing for continuity. Political thinking revolves around the order of freedom in a democratic constitutional state.[16] The Corona virus put this thinking to a stress test. The democratic mode of governance was extremely challenged, since it was not only a matter of efficient problem-solving mechanisms but, much more existentially, of the survival of citizens. The risk decisions[17] were not only under special conditions of uncertainty, but were also structurally designed as a dilemma: Freedom or health? These conflicting goals alarmed. Coronacracy is the answer to these challenges. It requires no special foresight that crises will continue to shake our political system. Since February 2022, for example, we have been experiencing in our European neighborhood what it means to enforce interests by force. The Russian war of aggression against Ukraine once again has a disruptive quality.

Responding to this appropriately politically means using complex policy management, which includes formal and informal components among top individual actors as well as a broad social science focus for analytical tools.

---

[12] Coalition Agreement (2022).

[13] Delhees et al. (2008).

[14] Korte and Fröhlich (2009), Korte (2001).

[15] Florack et al. (2021).

[16] Kielmansegg (2013).

[17] Risks in this context of multiple and polycrises are always systemic crises that further complicate the weight on transformative processes.

3 Appropriate Policy Management: Adapting, Curating, Prioritizing, Swapping

How does policy management respond to these transformations—in a range from disruptive, disruptive shock to incremental, nuanced, slow policy change? Transformative governance as change-driven policy management follows hybrid tracks because different arrangements are required in different policy fields (from health care to military defense). Four varieties can be distinguished:

3.1 Customize

When it comes to environmental and climate protection, there is no counterpart, because earthquakes or global warming cannot be negotiated in principle.[18] The new paradigm of adequate government action geared to this would be adaptation instead of negotiation.[19] Adaptation would consist of implementing climate protection transformations politically. This can be done via incentives as well as via regulations or bans. The so-called Easter package from the Federal Ministry of Economics and Technology on the energy transition is a good example of this.

3.2 Curate

In the case of health and Corona, another type of transformative governance becomes apparent: curation. In principle, the virus is not visible as a negotiating partner either, but in social interaction it potentially comes to infection. How do you negotiate life-saving isolation measures? Conceptually, it is not adaptation, as in the case of climate protection, but rather curated governance (of the federal government) that offers itself as a possible answer.[20] Under the primacy of politics, it rapidly transforms transparent information processing into sorted and explained political decisions by crisis pilots. Curated governance does not come as a guiding suggestion, as in "nudging."[21] The variants of lockdown (decelerate everything, decouple everything, deconstruct everything) were state-imposed, not a suggestion. Curated governance has more to do with crisis-driven, appellative order. It uses a space of possibility.[22] It presupposes design knowledge (being able to learn situationally) and combines that with a sense of possibility (leading with confidence and deciding expeditiously). In this way, top-level politics advances to become a *hermeneutic of*

[18] Welzer (2022): 10.

[19] Bude (2022).

[20] Florack et al. (2021), Korte (2022).

[21] Derrig (2020).

[22] Korte (2019a).

resilience. Transformative decision-making in the mode of curated governance also always has moments of diffusion of responsibility in the context of the Corona pandemic. This can be observed, among other things, in the debate about mandatory vaccination. So far, political actors have not wanted to take responsibility for a clear yes or no; instead, for very different political motives, they have stuck to "urgent recommendations."

3.3 Prioritize

When it comes to war and peace, a radically twisted dialectic is evident in the combination of moral and realpolitik: talk and armament, peacefulness and deterrence, cooperation and militancy. In governmental action, the counterpart is not only visible, but existentially threatening. Here, transformative governance operates in the mode of extreme prioritization: hierarchy for leapfrog innovation. The chancellor used the government statement in the special session of the Bundestag on the war in Ukraine for this purpose, as will be shown below. Under existential seriousness, he decided the paradigm shift in German foreign and security policy.

3.4 Swap

Unlike traditional coalitions with two partners and a clear hierarchical structure, the traffic light parties do not look for intersections or lowest common denominators in the game of swapping—as dilatory formula compromises or quantitative distribution compromises.[23] The compromise emerges visibly in comprehensive legislative packages in which, as became visible in the so-called Easter package on energy security, offers are made to all three partners and their constituencies. In the day-to-day bartering of government, however, one sees difference models more frequently. They describe compensations that work through barter because they endure difference. They are therefore stronger than mere compromise. Those who endure difference bridge contradictions. If commonality remains the main goal, de-differentiation can secure more unity in difference via trust-based and departmental exchange policies. In this way, each of the three traffic light partners can also score points in public.

The mode of change in the context of transformations in times of crisis permanence is multifaceted, as the four varieties of transformative governance show. The mode combines various components, which are described below. There is much to suggest that in the classical understanding of policy research, the policy type, the particular policy field, determines the primary type of policy management.[24]

[23] Günther (2006).

[24] Lowi (1972): 299.

4 The Agents of Change

Our categories of classification and assessment of transformation, change, and policy change are guided by theoretical frameworks and empirical findings on policy change.[25] The scope and speed of policy change vary: From incremental to comprehensive to rapid. The phases of transformation and the respective extent depend on institutional and individual criteria, depending on the policy field.[26] What remains important in many studies on the conditional factors for success is not only to include civil society in the change process, but also to let it participate actively as an "agent of change."[27] In reality, this is often difficult. After all, change is usually an imposition and triggers countervailing forces or routine behavior. Both of these factors - the trigger of resistance and routine behavior - prevent people from dealing appropriately with the necessity of the structurally new. How does politics convey that transformation should always begin with the individual?

The analysis of Corona politics shows the difficulties of using appropriate routines of thinking about changes, turns, transformation, and change. Harald Welzer spoke of a "stagnation of the future," of "frozen situations," and "still images."[28] This is appropriate extra-ordinary metaphor. If one does not follow any thinking routines after that, then even the professional anchor points do not offer any interpretative support to develop conclusions for the future from the observed change. The situation is similar with the epochal break caused by the war scenario in Europe: First shock rigidity, then the emergence of the thought pattern of a global process of modernization. The belief in the automatism of modernization in the Western liberal sense as a pattern of progress is stopped by tanks. Andreas Reckwitz offers ways out in the project of modernity of a transformed thinking "as a normative and strategic project conscious of its own weaknesses, in the knowledge of its opponents."[29] This means to use the uncertainty strategically for new defensibility, when the caesura softens a linking to traditional world views.

What we know in political science as anchor points about abrupt policy changes follows a logic of appropriateness. It is oriented toward found routines, habits, and rules. Policy researchers speak of path dependence, historians of lines of tradition, government researchers of styles, practices and instruments of policy management. But the multicrises have paradigmatic features. Previous patterns of explaining policy management have reached their limits. In this respect, it is worth recalling once again systematic approaches to how and through what politics can change.

From a *neoinstitutionalist* perspective, it can be shown how institutions influence political decisions and how the actors involved in them change.[30] A *rational-choice*

[25] Rüb (2014), Korte (2021).

[26] Leggewie and Welzer (2009).

[27] Ibid: 149.

[28] Welzer (2020).

[29] Reckwitz (2022, 2017).

[30] Schleicher (2021), Florack (2013), Korte and Florack (2022).

approach favors the calculated, rational-personal decisions of actors. *Historical institutionalism*, on the other hand, tends to see the dominance of the institution shaping routinized practices and systems of rules. Path dependence limits the number and content of decisions in this respect. Without windows of opportunity and external shocks, little changes. *Sociological institutionalism*, on the other hand, focuses on social role patterns, in which actors act according to learned patterns of action and each strive for certainty of expectation. In *Discursive Institutionalism*[31], the causal meaning-charging of change is all about discourse and ideas. With communication, political change and transformative processes are achievable. Political institutions change through the transfer of ideas. All neoinstitutionalist approaches helpfully serve to systematically structure transformational processes. They will only be named at this point of the essay to show the possibility space for transformative governance. Exemplary Discursive Institutionalism will be recalled below, using the example of Chancellor Scholz's Ukraine government declaration. For it is the agents of change who resort to the varieties of action in transformative policy management despite systemic and structural constraints.

5 Resilience as a Resource for Change

Changes and transformations remain necessary in liberal societies in order to be able to act in a way that is suitable for globalization. I had already argued in this way. Types of transformation give us a sense of what might change as a result. With the concept of resilience as a political science category, we again expand the interpretive space of transformations, using resilience as a mechanism for resilience.

Resilience research concepts describe processes and components of a society's resilience to cope with and manage complexity.[32] Resilience research thus measures the readiness to cope with change. Politically and culturally, the Germans are said to be world champions when it comes to resilience.[33] They clearly have the ability to cope well with major changes and crises. Otherwise, the internationally high welfare and prosperity gains of the prosperous German competitive society could hardly be explained. Specifically, the prospect of change triggers passion. But neither the prospect nor the ability to change equates to a willingness to embrace change swiftly—and ideally—to plan ahead.

Resilience lends itself to policy management as a complexity management strategy.[34] How does one prepare as a government for crises that one knows will come, but does not know exactly when they will come and how they will manifest themselves? How do you arm yourself against the unpredictable and at the same time the inevitable? *Policy management* combines the controllability of the political

[31] Schmidt (2010).

[32] Endreß and Maurer (2015).

[33] Druyen (2018).

[34] Korte et al. (2022).

system with the controllability of key political actors.[35] Governance as a form of political management uses resilience to cope with crises. In doing so, top actors can refer to resources as resilience-enabling and resilience-influencing variables.[36]

The war shows once again that we should have taken precautions for national defense. It would be resilience-enhancing, for example, to have a common European missile defense system. With catching up (after the initial rigidity of shock), we are learning every day what we apparently lack in terms of external security. This also applies to the area of energy security, where the manifold dependencies—especially on Russian fossil fuels—are obvious. In a figurative sense, however, it is not only about buffers, stores, and reserves of our society that would have to be filled, but resilience-shaping abilities for precaution[37]. Markus Brunnermeier writes:

> A society as a whole is resilient when all or at least most people have the ability to respond in a way that can bounce back.[38]

Back to one's own actions, to sovereign decision-making, to political management that makes change possible. Robustness means withstanding shock, like an oak tree in the wind. Resilience, on the other hand, according to this view, means yielding rather—like reeds in the wind that do not break. The oak, to use the metaphor further, loses branches in the storm. Transformation in the sense of sustainable change should therefore focus more on flexible "yielding."

Ivan Krastev, in turn, refers to resilient communities that are characterized by the fact that power lies in society and not simply in the capabilities of the state.[39] Thus, the concept of "freedom energies" coined by Finance Minister Christian Lindner (FDP) also makes sense: renewable energy as a resource of democratic resilience owned by citizens, by communities. In this respect, transformation in a resilient society works not only with precautionary determination and foresight, but also with the individual capacities of citizens to engage in change. Transformative governance is resilience-oriented and takes the form of adaptation.

---

[35] Korte and Fröhlich (2009).

[36] Endreß and Maurer (2015), Jage-Bowler (2020).

[37] von Thadden (2022).

[38] Brunnermeier (2021): 89.

[39] Krastev (2022).

6 On the Contours of a Political Rationality

The decision-making burden for the politician is tantamount to a coping task.[40] This is perpetual, regardless of transformation tasks. Coping refers to a multitude of political interdependencies.[41] The decision-making action of the individual actor interacts with actions of other actors. Interdependence coping contains various components in the field of political rationality[42]; two of them—the substance of the political and the situation assessment of the actor will be presented. From this, one can see the degree of informality required and the sociological context.

7 The Fabric of the Political

In a democracy, making and enforcing collective decisions to solve problems succeeds only with legitimized power. It weaves the fabric of the political.[43] It does not simply exist, but emerges—if one follows Hannah Arendt's associative-communicative approach—only in togetherness, in interaction, concretely in action and speech. The decision-making ability of actors is at the center of this approach, not so much the sanction-based decision-making competence. Action, in which one joins with others to pursue a common concern, to take responsibility, generates power. Power as a potency is thus first realized in political action.[44] The political choice of decisions—certainly by means of heuristics—includes power. The subject matter of politics that has to be dealt with deviates significantly from the picture originally sketched by Max Weber. Politicians are not thick-board drillers who proceed with slow drilling movements of hard boards with passion and a sense of proportion; at least not primarily. The material of politics, to stay with the image, is not a board. The material is fluid and therefore always takes on different forms, knows neither beginning nor end. Navigating this fluid ground, reacting to the unexpected strategically, perhaps experimentally, and not losing sight of majorities in the process is much more demanding than slowly drilling through thick boards.

Political power is geared toward not only remaining empowered to make decisions in the cooperation of a majority, but also to be able to act in a way that is relevant to decision-making. The ability to reach agreement is part of the tools of the political trade in a democracy. In politics, however, majorities are always fragile constructs. The constitution of fluid communities, which must therefore always be

---

[40] Cf. on this argumentation Korte, Karl-Rudolf (2022b). Political mechanics: On decision heuristics in politics. In Heuristics of Political Decision-Making, ed. Karl-Rudolf Korte, Gert Scobel, and Taylan Yildiz, 298–320. Berlin: Suhrkamp.

[41] On the concrete background of interdependence management from a sociological perspective, see Schimank (2005), pp. 122–155.

[42] For a European perspective, see Sturm (2018), especially pp. 77–86.

[43] For a comprehensive classification of the concept of the political, see Rüb (2020).

[44] Robert Habeck writes about this in reference to the classic: "Politics is power as a profession," cf. Habeck (2021): 330.

kept in view on a day-to-day basis in order to implement purposeful solutions to problems, aims for the "best possible minimum."[45] What is meant by this is that political decision selection is not about the optimum or the maximum, but precisely about the balance of both. The heuristic to be applied situationally to ensure the best possible balance has an important reason. Ideally, it enables a government not only to consistently mobilize the necessary parliamentary majorities, but also to increase its chances of re-election. After all, in the decision-making stream of politicians, it is important to consider not only solving a problem, but solving it in such a way that the outcome does not minimize options for re-election. Politicians are characterized by precarious employment: unsecured, episodic, fuzzy. Without job protection, they must do everything possible to minimize career risks.

Politics is the enforcement of collectively binding norms of action. This is done by organizing social interaction through communal rules.[46] Ideally, the solution or overcoming of problems should be politically rational and increase the chances of re-election. Political power determines the form and content of these collectively binding decisions. Issues of fact and issues of power must be reconciled in the sense of decision-driven political management.

The fabric of politics further includes the fact that all political actors act purposefully. They are extremely aware of their interdependence. The actions of the actors are interconnected. As one congressman put it, "Politics thrives on seeing reactions in others—annoyance, outrage, enthusiasm, disinterest."[47] The outcome of one actor's decision depends on the decisions of other actors. This can lead to functional friendships in one's own partisan political camp. *Do ut des*—I give so that you give—describes the core of relationships among party friends.[48] Behind this are political networks that are designed for commitment and lasting reciprocity, for mutual benefit.

They act and react reciprocally, thus trying to anticipate the steps of other top players. They always act intentionally. The intentional refers to the political context of the action, in which every political decision—even the non-action, the waiting, the missing, the postponing, the refused decision—can be interpreted as an intentional and at the same time anticipatory interaction. This constitutes the essence of political space. Every decision (verbal, non-verbal, visual, analog, digital), consequently every design, is interpreted. The perpetual and professional handling of ambiguity (in principle ambiguous) and contingency (everything that is could also

[45] Thus formulated by Peter Graf Kielmansegg with reference to statements by Eugen Kogon, in Cavuldak (2020): 62.

[46] For background, see also Maier and Blum (2018): 141.

[47] Thus the politician of the LEFT Jan Korte MdB, "Politik ohne Anfassen," in: *Die Zeit* 1, (2021): 8.

[48] With a large number of specific examples in a study of parliamentarians, see Dausend, and Knaup (2019): 389.

be different from the actor's point of view) remains the yardstick for successful power ambitions of individual actors.[49]

Without the awareness of both, no decision-maker succeeds in his or her own complex form of reality processing. This is what makes politics specific and provides the material for the fabric of the political. Political power changes depending on the constellation. As a rule, it does not depend on formal criteria of organizational charts, but largely on projections, on conjectures, on—as explained above—acting with others. Political power is therefore often also a power of the moment assumed and constructed by third parties—latent and potential at the same time. The French writer André Malraux aptly formulated on this interactionist understanding of power: "You can't flirt with power, you have to marry it."[50]

From this follow quite typical practices in decision-making situations[51]. Some central practices are listed below:

- Political communications make political decisions public. They are the continuous effort to obtain a political mandate. In order to retain the scope for decision-making, the political actor cultivates a *sovereign vagueness* in communicative terms. Whoever formulates in a clichéd and inauthentic manner can still be sure of his allegiance at the end of the sentence. At least he escapes the trap of polarization without offensive clarity. Political language must suggest coherence in decision-making situations. Counterfactual formulations help in this. Even pending disparate, paradoxical, contradictory decisions must be contained linguistically. The back doors of power remain open. Language promotes the art of illusion: to be able to clown around in an elegant and educated manner.
- *Time strategies* also help to test one's own allegiance. In the flow of events, the weights can be shifted carefully. Deceleration not only saves time, but also increases the possibilities of winning over one's own followers, both publicly and non-publicly. At the center is the "management of attention,"[52] which can change political positions. Dilatory decision-making is one of the practices surrounding time management. It extends action into legislative periods. The decision is made now, but its consequences are postponed. Politicians have less and less time to make decisions with ever longer impact periods and implementation phases (for example, in climate policy). An intergenerational agenda-setting distributes with a dilatory approach, positive prospects today, at the price that later generations will have to bear the burdens. The "present shrinkage" makes it easier to decide in the now, by externalizing the consequential costs.

[49] On the concepts of ambiguity and contingency, also as applied to the field of design decisions, see Korte (2019a): 41–75.

[50] quoted from Dausend and Knaup (2019): 54.

[51] Korte (2019a), pp. 41–75, Korte and Fröhlich (2009), pp. 240–270, Schmid (2011), pp. 324–344; Sturm (2018), pp. 41–60, Eckert et al. (2013), Lettrari (2019), Rüb (2022).

[52] Sarcinelli (2014).

Democracies also weaken for this reason when it comes to responsibility for the future.

- For the upcoming decisions, the possible *excitement thresholds* must be *reduced* so that voters also become accustomed to them in the long term. This applies all the more when unpopular decisions are involved. Moderating a conflict in advance can succeed if one can show the light at the end of the tunnel (time limits). Acclimatization effects can also be achieved through small-scale decision-making above the shadow of major events.
- *Opponents* in one's own camp must be *integrated*, kept busy, pushed away or, in the worst case, isolated. For every decision, the question arises as to how one's own majority feels about it. It is not the observation of opponents that is decisive, but the orchestration of the supposedly own supporters. The focus here is on the day-to-day integrative management of majorities from very different communities of interest.
- The practices of power in decision-making situations include political *barter and countertrade*: quid pro quo—this for that! The larger and more disparate the decision package, the more effectively personnel and also factual issues can be accommodated with one another in an exchange to reach a compromise.[53] In decision-making situations, it is also power-preserving for politicians to exclude difficult problems or—in the sense of an escape—to migrate to other decision alternatives.
- Rule deviations in the sense of a new or reinterpretation of existing rules are also used.[54] They serve situational elasticity without formally violating rules. The so-called useful illegality in political organizations occurs publicly.
- The *power of the informal* complements formal measures that safeguard procedures. In the informal discussion formats, it is not only knowledge and positions that expand, but above all the anticipated playing out of possibilities. Thomas de Maiziere calls this the "pre-effect of the formal procedure."[55] In terms of systems theory, "anticipations... are excursions into the uncertain."[56] Informality reveals a practice that professionally anticipates uncertainty of expectations in order to remain capable of making decisions

The listing of central practices characterizes the substance of the political. It is part of political rationality, which a politician should have at his disposal when he has to make decisions. However, interdependence management within the framework of political rationality also includes the decisive individual and mostly *informal* assessment of the situation.

[53] You could also describe it casually as "distributing balls."

[54] On this subject in detail Kühl (2020).

[55] de Maizière (2019): 41.

[56] Luhmann (2017): 461.

8 The Situation Assessment of the Actor

In the momentum of decision-making imposition, the actor uses the cognitive shortcuts. What profiles the specifics of the momentum? The answer: it is essentially the political situation assessment. What is the political situation at this moment? The answer directs the focus to the politician's information management, how he or she concretely arrives at the situation assessment. Information management is important for the top actor because information is one of the key power resources[57]. Information can refer both to policies—in the sense of a problem-solving chain of reasoning—and to indicators that secure power: What is the power-political value of information, who is for it, who is against it? The maxim is: it is not how one makes decisions, but how one prepares them in a power-preserving way that is important.

Questions of fact are always linked to questions of power. The goal should be to keep the diversity of information permeable all the way to the top player. Only those who secure the long-term independence to use *alternative sources of advice to* prepare decisions can use the resource of information to stabilize power. Political power always means having *decision-making alternatives.* Alternative sources of advice besides the prepared written and oral procedures of one's own administration can be personal advisors as well as telephone calls with important party members. It is also always necessary to weigh the factual rationality of the planned measure against the political rationality of mediation and enforcement.

In the abstract, one can imagine this permanently ongoing information management of policy advice from the inside in a *three-stage strategy process* that combines various partial rationalities:

1. Factual rationality: Which objective societal problems need to be solved? Which policy field-specific concepts are suitable to solve the problems (policy output and impact)? What impact do these concepts have on other policy fields (policy outcome)?
2. Mediation rationale: Are the necessary legislative powers available? Are sufficient administrative and financial resources available? How is the goal of the legislation operationalized administratively (implementation knowledge)?
3. Rationality of enforcement: How can majorities be organized for the targeted legislation in the party and parliament? What impact will the legislative project have on re-election? How must the project be communicated to mobilize public approval and support? What reactions does the legislation elicit from important interest groups?

In concrete terms, *informal information management* has the function of *reducing complexity from the* point of view of the top actor. The decisive factor is where the

[57] On the information situation in political organizations cf. Korte (2019b): 129–143. Using the example of Kohl's chancellorship, cf. Korte (1998). Using the example of the Office of the Federal President, cf. Korte (2019a): 101–122.

information comes together and when. Ideally, it is bundled at the top actor (information pyramid). In reality, this usually takes place via persons with brokerage power in the immediate environment of the top actor. In order to coordinate the daily work, all top actors used to schedule a morning situation meeting ("Morgenlage"). This situation meeting is not a decision-making body, but an information round. It is exclusively related to the respective actor. The personal character of this meeting dominates. The situation serves to classify the political situation of the day: What issue is at stake? How explosive is the issue that the situation already has to deal with it? In this circle, however, decisions are pre-drawn. This is because the selection of information sets the *filter for the* actors' *perception of reality* and, at this moment, also pre-structures the political *construction of reality*. Every reassessment of the situation by those involved is already a decision. The task of these informal circles and situation meetings is to form a picture of the political situation and to develop patterns of reaction to it.

Political situation analysis is the foundation of any political decision. Political decisions are decisions adapted to the situation. Components of such a situation analysis should be: problem interpretation and problem identification, information processing, risk assessment, benefit measurement, preference order of the actors (who is to be involved, etc.). Internal memos transfer situation assessments into a three-step process: a. Own *interest definition* (What is our agenda? What is at stake?); b. *Goal definition* (What do we want to achieve with which issues and how? What will other key protagonists want?); c. Assessment of *design opportunities* (Where are our strengths and where are our weaknesses?).

Policy advice from within means clever and efficient *knowledge management* with the resources available to an administrative apparatus. This requires a permanently changing flow of information. Information is a power resource. It is not the exclusivity of information that is important, but rather the earliness and speed of its access. If information is so important as the basis for a political assessment of the situation, then the question arises as to the weaving pattern of a situation analysis and problem interpretation. What criteria does a top political actor use to create a risk assessment? How does the communicative apparatus—also in the sense of an early warning system—remain sensitive to the risks and opportunities of political and social processes and events? The information channels suggest not only transparency for the basis of decision-making, but also a form of complete, unlimited rationality. But that would be an illusion, because there can be no such thing, certainly not in politics, where every decision depends more on perception than on the supposed reality behind it. But the information paths, which are worked out transparently, also offer the possibility of documenting a decision path, which can be important for the legitimization of the decision for the actor and his team.

Barack Obama describes this as an important background to his decision-making ability:

> With a firm decision-making process, I was able to put my ego aside and really listen to follow the day's facts and logic as best I could, weighing everything in line with my goals and principles. This allowed me to make difficult decisions and still sleep at night, because at

> least I knew that no one in my position could have made the decision better with the same information as a basis. A good process also meant that I could allow each member of the team to feel co-responsible for the decision.[58]

The assessment of the situation in situational permanence creates a reality that constitutes the momentum of decision-making. The always uncertain and unclear information situation based on limited rationality (incomplete, inaccurate, random) paints a picture as a snapshot for the actor. It must be constantly repainted. If this accompanying process is missing, then decisions are also made, but these are less power-securing.

9 The Powerful Transformational Impositions

Transformation produces costs: Individual, collective, private, public—and this in all areas and in all policy fields. How high are the prices? Who is prepared to pay them, when and for what? In a democracy, the answer is initially not economically relevant, but politically. For it describes elementary decision-making impositions of politics, completely independent of the extent of the transformation processes. Anyone who deals with mechanisms of transformation should know the substance of the political. For those who disregard it lose majorities to solve problems or to be re-elected.

In a democracy, the production and implementation of collective decisions for transformation processes succeeds only with legitimized power. It weaves the fabric of the political.[59] It does not simply exist, but—if one follows Hannah Arendt's associative-communicative approach—only emerges in togetherness, in interaction, concretely in action and speech. When this happens in the context of governance, it can again be used as evidence for Discursive Institutionalism. The decision-making capacity of actors is at the center of these approaches, not so much the sanction-based decision-making competence. Action, in which one joins with others to pursue a common concern and to take responsibility, generates power. Power as potency is thus first realized in political action. Political decision-making for processes of transformation includes power.

Political power is geared to not only remaining empowered to make decisions in the company of a majority, but to be able to act in a way that is relevant to decision-making. In democracy, the ability to reach agreement is part of the tools of the political trade. In politics, however, majorities are always fragile constructs. The constitution of fluid communities, which must therefore always be kept in view on a day-to-day basis in order to implement purposeful solutions to problems, is aimed at the "best possible minimum."[60] What is meant by this is that political decision

[58] Obama (2020): 417.

[59] Korte (2022): 296–302; Habeck (2021): 330.

[60] Kielmansegg (2020): 62.

selection is not about the optimum or the maximum, but precisely about the balance between minimum and maximum—between, for example, renewable energies and fossil fuels. The heuristic to be applied situationally to ensure the best possible balance has an important reason. Ideally, it enables a government not only to consistently mobilize the necessary parliamentary majorities, but also to increase its chances of re-election. After all, in the decision-making stream of politicians, it is important not only to solve a problem, but to solve it in such a way that the result does not minimize the options for re-election. Thus, it is not enough to introduce renewable energies across the board as a transformation gain if, at the same time, energy prices are socially overburdening or energy security cannot be guaranteed. How then should one be re-elected? Political power determines the form and content of collectively binding decisions. Factual and power issues must be reconciled in the sense of decision-driven political management. Anyone who follows these traces of political rationality for political actors recognizes the dilemma of transformation processes. Changes require democratic majorities and, ideally, long-term time corridors that point far beyond a legislative period. In this respect, elections are important for recognizing the momentum for transformations. It is therefore worth looking back at the constellation of the 2021 super election year in the Federal Republic of Germany.

The super election year followed a non-ordinary logic.[61] It remained peculiarly unique. The basic mood oscillated between an enthusiasm of the positive ("in solidarity we defeated Corona") and the melancholy of the cautious and disappointed ("nothing works here"): Worriedly content or content in discomfort? This ambiguous basic mood nevertheless mobilized more voters again than in 2017, which was surprising given the lack of maximum issue polarization in the election campaign. Higher voter turnout spoke for an overall successful election campaign by the parties, which was able to mobilize sufficiently.

Since Election Day, the elected parties have been sorting themselves out in the setting of the Bundestag in a multipolar, multi-party system. The Bundestag is less polarized than in the last legislative period. Radicalization in the party spectrum was reserved for the AfD, which had little to contribute to Corona politics in a solution oriented way. The political center has continued to expand and differentiate, presenting itself as a pronounced electoral fragmentation. Even the Danish-Frisian minority has re-entered the Bundestag with its own party. Segmentation in the sense of alliances that are difficult to overcome has diminished, especially since the Greens are now located in the electoral spectrum as a post-campaign party. Election day confirmed the pattern of voting behavior in Germany: moderate, centrist and often mediocre.

Corona politics posed obvious mobilization challenges in the super election year 2021. Future-proofing played a major role in voter market motives: How do we create a resilient democracy? This means much more than just pandemic prevention. After all, the need for repair of the post-pandemic state was particularly noticeable

---

[61] Korte et al. (2022).

during the pandemic. In this respect, the longing for a cleverly protective, a guiding precautionary state grew. For the security germans, security is multifaceted: external, internal, social, cultural, and health-related.

The election campaign nevertheless came across as maximally unconditional. But voters sensed that unconditional provision of services of general interest would come at a price. The key resources for building a resilient democracy played an indirect role in the election campaign, but advanced to become a political driver in the coalition negotiations. The government's program, as stated elsewhere, is ambitious for change. All three traffic light parties want change. They also feel legitimized to do so by the election results. How they succeed in balancing interests will depend not only on the actors' ability to deal with conflict, but above all on their ability to communicate. A successful inclusive transformation requires intensive political communication. The war scenario has fundamentally changed the dynamics of the traffic light plans, but not their focus on the transformation necessities.

Those who do not take the trouble to explain transparently to the population why changes also bring improvements will lose their majority.[62] Social democratic, liberal, and green narratives can push for the desire for change that keeps social and societal peace in Germany at a high level, especially under new security policy conditions. Political leadership must then make even the supposedly unpopular acceptable to the majority if it serves the goal of increasing prosperity for all in the long term. In this context, the probability of disruptive changes is lower than the continuation of routines. This is where the dangers of losing future viability lie.

10 Conclusion

Major conflicts over distribution lie behind major transformations. It is not the political colors in party competition that are central here, but the common will to change. Resilience-enabling aspects are not to be found in the ideological orientation of the parties, but in their potential for communicative interaction. Whether the changes are then called progress, modernization, reform, or transformation is not as decisive as the openly communicated road map. With the transformation narrative as an important policy driver, the Berlin traffic light as a learning community can advance innovations without having to calculate who gains more or loses more at which step. The impositions follow the motto: "Tackle and do." This can be unconventional and unexpected.

Transformative governance in crisis permanence must make use of different modes and mechanisms, depending on the policy field and the dramaturgy of the day. Whether this involves adapting, curating, prioritizing, swapping or even hybrid forms must be demonstrated in concrete case studies. The ecological-social transformation will challenge the traffic light in coronacracy and with a view into the abyss of war.

[62] Kimhi and Eshel in (2022) and Korte in (2024).

The conclusions for us as voters are ambivalent. Because the ampelig governance we have experienced seems obtrusively unfinished. There is great resistance among the population to change. The masters of the diffuse and the non-competent stand alongside ministers who reveal their day-to-day weighing of goods by the minute. This is exhausting and does not fit the patterns of political management of a chancellor's democracy. There is much to suggest that the political system is evolving from chancellor democracy in the form of the current three-party coalition into a perspective of multicentric, collaborative governance—driven by the resource curse of transformational pressures. It is too early to judge whether this is due to the actors, the tripartite constellation, or the transformation task. The transformation should be communicatively inclusive, politically participatory and always oriented toward social solidarity. However, anyone who wants to transform a growth society into a climate-neutral sustainability society also needs new models of democratic decision-making. New "we-ness" does not arise automatically, especially when scarcity also brings distribution issues to a head.

The "ampelig" patriotism of change is highly ambitious in this respect, if the "Great Transformation" serves as a model. The appropriate policy management of the top actors makes use of transdisciplinary instruments, styles, and techniques, as has been shown, with a focus on the social sciences. Politicians act powerfully in balancing formal and informal mechanisms.

References

Brecht, B. (2020). *Geschichten vom Herrn Keuner*. Suhrkamp.

Brunnermeier, M. (2021). *Die resiliente Gesellschaft*. Aufbau Verlag.

Bude, Heinz. 2022. Kein Stein wird auf dem anderen bleiben. Verfügbar unter: https://www.sueddeutsche.de/kultur/ampelkoalition-wahlversprechen-heinz-bude-1.5522062?reduced=true. Accessed August 7, 2022.

Cairney, P. (2011). *Understanding public policy*. MacMillan Education.

Dausend, P., & Knaup, H. (2019). *Alleiner kannst du gar nicht sein. Unsere Volksvertreter zwischen Macht, Sucht und Angst*. dtv.

Delhees, S., Korte, K.-R., Schartau, F., Switek, N., & Weissenbach, K. (2008). *Wohlfahrtsstaatliche Reformkommunikation*. Nomos.

de Maizière, T. (2019). Regieren. In *Innenansichten der Politik*. Herder.

Derrig, R. (2020). Was Menschen dazu bringt, auch strenge Regeln zu befolgen. Die Pandemie zeigt Grenzen der Nudging-Theorie auf. *WZB Mitteilungen, 168*, 66–72.

Druyen, T. (Ed.). (2018). *Die ultimative Herausforderung—über die Veränderungsfähigkeit der Deutschen*. Springer VS.

Eckert, G., Novy, L., & Schwickert, D. (Eds.). (2013). *Zwischen Macht und Ohnmacht. Facetten erfolgreicher Politik*. Springer VS.

Endreß, M., & Maurer, A. (Eds.). (2015). *Resilienz im Sozialen*. Springer VS.

Florack, M. (2013). *Transformation der Kernexekutive*. Springer VS.

Florack, M., Korte, K.-R., & Schwanholz, J. (Eds.). (2021). *Coronakratie. Demokratisches Regieren in Ausnahmezeiten*. Campus Verlag.

Günther, K. (2006). *Politik des Kompromisses*. Springer VS.

Habeck, R. (2021). *Von hier an anders. Eine politische Skizze*. Kiepenheuer & Witsch.

Jage-Bowler, F. (2020). Das gesellschaftliche Immunsystem stärken Elemente eines positiven Risikowissens. *WZB Mitteilungen, 168*, 79–81.

Kielmansegg, P. G. (2013). *Die Grammatik der Freiheit.* Nomos.
Kielmansegg. (2020). Peter Graf mit Bezug auf Aussagen von Eugen Kogon. In A. Cavuldak (Ed.), *Peter Graf Kielmansegg im Gespräch* (p. 62). Übungen im politischen Denken.
Kimhi, S., & Eshel, Y. (2022). Politische Psychologie und Resilienz. In S. Zmerli & O. Feldmann (Eds.), *Politische Psychologie* (pp. 453–462). Nomos.
Kollmorgen, R., Merkel, W., & Wagener, H.-J. (Eds.). (2015). *Handbuch Transformationsforschung*. Springer VS.
Korte, K.-R. (1998). Deutschlandpolitik in Helmut Kohls Kanzlerschaft. In *Regierungsstil und Entscheidungen 1982–1989*. DVA.
Korte, K.-R. (2001). Regieren. In K.-R. Korte & W. Weidenfeld (Eds.), *Deutschland-Trendbuch, Fakten und Orientierungen* (pp. 515–546). Springer VS.
Korte, K.-R. (2019a). *Gesichter der Macht. Über die Gestaltungspotenziale der Bundespräsidenten*. Campus.
Korte, K.-R. (2019b). Politikberatung von innen. In S. Falk, M. Glaab, A. Römmele, H. Schober, & M. Thunert (Eds.), *Handbuch Politikberatung* (pp. 129–143). Springer VS.
Korte, K.-R. (2021). Bundestagswahlkampf in Zeiten der Pandemie. *Aus Politik und Zeitgeschichte, 47–49*, 17–21.
Korte, K.-R. (2022a). Transformatives Regieren in Zeiten der Krisenpermanenz. In *dms. Der Moderne Staat, H.2* (pp. 413–429).
Korte, K.-R. (2022b). Politische Mechanik: Über Entscheidungsheuristiken in der Politik. In K.-R. Korte, G. Scobel, & T. Yildiz (Eds.), *Heuristiken des politischen Entscheidens* (pp. 298–320). Suhrkamp.
Korte, K.-R. (2024). *Wählermärkte: Wahlverhalten und Regierungspolitik in der Berliner Republik.* Campus Verlag.
Korte, K.-R., & Florack, M. (Eds.). (2022). *Handbuch Regierungsforschung*. Springer VS.
Korte, K.-R., & Fröhlich, M. (2009). *Politik und Regieren in Deutschland.* UTB.
Korte, K.-R., Scobel, G., & Yildiz, T. (Eds.). (2022). *Heuristiken des politischen Entscheidens: Zwischen Komplexität und Kunstfertigkeit.* Suhrkamp.
Krastev, I. (2022). Jetzt beginnt eine neue Geschichte. *Die Zeit*, 57.
Kristof, K. (2020). *Wie transformation gelingt.* Oekom Verlag.
Kühl, S. (2020). *Brauchbare Illegalität. Vom Nutzen des Regelbruchs in Organisationen.* Campus.
Leggewie, C., & Welzer, H. (2009). *Das Ende der Welt, wie wir sie kannten.* S. Fischer Verlag.
Lettrari, A. (2019). *Politische Hochleistungsteams im Deutschen Bundestag. Professionelles Management in Abgeordnetenbüros in Zeiten hyperkomplexer Anforderungen.* Nomos.
Lowi, T. J. (1972). Four systems of policy, politics and choice. *Public Administration Review, 32*(4), 298–310.
Luhmann, N. (2017). *Systemtheorie der Gesellschaft.* Suhrkamp.
Maier, D., & Blum, C. (2018). *Logiken der Macht.* Nomos.
Nassehi, A. (2021). Mut brauchts. *Süddeutsche Zeitung*, 6.
Obama, B. (2020). *Ein verheißenes Land* (2nd ed.). Penguin Random.
Polanyi, K. (2021). *The great transformation.* Suhrkamp.
Reckwitz, A. (2022). Der Optimismus verbrennt. *Die Zeit*, 47.
Reckwitz, A. (2017). *Die Gesellschaft der Singularitäten.* Suhrkamp.
Rosa, H. (2021). *Unverfügbarkeit.* Suhrkamp.
Rüb, F. W. (Ed.). (2014). *Rapide Politikwechsel in der Bundesrepublik.* Nomos.
Rüb, F. W. (2020). *Das Jahrhundert der Politik.* Nomos.
Rüb, F. W. (2022). Mikropolitologie. Auf dem Weg zu einem neuen Konzept? In K.-R. Korte & M. Florack (Eds.), *Handbuch Regierungsforschung* (pp. 569–580). Springer VS.
Sarcinelli, U. (2014). Von der Bewirtschaftung der Aufmerksamkeit zur simulativen Demokratie? Politische Visionen—Von Platon zum Global Village. *Zeitschrift für Politikwissenschaft, 24*(3), 329–339.
Schimank, U. (2005). *Die Entscheidungsgesellschaft.* Springer VS.

Schleicher, K. (2021). *Von alternativen Paradigmen zur umfassenden Transformation*. Springer VS.
Schmid, J. (2011). Mikropolitik. Pluralismus mit harten Bandagen? In N. C. Bandelow & S. Hegelich (Eds.), *Pluralismus, Strategien, Entscheidungen*. Springer VS.
Schmidt, V. (2010). Taking ideas and discourse seriously: Explaining change through discursive institutionalism as the fourth 'New Institutionalism'. *European Political Science Review, 2*(1), 1–25.
Sturm, R. (2018). *Wie funktioniert Politik. Die Beweggründe des Politischen in den Nationalstaaten und in der EU*. Nomos.
Thadden, E. v. (2022). Wärme pumpen. *Die Zeit*, 49.
Weber, M. (2010). *Politik als Beruf*. Duncker Humblot.
Welzer, H. (2020). Im leeren Raum der stillgelegten Zukunft. *Futur Zwei, 13*, 9–11.
Welzer, H. (2022). Machen. *Futur Zwei, 19*, 9–11.

Korte, Karl-Rudolf, Prof. Dr. phil., received his doctorate (Dr. phil.) from Johannes Gutenberg University in Mainz and his habilitation (Dr.rer.pol.habil) from Ludwig Maximilians University in Munich. In 1998, he received the award of the University of Munich for the best habilitation thesis. Substitute professorships in Tier, Cologne and Munich. The 64-year-old family man and grandfather has been a professor of political science at the University of Duisburg-Essen in the field of "Political System of the Federal Republic of Germany" since 2003. He was able to turn down calls to the universities in Bonn in 2008 and Munich in 2013. In 2006, he was named "Professor of the Year" in the category of humanities, social and cultural sciences by the professional magazine UNICUM-Beruf. Since its foundation in 2006, he has been the director of the NRW School of Governance (University of Duisburg-Essen). From 2010 to 2017, he was Dean of the Faculty of Social Sciences; from 2013 to 2015, Chairman of the German Political Science Association. He is co-editor of the "Zeitschrift für Politikwissenschaft. Journal of Political Science." Since 2018, he is a full member of the European Academy of Sciences and Arts (Salzburg). Since 2021, he has been vice president of the German Association of Universities. He chairs the internationally staffed scientific advisory boards of the Federal Chancellor Helmut Kohl Foundation and the Haus der Geschichte NRW. Korte has been known to a broad public for many years through regular appearances and trenchant analyses on ZDF, Deutschlandfunk, WDR, Phoenix and SWR. Recent publications: "Gesichter der Macht. On the Shaping Potentials of the Federal Presidents" 2019 (Frankfurt/New York, Campus); "Coronacracy. Demokratisches Regieren in Ausnahmezeiten" 2021 (Frankfurt/New York, Campus); Heuristiken des politischen Entscheidens 2022 (Berlin/Suhrkamp Verlag); and "Regieren in der Transformationsgesellschaft" (Wiesbaden/Springer VS). In January 2024, Campus will publish his monograph: "Wählermärkte. Electoral Behavior and Government Policy in the Berlin Republic." Contact: krkorte@uni-due.de; Insta: krkorte; www.nrwschool.de; www.karl-rudolf-korte.de.

The Art of Listening in Transdisciplinary Management

Claudia Pölderl

1 Introduction

"*The word listen contains the same letters as the word silent.*" philosophizes pianist and author Alfred Brendel. Management has little to do with silence: it is about communicating, negotiating, goal setting, performance evaluating, motivating, to name just a few aspects of management.[1] The primacy of an action orientation underlies most entrepreneurial activities.[2] In these activities, an active role of language is central. This one-sided understanding of management leads to rash and therefore non-optimal decisions. These can cause massive damage to the economy, society and nature. Therefore, the suppressed half of responsible leadership[3] is indispensable for minimize such risks: the art of listening. In this way, the myriad impacts of management decisions can be specifically identified from the perspective of those affected and potentially avoided before they occur.

But how would management practice change if the predominant expressive side of language gave way to the other side of language, to listening? This is not just about applying new techniques or tools to improve managerial effectiveness. Today's management and leadership are anchored in a culture of Western thinking tradition that is based on a "*system of knowledge that tends to ignore listening*

[1] Birkinshaw (2012): 38.

[2] Steiger and Vey (2003): 58.

[3] In this article, management and leadership are used synonymously, following Alvesson and Sveningsson (2003:1435), who conceptualize leadership as the extra-ordinarization of the mundane.

C. Pölderl (✉)
School of Management and Law, Zurich University of Applied Sciences (ZHAW), Winterthur, Switzerland
e-mail: claudia.poelderl@zhaw.ch

© The Author(s), under exclusive license to Springer Nature Switzerland AG 2025

M. Schüz (ed.), *Transdisciplinary Thinking and Acting*, Transdisciplinary Management of Ecological and Social Crises,
https://doi.org/10.1007/978-3-031-70680-6_14

processes".[4] For this reason, a transdisciplinary exploration of listening is worthwhile. Corradi Fiumara, an Italian language philosopher, points out that "*there is a need, however, to look further into the possible ways of understanding such a pivotal word in the west as* logos. *Perhaps we could start out by admitting that there could be no saying without hearing, no speaking which is not also an integral part of listening, no speech which is not somehow received*."[5] This article therefore explores the possibilities of listening management practice. It starts with a deep investigation into how listening can be understood through looking beyond a mere mechanistic understanding of hearing through the ears. It then portrays and dis-cusses listening in management and leadership, and what gets in the way of turning to that other side of language. The article concludes with transdisciplinary thoughts on what is beyond listening.

2 Considering Listening

All actual life is encounter.
Martin Buber

In the 1980s, management literature increasingly emphasized the role of listening. In "Search of Excellence", Tom Peters notes that excellent companies are not only better on service, quality, reliability, and finding a niche. They are also better listeners.[6] On a personal level, highly effective people seek first to understand and then to be understood. The problem is, to cite Steven Covey, that *"Most people do not listen with the intent to understand; they listen with the intent to reply"*.[7] In order to understand why a skill such as listening, although postulated as desirable in business practice, quickly reaches its limits in implementation, the following article will take a closer look at the fundamental lack of anchoring listening in our Western culture. It will become clear that listening can be taught and applied as a technique. If listening is also understood as an attitude and as a community affair, a cultural change is needed in organizations.

2.1 The Silence Around Listening

In her philosophical analysis on the other side of language, Corradi Fiumara approaches the missing place of listening through the consideration of the Greek term *logos* and its verb *legein*. The meaning of the latter has been drastically limited

---

[4]Corradi Fiumara (1995): 1.

[5]Corradi Fiumara (1995): 1.

[6]Peters and Waterman (2008).

[7]Covey (2004).

to the expressive aspect, to "saying". If only this one part is taken into account, logos is mainly used as the ability to order and explain. A rationality, however, which only listens to what is said and conceals what is heard, turns into a rationality which suppresses itself.[8]

Heidegger describes the meaning of *legein* beyond the expressive element as follows:

> No-one would want to deny that in the language of the Greeks from early on legein means to talk, say or tell. However, just as early and even more originally, legein means what is expressed in the similar German word legen: to lay down, to lay before. In legen a "bringing together" prevails... However, legein...means just this, that whatever lies before us involves us and therefore concerns us.[9]

Thus, the verb *legein* means not only to speak, but also to *gather*. Gathering means to bring something together, that can be people, ideas, or things. In this original sense, *logos* thus also implies a fundamental belonging, or relationship—a meaning that had disappeared from the beginning in the development of Western thought. Heidegger recognized the primacy of listening in creating meaning and in developing our relationships with one another. Listening enables us to maintain meaningful relations with family, friends, and professional associates.[10] Heidegger formulates what he means by listening as follows. He writes:

> *So long as we only listen to the sound of a word, as the expression of a speaker, we are not yet even listening at all. Thus, in this way we never succeed in having genuinely heard anything at all. But when does hearing succeed? We have heard [gehört] when we belong to [gehören] the matter addressed.*[11]

This listening, as Heidegger describes it, is the opposite of the listening of a "have-man" as it was described by Erich Fromm. The latter is identified with his opinion. It is a matter of presenting more accurate arguments in order to be able to convince the counterpart or to be able to defend his own position well. To leave one's own position would mean a loss, since one's own opinion is understood as a possession.[12]

Much rather a similarity of hearing with a "being-man" is to be discovered. The mode of existence of being means aliveness and an authentic relatedness to the world. The conversation is less an exchange of goods (like information or knowledge), but it becomes a dialogue in which it is no longer about who is right, and in which new things are allowed to emerge.[13] Fig. 1 demonstrates how Chinese art expresses the relatedness of listening. The Chinese signs mean "Hear all parties. Listen to both sides and you will be enlightened".

---

[8] Corradi Fiumara (1995): 12.

[9] Heidegger (1984): 60f.

[10] Heidegger (2007): 204.

[11] Heidegger (1984): 66.

[12] Fromm (1991): 42.

[13] Fromm (1991): 35.

Fig. 1 The art of listening according to the Chinese artist Jenny Xu. (Source: Private property of Mathias Schüz)

The juxtaposition of listening from the perspective of Fromm's forms of human experience, of being and having, is not meant to be understood as a polarity. Rather, it shows the different manifestations and dimensions of listening. A more nuanced account of this can be found in Michael Purdy, an American communication scholar. For him, listening received no attention in Western history for a long time because it was conceived as something that happens automatically. Purdy examines different forms of listening by placing them in the context of different structures of consciousness.

2.2 *Listening Across Levels of Consciousness*

Jean Gebser's phenomenology of culture can be understood as a description of distinct ways of being in the world, and constructing it emanating from different structures of consciousness. These five structures are called archaic, magic, mythic, mental, and integral, and they are all present as part of everyday experiences.[14] In his descriptive research, Purdy relates different modes of listening to these structures of consciousness:

In an archaic mode of consciousness, listening (and speaking) were not necessary, as there was not a sense of separateness.[15] Sympathetic listening and empathetic

[14] Gebser (1985).

[15] Purdy (2000): 51.

listening are allied with magic and mythic consciousness, and underlie all communication and listening processes as a felt connection, which is less conscious but accessible if one is trained to do so. The predominant mode of listening in most of contemporary western world relates to the mental structure of consciousness and is called "rational listening", a way of listening as *thinking*. This perspective is reflected in cognitive psychology, which understands communication in general and therefore listening as well, as a product of a thinking individual. Research of listening embodies this position through its extensive rational analysis. To theorize about listening, researchers develop constructs. They break listening into relevant components or types, examine the stages in the listening process (such as attention, perception, interpretation, constructing meaning, and responding) or identify characteristics of good and poor listeners. Another aspect of the mental consciousness mode is its duality: the speaker and the listener are understood as separate and exclusive components of communication.[16] This conception of listening, imbued with rationality, is strongly reminiscent of the one-sided logos, as discussed earlier.

Corresponding to the fifth, integral mode of consciousness, listening is denoted as "conscious listening". Integral in this context means that all previous levels of listening are accessible. As a main difference to rational listening, it goes beyond the individual's meaning and understands *listening "as part of a cultural field*".[17] This perspective thus not only takes into account the own, individual, but places listening at the service of the community, be it the family, the neighbourhood, or the company. Here, therefore, an essential transition takes place from the personal *I* to a (collective) W*e*.

3 Listening in Transdisciplinary Management

3.1 The Need for Management to Become Transdisciplinary

Management, and equally leadership, aims to drive people towards a common goal.[18] In the context of corporate organizations, it has to be acknowledged that organizational leaders produce quite often catastrophic consequences in ecological, social, and spiritual respect.[19] Even in the companies itself, according to Gallup surveys, only 15% of employees engage at the workplace. Additionally the current working world is subject to substantial changes, caused by developments and megatrends such as digitalization, demographic change, climate change, globalization, changing values, and the question of the preservation of democracy.[20] This

[16] Purdy (2000): 53.

[17] Purdy (2000): 53.

[18] Northouse (2016).

[19] For example, Reybrouck (2018).

[20] Hackl et al. (2017).

raises questions about new adequate forms of management, leadership, and cooperation to be able to deal with this demanding disruption, as one main task of organizations is to adapt effectively to uncertainty.[21]

The global crisis on ecological, economical, and social dimensions provokes a basic rethinking of the purpose of corporations.[22] To think of making profit as the only responsibility of companies, as Nobel prize winner Milton Friedman formulated in the 70s, seems not be adequate anymore.[23] The economy shall serve the human beings—and with them all living beings—and not vice versa.[24] And although organizations do not want to have the whole person with all personal concerns in the work environment, but mostly only employees as function owners, it is time to ask how organizations can be designed to be more in line with the needs of their members, as a call to "humanize organizations".[25]

On an individual level, this could mean shifting the motivational system from egosystem to ecosystem, transcending self-interest and starting to ask questions about what one wants to contribute and what one wants to give..[26] For creating a future that is highly unpredictable, collective and collaborative capacities such as the ability to decide in groups will be central.[27] It is *joint* thinking and problem-solving that will enable the members of an organization to be able to meet the world's complexity.[28]

Transdisciplinary management would therefore set up conditions to foster and cultivate "the emergence of a new type of intelligence, founded on an equilibrium between analytic intelligence, feelings, and the body" and "the open totality of the human being".[29] And as transdisciplinary education revalues the role of intuition, imagination, sensibility, and the body in the transmission of knowledge (Article 11 of the Charter of Transdisciplinarity), I would claim the same for transdisciplinary management.

3.2 *Bohm Dialogue: The Collective Dance of the Mind*

David Bohm, like Basarab Nicolescu (transdisciplinary thinker and pioneer and founder of the International Center for Transdisciplinary Research and Studies), was a physicist, more precisely a quantum physicist with a philosophical mind. If

[21] Williamson (1981).

[22] Fink and Moeller (2018), Laloux (2014), Quinn and Thakor (2018).

[23] Friedman (1970).

[24] Devivere (2021), Nicolescu (2002).

[25] Matthiesen et al. (2022).

[26] Brown et al. (2012), Crocker (2011).

[27] Samochowiec (2020).

[28] Knapp (2013), Scharmer and Kaufer (2013).

[29] Nicolescu (2002): 37f.

one goes back in time a few centuries, then one can understand it as a great liberation when man, the subject, understands himself as separate, distinct from the environment. For only this enabled science the necessary distance from the "objects of investigation".[30] Thinking and analysing have been sharpened and trained to such an extent in the last centuries that other ways of knowing the world have been pushed strongly into the background. Bohm saw himself limited in his work as a researcher by the individual, discipline-oriented silo-thinking and the concurrence in Academia, and developed a way that should make it possible to look behind and beyond. The perspective sought was one that went beyond the self, the ego. Gebser describes the aperspectival world as follows:

> "The designation 'aperspectival', in consequence, expresses a process of liberation from the exclusive validity of perspectival and unperspectival, as well as pre-perspectival limitations. Our designation, then, does not attempt to unite the inherently coexistent unperspectival and perspectival structures, nor does it attempt to reconcile or synthesize structures which, in their deficient modes, have become irreconcilable".[31] Bohm developed a way of dialogue with which this aperspectival world becomes accessible and tangible.

4 On Dialogue

The word "dialogue" originates from the Greek διάλογος (dialogos). It is made up of the components διά (dia: "through, inter") and λόγος (logos: "word, meaning, sense"). This creates an understanding of "dia-logos" as "that which comes to flow through words", or the "flow of sense".[32] In this respect, the spoken word is necessary to investigate on the flow of sense that emerges in course of a dialogue. However, it is not only what is said that gets attention. In suspending the dialogue participants listen to the others as well as to themselves. They listen above all to their reactions, impulses, feelings, thoughts, opinions. It is like listening to the rustling of the leaves, unfiltered, unbiased. And then, this listening is made visible, audible, so that the others in the group can also perceive it.[33] Dialogue is therefore a practice of presence, which serves to recognize the opinions and assumptions with which all humans are continually identified.[34] Isaacs summarizes the four practices required to practice dialogue as follows[35]:

- **Voice**: Telling the truth as far as you know it, telling who you are and how you think.

[30] Nicolescu (2010): 19.

[31] Gebser (1985): 2.

[32] Hartkemeyer et al. (2016): 56, Rohn and Sutrich (2014): 147.

[33] Bohm et al. (1991): 6.

[34] Hartkemeyer et al. (2016): 61.

[35] Isaacs (2011): 328.

- **Listen**: Listen without resistance and coercion. Listening means to develop inner silence.[36]
- **Respect**: This means to be aware of the integrity of another's position and the impossibility of fully understanding it.
- **Suspend**: This is about observing emerging thoughts and feelings without necessarily having to act on them, and present them in a way that others can grasp them.

By its nature, dialogue is not a discussion, not a debate nor a salon, nor a T-group or sensitivity training.[37] Bohm characterizes the dialogue as "*a mean to explore the field of thought. It is something new to our culture*".[38] It is meant to remove the fragmentation of reality by occidental rational-analytical thinking.[39] The fragmentation of our world is a result of our thinking, which divides and splits everything. This way of thinking has consequences, but thinking says: I didn't do it![40] The impact of an overemphasized *logos* as described before cannot be underestimated:

> We became aware that we live in a world produced almost entirely by human enterprise and thus, by human thought. The room in which we sit, the language in which these words are written, our national boundaries, our systems of value, and even that which we take to be our direct perceptions of reality are essentially manifestations of the way human beings think and have thought. We realize that without a willingness to explore this situation and to gain a deep insight into it, the real crises of our time cannot be confronted, nor can we find anything more than temporary solutions to the vast array of human problems that now confront us.[41]

People spend a tremendous amount of time in organizations, and there are many different ways to think about organizations.[42] What happens when the unspoken, the unfelt, the unthought, is allowed to emerge into the collective space of the organization or team?

5 Dialogue and Listening in Organizations

There are some reasons why organizations prevent genuine dialogue in the Bohmian sense. The circle in which dialogue takes place, as an expression of an attitude of open mind and heart and a free exchange of what someone thinks and feels, is in contrast to the asymmetrical forms of relationships and power imbalances typical in organizations.[43] Nevertheless, dialogue has taken its place in organizations,

---

[36] Isaacs (2011): 85.

[37] Hartkemeyer et al. (2016).

[38] Bohm et al. (1991).

[39] Hartkemeyer et al. (2016): 59.

[40] Bohm (2002): 38f.

[41] Bohm et al. (1991): 1f.

[42] Morgan (2006).

[43] Ellinor and Gerard (2000), Isaacs (2011), Rautenberg (2010).

primarily also due to the challenges for companies already mentioned. There are various areas in which Dialogue is used as an approach, for example strategy development, team and organizational development, and organizational learning.[44] Dialogue is especially useful to enhance creativity, improve conversation culture, and support open communication.[45]

Companies, for example, use dialogue to work on strategic directions. A Swiss rehabilitation centre with 600 employees and two locations used dialogue in their strategic task force, counting 32 members of executive management. The dialogue allowed dissenting opinions to be expressed without fear of consequences. It was a safe vessel to address even sensitive issues. The conversation did not get stuck in an either-or of strategic options, but could develop into a both/and view. The various perspectives that spoke for or against a new location could be addressed openly, thus contributing to a differentiated decision that was not foreseeable. The introduction of dialogue in the strategic task force but also for executive managers contributed to a change in the communication culture of the company. Communication is experienced in a more appreciative way, and mutual understanding has been improved through the openness of the dialogue. Employees are listened to and have more confidence in expressing their opinions.[46]

In addition to strategic dialogue, companies can also conduct generative dialogue, which can delve into questions such as "What do we understand by leadership in our organization?", "Are we satisfied with the way we work together?", or "What does digitization mean for our work area?"[47] At first glance, the questions don't seem any different than in usual meetings or team retreats. However, the format of the Dialogue makes a depth of shared exploration possible that is, on the one hand, very personal, and at the same time connects to the shared potential of the participating people.

6 The Art of Listening: Requirements and Limits

Listening as it has been argued in this article is not only understood as a one-directional experience of receiving and processing information as a thinking individual.[48] In listening to what is expressed and how it feels, it opens a space for a collective evaluation of a possibility. In this very sense, listening becomes a

[44] Senge (2006), Steiger and Vey (2003).

[45] Ehmer (2004).

[46] Gut (2018).

[47] Many concrete experiences about the introduction of dialogue in companies can be found in Isaacs (2011).

[48] For a recent review on the conceptualization of listening at work, see Kluger and Itzchakov (2022).

community affair; it is an interactive process that involves the other individuals.[49] Being related in listening to each other involves "*conversation partners being sensitive to the unfolding, dynamic understanding emerging out of what is being mutually shared through conversation*".[50] Listening, then, means being open to something new that emerges out of the shared space. It requires a certain attitude, to be open and tolerant to meet the unexpected.[51] Embarking on something new also means admitting uncertainty, especially facing fear when the familiar has to be left behind.[52] In addition to this willingness to face fear, listening and dialogue in particular require individuals to have the courage to be authentic, open, and eager to discover. Research on psychological safety clearly demonstrates that employees need psychological safety in their team to be able to express themselves openly in an uncertain, complex world.[53] Trusting relationships are conducive as well.[54] However, these cannot be assumed in the working world—they would evolve, however, if a group engages in a dialogue process.

In addition to this fundamental need for openness and overcoming fear, there is another significant hurdle to overcome if a more conscious way of listening and dialogue is to be introduced into a company: Listening is powerful because it affects power structures. This already starts with a very little thing: the seating arrangement in a dialogue is in a circle, where everybody sits equally next to each other, and everybody has the same right to take the "talking stone" that is used to slow down the process. For the success of the dialogue though, it is imperative that insights from the dialogue are not instrumentalized against participants and used in power games.[55]

The art of listening therefore concerns fundamental aspects such as dealing with uncertainty and power structures, and engaging in it also means engaging in a cultural change within the organization.[56] Isaacs describes the potential of such a change: «*The power that respects no one but includes all, demands the best in man and awakens great creativity, is love. Through dialogue, the power of love can be released, not in the sentimental or moralistic sense, but in the sense of true creativity*».[57] He describes in detail the results that these processes can lead to in companies, using examples from companies such as Volvo, IKEA, Ford, and Shell.

---

[49] Purdy (2000).

[50] Hinz et al. (2022): 78.

[51] Koivunen (2006).

[52] Scharmer (2022): 42.

[53] Edmondson and Lei (2014): 27.

[54] Steiger and Vey (2003): 33f.

[55] Hartkemeyer et al. (2016): 161.

[56] This is also described by Mölleney and Sachs (2019) in their approach "Beyond Leadership", which enables an encounter at eye level, and brings the relationship aspect back into the working world.

[57] Isaacs (2011), p. 319f.

7 Finally: Listening Is Connecting

I am because we are.
Zulu proverb

Conscious listening, joint dialogue, or presencing—these examples illustrate the movement of listening that goes beyond an individual, mechanistic understanding of listening and opens up a community field. Enabling this kind of listening at different levels of consciousness is definitely a task of management for sustainable and viable organizations. In this sense, viable does not only mean that the organizations meet environmental and market requirements successfully, but in particular that they are designed in accordance to human beings. The man himself could become part of the solution if he explores his deep dimensions and recognizes his fundamental belonging. Heidegger says: "Being-with others develops in listening to one another".[58] In listening, we cannot separate from the world.[59] Changing the quality of listening means changing the relationship to the own experience.[60]

And that can mean a lot—more than one could say. Or: Listening is . . . when the silence is speaking to you.
Claudia Pölderl

References

Alvesson, M., & Sveningsson, S. (2003). Managers doing leadership: The extra-ordinarization of the mundane. *Human Relations, 56*(12), 1435–1459. https://doi.org/10.1177/00187267035612001

Birkinshaw, J. (2012). *Reinventing management: Smarter choices for getting work done*. Jossey-Bass.

Bohm, D. (2002). *Der Dialog: Das offene Gespräch am Ende der Diskussionen* (3rd ed.). Klett-Cotta.

Bohm, D., Factor, D., & Garrett, P. (1991). Dialogue: A proposal.

Brown, S. L., Brown, R. M., & Penner, L. A. (Eds.). (2012). *Moving beyond self-interest: Perspectives from evolutionary biology, neuroscience, and the social sciences*. Oxford University Press.

Corradi Fiumara, G. (1995). *The other side of language: A philosophy of listening*. Routledge.

Covey, S. R. (2004). *The 7 habits of highly effective people: Restoring the character ethic*. Free Press.

Crocker, J. (2011). Safety in numbers: Shifting from egosystem to ecosystem. *Psychological Inquiry, 22*(4), 259–264. https://doi.org/10.1080/1047840X.2011.624978

Devivere, B. v. (2021). *Sinn und Arbeit: Antworten zur Sinnsuche im 21. Jahrhundert—Viktor Frankl und andere*. Springer.

[58] Heidegger (2007): 206.

[59] Koivunen (2006): 100.

[60] Scharmer (2022): 4.

Edmondson, A. C., & Lei, Z. (2014). Psychological safety: The history, renaissance, and future of an interpersonal construct. *Annual Review of Organizational Psychology and Organizational Behavior, 1*(1), 23–43. https://doi.org/10.1146/annurev-orgpsych-031413-091305

Ehmer, S. (2004). *Dialog in Organisationen: Praxis und Nutzen des Dialogs in der Organisationsentwicklung; eine Untersuchung*. Univ. Press.

Ellinor, L., & Gerard, G. (2000). *Der Dialog im Unternehmen: Inspiration, Kreativität, Verantwortung*. Klett-Cotta.

Fink, F., & Moeller, M. (2018). *Purpose Driven Organizations: Sinn—Selbstorganisation—Agilität*. Schäffer-Poeschel Verlag.

Friedman, M. (1970). The social responsibility of business is to increase its profit. *New York Times*.

Fromm, E. (1991). *Haben oder Sein: Die seelischen Grundlagen einer neuen Gesellschaft* (20th ed.). Dt. Taschenbuch-Verl.

Gebser, J. (1985). *The ever-present origin*. Ohio University Press.

Gut, D. (2018). *Dialog-Kultur mit schöpferischen Gesprächen zu neuen Impulsen*. SPEKTRAmedia.

Hackl, B., Wagner, M., Attmer, L., & Baumann, D. (2017). *New Work: Auf dem Weg zur neuen Arbeitswelt: Management-Impulse, Praxisbeispiele, Studien*. Springer Gabler. https://doi.org/10.1007/978-3-658-16266-5

Hartkemeyer, M., Hartkemeyer, J. F., Hartkemeyer, T., & Hüther, G. (2016). *Dialogische Intelligenz: Aus dem Käfig des Gedachten in den Kosmos gemeinsamen Denkens (Zweite Auflage)*. Info 3-Verlagsgesellschaft Brüll & Heisterkamp KG.

Heidegger, M. (1984). *Early Greek thinking*. Harper & Row.

Heidegger, M. (2007). *Being and time*. Blackwell Publishing.

Hinz, J., Stephens, J. P., & Van Oosten, E. B. (2022). Toward a pedagogy of connection: A critical view of being relational in listening. *Management Learning, 53*(1), 76–97. https://doi.org/10.1177/13505076211047506

Isaacs, W. (2011). *Dialog als Kunst gemeinsam zu denken: Die neue Kommunikationskultur in Organisationen* (2nd ed.). EHP.

Kluger, A. N., & Itzchakov, G. (2022). The power of listening at work. *Annual Review of Organizational Psychology and Organizational Behavior, 9*(1), 121–146. https://doi.org/10.1146/annurev-orgpsych-012420-091013

Knapp, N. (2013). *Kompass neues Denken: Wie wir uns in einer unübersichtlichen Welt orientieren können* (3rd ed.). Rowohlt-Taschenbuch-Verl.

Koivunen, N. (2006). Auditive leadership culture. Lessons from symphony orchestras. In D.-M. Hosking & S. McNamee (Eds.), *The social construction of organization*. Liber & Copenhagen Business School Press.

Laloux, F. (2014). *Reinventing organizations: A guide to creating organizations inspired by the next stage of human consciousness*. Nelson Parker.

Matthiesen, K., Muster, J., & Laudenbach, P. (2022). *Die Humanisierung der Organisation: Wie man dem Menschen gerecht wird, indem man den Großteil seines Wesens ignoriert*. Verlag Franz Vahlen.

Mölleney, M., & Sachs, S. (2019). *Beyond leadership*. Verlag SKV.

Morgan, G. (2006). *Images of organization*. Sage Publications.

Nicolescu, B. (2002). *Manifesto of transdisciplinarity*. State University of New York Press.

Nicolescu, B. (2010). Methodology of transdisciplinarity–levels of reality, logic of the included middle and complexity. *Transdisciplinary Journal of Engineering & Science, 1*(1), 10.22545/2010/0009.

Northouse, P. G. (2016). *Leadership: Theory and practice* (7th ed.). SAGE.

Peters, T. J., & Waterman, R. H. (2008). *In search of excellence: Lessons from America's best-run companies*. HarperCollins.

Purdy, M. W. (2000). Listening, culture and structures of consciousness: Ways of studying listening. *International Journal of Listening, 14*, 47–68.

Quinn, R. E., & Thakor, A. V. (2018). Creating a purpose-driven organization. *Harvard Business Review, 2018*, 78–85.
Rautenberg, M. (2010). *Der Dialog in Management und Organisation—Illusion oder Perspektive? Eine systemtheoretische Zuspitzung*. Carl Auer Verlag.
Reybrouck, D. v. (2018). *Kongo: Eine Geschichte* (4th ed.). Suhrkamp Taschenbuch Verlag.
Rohn, C., & Sutrich, U. (2014). Dialogue as shared social space in management and organizations. In *Leadership learning for the future* (pp. 145–162). Information Age Publishing, Inc.
Samochowiec, J. (2020). Future skills. Vier Szenarien für morgen und was man dafür können muss. GDI.
Scharmer, C. O. (2022). *Essentials der Theorie U: Grundprinzipien und Anwendungen* (2nd ed.). Carl-Auer-Systeme Verlag.
Scharmer, C. O., & Kaufer, K. (2013). *Leading from the emerging future: From ego-system to eco-system economies*. Berrett-Koehler Publishers, Inc.
Senge, P. M. (2006). *The fifth discipline: The art and practice of the learning organization*. Doubleday/Currency.
Steiger, R., & Vey, K. (2003). *Chancen und Grenzen des Dialogs: Denkanstösse für Beruf und Alltag*. Huber.
Williamson, O. E. (1981). The economics of organization: The transaction cost approach. *American Journal of Sociology, 87*(3), 548–577.

Pölderl, Claudia is a PhD candidate at VŠE Prague University of Economics and Management and a research associate at the Institute für Organizational Viability at ZHAW School of Management and Law (Switzerland). She holds teaching assignments in leadership, business ethics, and human capital management at Bachelor, Master, and continuing education level. Research areas include leadership, self-organization, and transformation on individual and collective level. Claudia holds a master degree in HR and OB from Johannes Kepler University, Linz (A).

Transdisciplinarity in Organizational Learning: How Multi-Perspectivity and Self-Transcendence Lead to Co-creation

Birgit Feldhusen

1 In Search for New Learning Outcomes

Organizations serve us humans in achieving goals together. In the textbook on organization, these are defined as "*social entities* that *permanently* pursue a *goal* and have a *formal structure* with the *help* of which the *activities* of the *members* are to be *aligned* towards the pursued goal."[1] Meanwhile, almost every element of this definition seems to be in flux, if not in crisis. *Social interaction* is being reduced by digital communication or even displaced by artificial intelligence. In an increasingly volatile world, *goals* and strategies can rarely be of lasting *duration* and are subject to ever shorter half-lives. Actual goals are rarely transparent and often subject to particular—non-organizational—interests, thus they tend to lose their power in *aligning* and bonding employees. Established *structures* allow little flexibility in volatile times and are often *obstacles rather than help. Activities* required to achieve goals are difficult to coordinate or even unknown, as usual practices of action and decision-making have reached their limits. *Employees and leaders* are in short supply in many sectors, especially those kinds who see themselves as committed members of the whole organization.

As a result, we observe dysfunctionalities such as ethical scandals (e.g., VW, Wirecard), the instrumentalization of entire organizations for individuals' interest in power (e.g., Twitter), product developments that ignore the (basic) needs of consumers or harm them (e.g., food scandals, manipulative social media platforms), environmental damage as part of the value chain (e.g., genetic engineering, the chemical and oil industry), loss of employees and performance due to burnout,

[1] Kieser and Walgenbach (2010: 6).

B. Feldhusen (✉)
University for Continuing Education Krems, Krems an der Donau, Austria
e-mail: Birgit.feldhusen@donau-uni.ac.at

© The Author(s), under exclusive license to Springer Nature Switzerland AG 2025

M. Schüz (ed.), *Transdisciplinary Thinking and Acting*, Transdisciplinary Management of Ecological and Social Crises,
https://doi.org/10.1007/978-3-031-70680-6_15

strikes, dismissals, or migration from entire sectors (e.g., gastronomy, care) as well as younger generations that increasingly refuse the performance paradigm of previous generations.

It seems that well-learned patterns of action and value orientations within the economy, ecology, and society cannot give adequate answers to current tasks and challenges anymore. The smart answer to such situations was and is for individuals as well as for organizations: learning. In times of high volatility, complexity, and existential threat, learning even becomes a survival strategy. Ongoing adaptation and permanent learning are the basis for sustainable capacities to act. Real learning always changes the learners themselves—in this case, the organization:

> *Organizational learning occurs when individuals in an organization experience a problematic situation and examine it on behalf of the organization. They experience a surprising mismatch between expected and actual action outcomes and react to it with a process of thought and further action; this leads them to modify their conceptions of the organization or their understanding of organizational phenomena and to reorder their activities so that outcomes and expectations match, thus changing the action guiding theory of organizations. To become organizational, learning [...] must be anchored in the images of the organization that exist in the minds of its members and/or the epistemological artefacts [...] located in the organizational environment.*[2]

Whether it is the learning organization,[3] dynamic capabilities,[4] or the current discussion on agility—the idea is always the same: to adapt to volatile, dynamic environments and to remain capable of acting through the ability and willingness to continuously renew oneself. However, learning outcomes that lead the way out of the current crises do show up very slowly or not at all. Therefore, it makes sense to generate a deeper understanding of how people can learn at a new level, grow beyond themselves individually and collectively, and thus find new solutions that will be actually implemented in everyday organizational life. What or how do organizations need to learn for acting appropriately in the face of the crises that have been mentioned? How do organizations have to change themselves for really serving people in achieving common goals, and specifically in overcoming the challenges ahead? What contribution can transdisciplinarity make here?

> *In order to explore organizational learning, we must rethink what we mean by organization. We must ask what an organization is that it may be said to learn.*[5]

Following this approach from Argyris and Schön, this article takes a transdisciplinary look at organizations to newly understand their nature and their capacity to act and learn for revealing possible starting points for new learning cultures. Following Nicolescu, key concepts of transdisciplinarity are 1) the existence of levels of reality, 2) the importance of subject–object interaction, 3) the existence of a hidden third (zone of non-resistance) beyond our experiences and models, and last but not least 4)

---

[2] Argyris and Schön (1996: 15 f).

[3] Senge (1990).

[4] Teece and Pisano (1994), Teece (2009).

[5] Argyris and Schön (1996: 6).

the insight that knowledge is "*forever open*."[6] In search of understanding what transdisciplinary organizational learning can mean, the article will establish connections between these key concepts of transdisciplinarity and theories of collective knowledge, learning and collective cognition. First, it derives an understanding of organization as a knowing collective mind that emerges through interaction and attribution of meaning from multi-perspectivity. Then, it shows different levels of learning and their characteristics and locates the zone of non-resistance in it. In the following, it integrates the two trains of thought into a map of transdisciplinary organizational learning. Finally, the article gives examples of implementation to show how transdisciplinarity in organizational learning can lead to new learning outcomes and thus contribute to overcoming current challenges.

2 Collective Knowledge: Multiple Perspectives in Coherent Interaction

2.1 *Subjectivity of Knowledge and Capacity to Act*

In knowledge management, organizational learning, and the sociology of knowledge, we can find an understanding of knowledge as *capacity to act.*[7] Combining fundamental propositions from epistemology,[8] learning theory,[9] knowledge management,[10] organizational learning,[11] organizational behavior,[12] sociology of knowledge,[13] and psychology,[14] we find that knowledge, resp. the capacity to act at a given point of time is the result of previous experiences that have been coherently connected by sense-making:

> *Making sense is a circular cognitive process that may start with some initially incomprehensible sensation, which then proceeds to imaging hypothetical contexts for it and goes around a hermeneutic circle during which features are distinguished—in both contexts and what is to be made sense of—and meanings are constructed until this process has converged to a sufficiently coherent understanding.*[15] *At that point observers are in harmony with what*

[6] Nicolescu (2010: 17).

[7] Churchman (1971), Argyris (1993), Stehr (1996).

[8] Krippendorff (1984).

[9] Bateson (1983), Kolb (1984).

[10] North (1998), Nonaka and Toyama (2005).

[11] Argyris and Schön (1978).

[12] Weick and Roberts (1993), Tsoukas (2000).

[13] Stehr (2001).

[14] Seligman et al. (2013).

[15] Krippendorff (1984: 13).

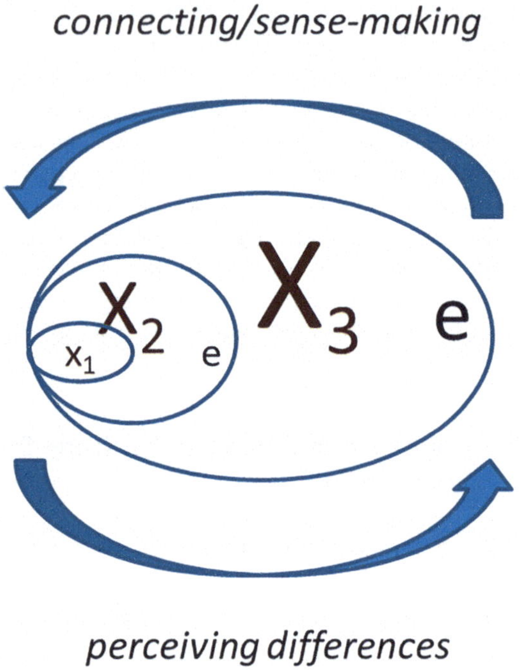

Fig. 1 Knowledge as basis and result of a cognitive hermeneutic circle

> *they believe they observe. The nervous system is organized or organizes itself such that it computes a stable reality. This reality is located neither inside nor outside the observing organism but resides in the ongoing process of drawing distinctions and formulating relations.*[16]

Thus, a process representing a *cognitive hermeneutic circle* continuously *organizes* the individual's world of experience based on meaning, and continuously creates possibilities for action that "*make sense*." Given the continuity of new experiences, knowledge is permanently in flux. We rather must think of a continuous knowledge process (*knowing*) instead of knowledge as a static concept. The latter would only exist at a theoretically frozen point of time within the knowledge process.

Knowledge is thus both the basis and the outcome of knowledge processes (Fig. 1). Starting point is an existing set of knowledge x_i. This spans a certain scope or space of action for the individual, in which current values, experiences, and results of action are in line, resp. *make sense*. If the individual makes a *new experiences e* that is outside this space, she needs to connect this *difference* to the existing set of knowledge by sense-making. This is how the knowledge space—ergo the scope of action—is expanded to x_{i+1} and the individual learns. An example: Every morning, a person goes to catch the train at 8:05 am to get to work (x_1). One morning, she sees the train's rear lights, although she thought to be in time (*e*). Then her mind starts to search for possible explanations (watch was wrong or even broken,

[16] Krippendorff (1984: 27 f.).

train driver left ahead of schedule, . . .) until she considers a change of timetable and finds it confirmed on the Internet. The next morning, she will be at the station at the new departure time of 7:55 a.m. Moreover, she will be more attentive to timetable changes in the future (x_2).

Individuals encounter such sources of learning involuntarily or voluntarily and even by intention. While Argyris and Schön understand learning as a necessary reaction to mistakes (*learning imperative*[17]), Kolb assumes learning as a holistic, lifelong process that transforms human experience (*experiential learning*[18]). Senge even sees learning as a creative process that realizes human potential (*generative learning*):

> *Real learning gets to the heart of what it means to be human. Through learning, we re-create ourselves. Through learning, we become able to do something we were never able to do. Through learning, we reperceive the world and our relationship to it. Through learning, we extend our capacity to create, to be part of the generative process of life. There is within each of us a deep hunger for this type of learning.*[19]

Thus, experiences which create a difference to the known space of action are triggers and sources of learning.[20] These can be encounters with the outside (e.g., statements of a human being, accessible new facts, confrontation with new situations, disciplines, and/or crises) or with the inside inner observer or the higher self, e.g., working with the inner team, but also sensing attractive futures states in contrast to the present, *learning from the future*, see Sect. 3 below. It is the chance and ability to perceive differences and to make new connections (see lower arrow in Fig. 1) that determine the possible degree and extent of learning for an individual. Whether differences are perceived and how they are transformed depends on the existing knowledge and its permeability (i.e., a certain willingness to revise) at the time in question, which in turn has been shaped by previous experiences. The existing knowledge at any point of time is partly conscious but mostly unconscious and forms a multi-layered self-context, a kind of glasses, which we use when looking at the world.

Conceptualized in such a way, knowledge cannot be transmitted, shared, or stored in databases, but is immanent for the knower. Only data, artifacts, or communicative contents and gestures can be transmitted, exchanged, or stored, but these again can result in subjectively different knowledge. A book on Einstein's theory of relativity is initially an artifact. Knowledge is not generated until human consciousness focuses on the artifact, the data, or information and processes its content. One and the same book can generate different levels of knowledge; in the case of Einstein's book, a student of physics will generate a different knowledge from it than a person

[17] Argyris and Schön (1978: 2).

[18] Kolb (1984: 31ff.).

[19] Senge (1990: 14).

[20] Bateson (1983), Argyris and Schön (1978).

who has never dealt with physics before. This insight is accompanied by a new differentiation of tacit and explicit knowledge. Knowledge management in the 1990s classified texts, data, and information as explicit knowledge, and inner, implicit knowledge as tacit knowledge.[21] *Explicit* referred to the outer form of representation, i.e., through articulation and writing. In a more recent understanding *explicit* refers to the clarity of representation in consciousness, e.g., through awareness and reflection, independently from sharing this explicit knowledge with the outside.[22]

To summarize, learning changes knowledge as well as the learner himself. It uses the scope of existing spaces of thought and capacities to act (knowledge), but at the same time transcends them. *With respect to transdisciplinarity, we find here Nicolescu's position that knowledge is forever open.*[23] Constructivist epistemology speaks of knowledge as a subjective key through which the subject, in interaction with the world, unlocks this very world and becomes capable of acting within its subjective reality.[24] In this sense, knowledge *organizes* reality. If something disturbs a coherent reality, the individual must learn and reorganize. Neurobiologist Wolf Singer puts it this way: The brain arranges sensory impressions in such a way that the organism functions.[25] Knowledge manager Klaus North also describes knowledge as information which is processed and networked by consciousness, with the networking being context-dependent and dependent on individual expectations and experiences.[26] Nonaka points out that Socrates' definition of knowledge as *justified true belief* should be understood less in the sense of a *true* justified belief, than rather in a sense of a belief *justified as true*, i.e. the focus is on justified, not true.[27] Knowledge is justified as true by one's own subjective cognition. *With respect to transdisciplinarity, we find here a reference to Nicolescu's position on levels of reality.*[28] Thus, a first step on the way to new effective learning outcomes is an *intellectual modesty*[29] by recognizing that reality is nothing absolute but always a self-construct: Personal perspective and truth is only one of many perspectives that can be justified as truth. It is dialogue with other perspectives that let us grow and come to far-sighted perspectives and solutions.

---

[21] e.g., Nonaka and Takeuchi (1995).

[22] Virtanen (2013).

[23] Nicolescu (2008: 8).

[24] Glasersfeld (1984).

[25] Singer (2016).

[26] North (1998).

[27] Nonaka et al. (2008: 8 f.).

[28] Nicolescu (2008: 4 f.).

[29] Wüthrich (2022).

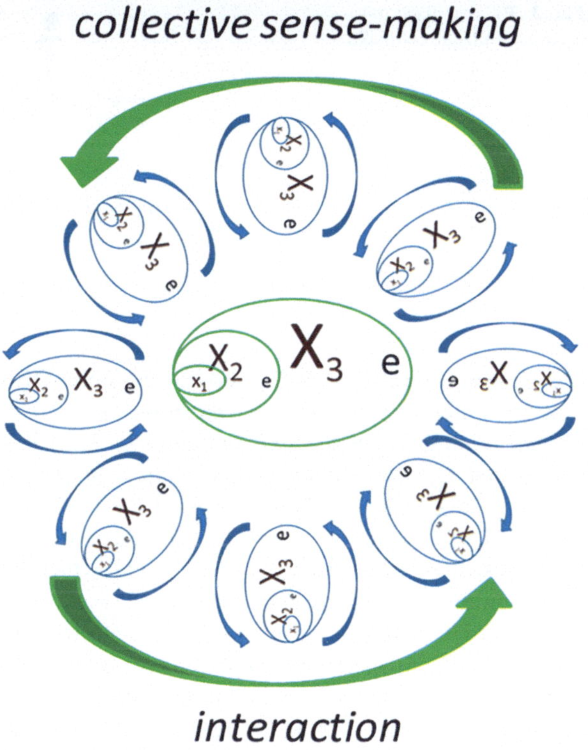

Fig. 2 Collective knowledge resulting from converging meaning formation

2.2 *Collective Capacity to Act: Sense-making in the Collective Mind*

Building on the understanding of knowledge as individual capacity to act which is organized through sense and meaning, we can understand collective capacity to act as collective knowledge organized through collective sense-making. This is exactly how organizational scientist Tsoukas describes organizational knowledge: individual knowledge that is organized by shared understanding about certain rules and cognitive categories, thus connected coherently.[30] If we stay with the image of the cognitive hermeneutic circle, individuals and their sense-making circles interact in an organization (Fig. 2).

Every utterance or gesture of a counterpart that is perceived as a *difference* forms an *experience e* for individual knowledge processes. Individual knowledge circles overlap and mutually impact each other in a kind of negotiation that can be seen as a *collective sense-making* circle: If this process comes to a state, where the sum of experiences and sense-making circles are coherently connected, and a shared sufficient understanding of the common space of experience has been reached,

[30] Tsoukas (2000 108 f.)

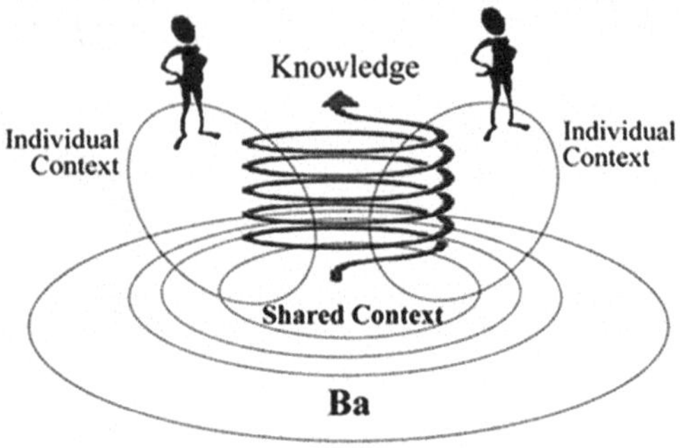

Fig. 3 Ba as a place of co-transcendence (Nonaka et al., 2000:14)

coordinated collective actions become possible. Organizational researcher Weick describes this as follows: "*Organizations [...] are talked into existence. [...] Organization emerges through sensemaking [...] In sensemaking, action and talk are treated as cycles rather than as a linear sequence.*"[31]

Knowledge manager Nonaka coined the term *ba* to describe the *space* where these processes of collective sense-making and knowledge creation take place.[32] *Ba* is a Japanese word for place, space, or field, which includes not only physical, but also cognitive, mental, or emotional space (Fig. 3). It describes a dynamic, shared context of interaction, interpretation, and dialectic, a place of "co-transcendence,"[33] where individuals meet and deeply engage with each other in a certain quality, so that they overcome their own limited perspectives and reach a higher state of knowledge together.[34] Nonaka refers to Nishida, a Japanese philosopher, who describes *basho* as a field of reflexive awareness. Depending on the degree of focused reflection, a *basho* can attain three different levels of quality, from unreflective *everyday basho* to *ultimate basho*, a joint level of consciousness that transcends the ego.[35] These levels correspond to 3 levels of learning, which will be explained in the next chapter.

The insight that cognitive processes also include "organism-external," social interaction is reflected in a wide variety of recent approaches to collective cognition: Extended Mind,[36] Group Cognition,[37] Enculturated Mind,[38] Collaborative Cognition.[39] They establish the proposition of 4E cognition (embodied, embedded, extended, enactive), which assumes that cognition, including knowledge and

[31] Weick et al. (2005: 409 f.,412).

[32] Nonaka et al. (2000: 14).

[33] Nonaka et al. (2008: 36).

[34] Nonaka et al. (2008: 27 ff.).

[35] Chia (2003: 970).

[36] Clark and Chalmers (1998).

[37] Theiner et al. (2010).

[38] Hutchins (2011).

[39] Kirchhoff and Newsome (2012).

learning, always emerges from interactions between brain, body, physical, and social environment. Mind thus presents itself not as an entity but as a space of activity, as a permanent process for coping with the world of experience.[40] *In terms of transdisciplinarity, we find here Nicolescu's position that reality is a result of the interaction between subject and object.*[41]

Like individual sense-making takes place in the individual mind and organizes reality, we can speak of a *collective mind* in collective sense-making, whose space or *ba* is spanned by interaction and organizes the common reality. Approaches that describe an organization as an emergent phenomenon of a collective mind, which bears knowledge and the ability to act, can be found in a wide range of disciplines:

- *Organizational behavior*: Weick and Roberts were leading in coining the term *collective mind*. When studying high reliability organizations, they recognized that mindful interactions of contribution, representation, and subordination generate a shared field with transindividual quality: "*To connect is to mind*."[42] These are the processes that generate shared meaning, which again calls organizations into existence: "*Sensemaking and organization constitute each other. [...] We need to grasp each to understand the other*."[43]
- *Knowledge management*: Erden and Nonaka describe how growing loops of social interaction give rise to a collective mind that bears the collective knowledge.[44] Stacey also understands the process of conversational interaction as a mind whose collective knowledge resides in the properties of interaction, such as ways of speaking. While interaction with myself constitutes individual mind, social interaction constitutes collective mind.[45]
- *Organization science*: Tsoukas describes organization as cognitive categories and rules of behavior through which a shared understanding has been achieved. The generation of understanding depends on the interpretations of each individual member. Organization is thus input and output at the same time: while existing rules of conduct are organization in the sense of input, interpretations of individuals and their resulting actions generate a new image every day, generating organization in the sense of output.[46]
- *Organizational learning*: Argyris and Schön understand the organization as an organism whose cells each have a specific, changing image of themselves in relation to the whole. Organization takes place through a continuous, coordinated intertwining and meshing of these individual images.[47]

[40] Weick and Roberts (1993).

[41] Nicolescu (2010: 29 f.).

[42] Weick and Roberts (1993: 374).

[43] Weick et al. (2005: 409 f.).

[44] Erden et al. (2008).

[45] Stacey (2000: 33).

[46] Tsoukas (2005: 381).

[47] Argyris and Schön (1996).

Forms and attributes of interaction (extent, attention, quality, technologies, novelty, etc.) determine the quality of collective knowledge, i.e. collective capacity to act. Distributed and reflective sense-making fosters a mindful, *witty collective mind*[48] and *collective intelligence*, where new, future-oriented knowledge and growth can emerge through plurality. Contrary, unreflective adaptation of individual cognitions to a group cognition promotes a *mindless group think.*[49] It destroys creativity, uniqueness, and independence. Erden and Nonaka have described different qualities of a collective mind depending on different *contributions of socialization*[50]:

- *Loose collection of people*: There is little or no relation to each other.
- *Collective action*: Shared memory and understanding, embedded routines/patterns of behavior, group culture, collective identity, group becomes a collective body and mind.
- *Phronesis*: Collectively grasp the essence of particular situations, decide heedfully, take actions guided by socially constructed common goodness (group values, culture, goals), "we-intention" exists: one for all, all for one.
- *Collective improvisation*: Make sense and react quickly and appropriately in complex, unpredictable situations through spontaneity and intuition, collective intuition, each member intuitively acts in a way that is consistent with the actions of others through full concentration and attention to others' actions over time. Outcome may range from incremental change of routines to radical creativity.

To summarize, collective knowledge resides in structures of human interaction. It is based on collective processes, but always has a subjective context. New collective knowledge can only emerge to the extent that interaction patterns are of a certain permeability, which in turn is based on the permeability of the individual sets of knowledge involved. Earlier, we derived multi-perspectivity in reflection as a first starting point on the way to new learning outcomes. Here we can add the joint reflection of interaction patterns and their underlying sense-making.

> *The degree and quality of learning within an organization can be regarded as a function of the different forms of reflection and of their interaction.*[51]

Reflecting on the shared understanding of rules and cognitive categories that organize joint action and span a corresponding *collective mind* or knowledge space *ba* include not only the way in which people work together, but also the meaning-giving *why* they work together. The forms of reflection mentioned by Jordan offer a third starting point and are addressed in the following chapter.

---

[48] Weick and Roberts (1993).

[49] Janis (1972).

[50] Erden et al. (2008: 13).

[51] Jordan et al. (2009: 470).

3 Levels of Learning and the Zone of Non-resistance

As described above, learning takes place through loops in a hermeneutic circle that connects perceived differences (new experiences, phenomena) with the existing knowledge space to an expanded knowledge space through sense-making. This construction of meaning toward a coherent understanding can take place at different levels. Bateson was the first to formulate the following levels of learning[52]:

- *Trivial learning (L0)* presents itself as non-learning. Differences in experience do not lead to change. Behavior and thought structures remain in habits and conditioned behavior.
- *Behavioral learning (L1)* leads to changed behavior within a known set of alternatives. The learner stays within a known context.
- *Context learning (L2)* leads to the recognition and differentiation of contexts, so that changed contexts open new alternatives.
- *De-identified learning (L3)* leads to a changed context for thinking itself, a correction of the system of sets of alternatives. Based on recognizing the nature of cognitive contexts and (self-)concepts it results in changed models about the person himself and his own ego identity, and consequently about the world he operates in.

Within his research on radical innovation, Peschl describes respective levels of reflection for learning levels L1–L3[53]:

- *Behavioral reflection.* Reflection of behavior: Adaptation, optimization in the pre-structured solution space.
- *Cognitive reflection.* Reflection of intellectual and cognitive domain: reframing, stepping out of one's own frame of reference and experience, looking holistically at the situation or problem.
- *Existential reflection*: Reflection of the fundamental domain in which the cognitive domain itself is embedded—the domain of existential self and wisdom: existential change, questions from the heart about deepest intention, purpose, shared vision, and common will.

If a certain level of learning does not bring the desired success, usually reflective thoughts step one level higher sooner or later (Fig. 4). A learning process that many FMCG companies have gone through in the last 30 years—at least up to level 2—may serve as an example: For a company, a certain split of advertising expenditure into various classical media such as print and TV works very well and delivers desired sales effects for a certain product. Knowledge and expected outcome coincide. Then, an unexpected experience occurs: Advertising impact analyses begin to show that spendings seem no longer bring expected sales effects. Learning 0 would result in continuing the advertising strategy, assuming that no correction is necessary

[52] Bateson (1983) (1972: 362ff.)

[53] Peschl (2007).

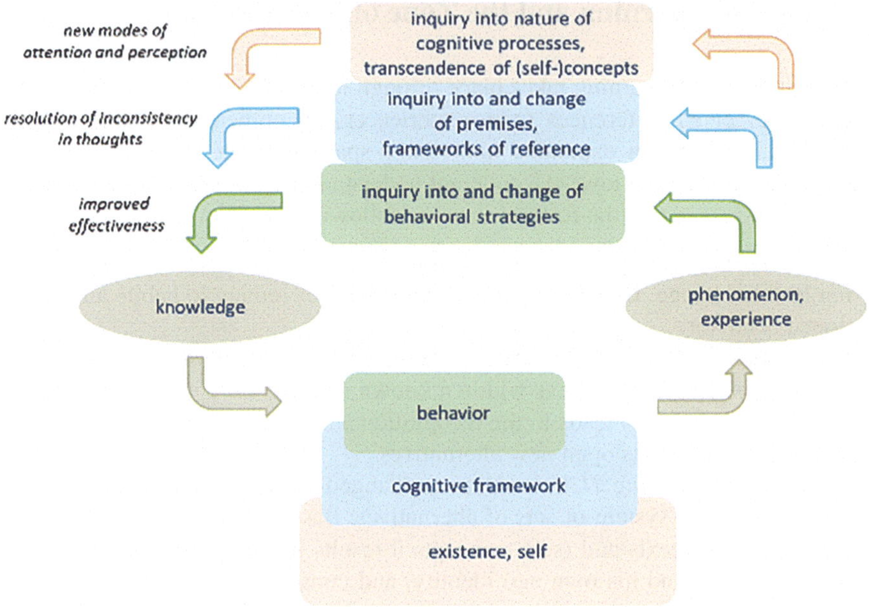

Fig. 4 Three levels of reflection and learning (Feldhusen, 2014: 92)

and that the usual effects would become visible again soon. Learning 1 would lead to a re-allocation of advertising spendings within the given media mix, e.g., shifting parts of the print and TV spendings to social media, as marketing people have realized a change in their target group's media use. Learning 2 would challenge the company's attitude that advertising need to be based on paid media and would explore further possibilities of brand presence, e.g., guerrilla marketing, brand ambassadors or influencers, or even question why products need to be brought to consumers by multi-million budgets and intensive communication efforts. Learning 3 would question how the organization originally came up with the idea of marketing this product and whether this makes sense at a deeper level for all parties involved: *"Who am I, thinking that I need to promote and sell this product?"*

Interestingly, predominantly L1 and the extension by L2 found their way into the theory of organizational learning, named *single-loop learning* and *double-loop learning*.[54] The third form of learning conceived there, *deutero learning*, does not refer to learning level 3, but to learning to learn, which makes learning processes themselves the object of L1/L2 learning. Learning level 3, the reflection of one's own ego identity, was initially omitted and thus avoided. It was Senge who, without explicitly mentioning L3, referred to this level when describing true generative learning of a learning organization: The deeper meaning of learning and the highest form of team learning would be *metanoia*, a Greek term combining meta = *beyond*

[54] Argyris and Schön (1978).

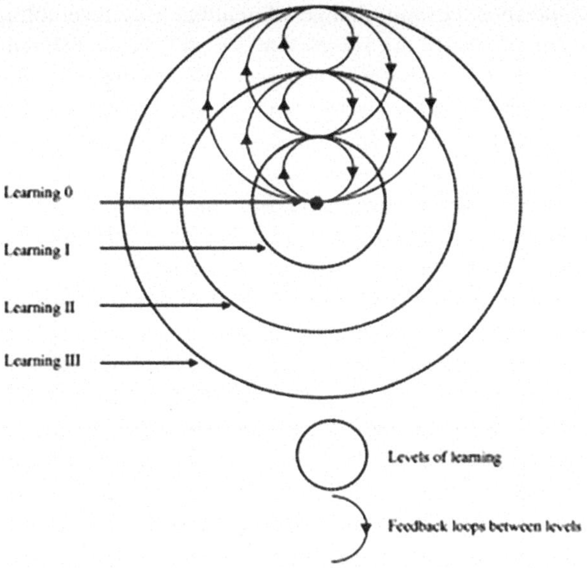

Fig. 5 Recursive hierarchy of Bateson's learning levels (Tosey et al., 2011: 296)

and nous = *cognizing/thinking mind* to describe a fundamental change or transcendence of the mind.[55] Isaacs was the first providing an appropriate treatment of L3 under the term *triple-loop learning*, denominating learning processes that—in addition to the two feedback loops on behavior (L1) and thinking frameworks (L2)—include the third level: the inquiry of and insight into the process of shaping human paradigms and therefore a new kind of attention and perception.[56] Starr and Torbert also take up the term and understand *triple-loop learning* as a shift in consciousness that requires total presence beyond mental thought processes.[57] Tosey et al. map these levels into a recursive hierarchy of learning loops (Fig. 5). Learning processes can take place on one, two, or three of the learning levels in parallel, whereby the learnings of each level influence each other. The third level, however, the existential reflection on the emergence of one's own identity up to self transcendence, forms a higher order of learning beyond cognitive abilities, which—interwoven with the other learning levels—takes the entire learning process to a new level.[58]

With respect to transdisciplinarity, we find here equivalents to Nicolescu's position of the "zone of non-resistance to our experiences, representations, descriptions, images and mathematical formulations" as the "hidden third between subject and object" which can be found between the different levels of reality but also beyond all levels.[59] While learning L1 and L2 constantly entails a certain broadening

[55] Senge (1990).
[56] Isaacs (1993: 30, 38).
[57] Starr and Torbert (2005: 87).
[58] Tosey et al. (2011:294ff.).
[59] Nicolescu, (2010: 17).

of perspective and change of reality, thus resembling a continuum, Nicolescu's levels of reality are stages that are subject to different rules. These are connected through the *zone of non-resistance* and at the same time are embedded in it. Thus, transdisciplinary learning can be understood as a learning process that involves learning level 3, i.e. includes existential reflection, taking the entire learning process and the learning outcomes to a new level: By diving into the realm beyond all levels of reality, a new level of reality can emerge.[60] So, a third starting point for new learning outcomes is the consideration of the third, existential level of learning, which brings the whole process to a new level.

Result of such transdisciplinary learning processes is what Nicolescu calls *transdisciplinary knowledge* or in vivo *knowledge,*[61] which aligns the internal and external world as well as mind, emotions, and body through a deeper understanding and integration of a value system. Real change and radically new solutions exactly require this consideration of the third level, so that integrative action, mind, and deepest intention can be brought into harmony with oneself and future potentials via all three learning levels. Peschl calls this process *individual cultivation* of a person or *radical-innovative change* of an organization. In such a mode, learning goes beyond processes of knowledge transfer or knowledge construction. It involves the deeper level of personal development and change, even beyond cognitive or personal skills and intellectual games.[62] Learners and their environment enter an active, dynamic process of mutual triggering, respecting, and deep understanding that can lead to mutual unfolding, a real change from within, a co-creation that realizes the deepest potential of all parties involved. Entering this level requires a radical openness and a deep willingness to suspension, receptivity, and redirecting,[63] thus a certain *non-resistance*. Scharmer describes knowledge emerging from these kinds of processes as *self-transcending knowledge*, or the *ability to see the coming-into-being of the new*.[64] For accessing this kind of knowledge, he proposes a technique called *presencing*, a combination of presence and sensing: changing one's own structure of attention in a way that one's own presence can be sensed, opening a space for engaging with our self and what we truly want to be. Non-essential aspects of the self can be dropped, and we can open up to new aspects of the highest possible future self.[65] Connecting these new future possibilities with the present, enabling new actions in the here & now, is what he calls *learning from the future*.[66] The more learning outcomes contain these aspects of going beyond oneself, the more likely it can be assumed that genuine, generative learning has taken place.[67]

---

[60] Nicolescu, (2010: 25).

[61] Nicolescu, (2008: 3).

[62] Peschl (2007: 138ff.).

[63] Peschl (2007: 143).

[64] Scharmer (2001: 137).

[65] Scharmer (2007: 163ff.).

[66] Scharmer (2007: 7 f).

[67] Kaiser and Peschl (2021: 537).

What does it mean for organizations to go to this existential level? How does a collective mind reflect the deeper level? Here, Schein's concept of organizational culture offers a well-known approach. Although there are differences between organizational culture and the concept of the collective mind these are not relevant within this context.[68] According to Schein, organizational culture emerges from the personal learning history of a team or organization:

> *The most useful way to arrive at a definition of [...] culture is to think in dynamic evolutionary terms, to think of culture as what the group has learned in its effort to survive, grow, deal with its external environment, and organize itself.*[69] *The culture of a group can be defined as the accumulated shared learning of that group as it solves its problems of external adaption and internal integration; which has worked well enough to be considered valid and, therefore, to be taught to new members as the correct way to perceive, think, feel, and behave in relation to those problems. This accumulated learning is a pattern or system of beliefs, values, and behavioral norms that come to be taken for granted as basic assumptions and eventually drop out of awareness.*[70]

According to Schein, there are three levels of cultural phenomena. These roughly correspond to the contents of the three levels of learning: 1) the level of artifacts and observable behavior, 2) the level of (more or less) conscious values and beliefs, and 3) the deepest level of basic unconscious assumptions. According to the five cultural dimensions of anthropology, the latter deal with assumptions about the human nature as such, about relations of humanity to nature, about relations of humans to other humans, ideas about truth and time, and about the nature of human action. For Schein, the level of basic assumptions is the actual, fundamental level of an organizational culture, thus determining identity. Existential reflection for transdisciplinary learning can and must therefore tackle this third level of organizational culture, e.g., as in the example mentioned earlier:

> *Who am I, thinking I have to market product x? What sense does this make for everyone involved and in the larger context?*

If the whole organization comes to new insights and new assumptions through this existential reflection, e.g., about the nature of human beings and their relationship to nature, and if these assumptions are not only reflected in single employees but are anchored in the entire common interaction space through organizational sense-making, radical changes and new organizational actions can emerge as a response to existential crises (see examples given in Sect. 5).

[68] Feldhusen (2018: 373).

[69] Schein (2016: 14 f.).

[70] Schein (2016: 6).

4 Transdisciplinarity and Transcendence: On the Way to Co-creation

Now, what does transdisciplinary organizational learning mean and where could be starting points? This chapter will try an answer by summarizing findings from the previous chapters. This paper described learning as a process that integrates differences via reflection to a broader or higher perspective, changing the learner himself. Learning uses existing ways of thinking and capacities to act (knowledge), and at the same time expands or transcends this self-context. Learning individually generated at learning levels 1 and 2 can expand contexts and bring up new individual perspectives and possibilities for action. This is where most learning processes in organizations (and education) take place. For radically new learning outcomes, we need additional new levels that go beyond:

Collective learning through multi-perspectivity goes beyond the individual by coherently combining different individual perspectives on reality into a new, overarching, shared perspective. Nonaka describes this as co-transcendence, transcending the individual context to the larger, collective space through encounter and deep engagement in a shared context ba:

> *Individuals interact with each other to reach out beyond the boundaries of their own existence and transcend the old self into a new self by acquiring a new context, a new view of the world and new knowledge, and as a result, change themselves, others, the organization and the environment.*[71]

Transdisciplinary learning with inclusion of Nicolescu's zone of non-resistance goes beyond the self by self-reflection on an existential level opening a higher level of reality. Scharmer describes this as self-transcendence, as dropping non-essential aspects of the self and opening to new higher aspects of the self through presencing, i.e. encountering the self beyond all cognitive structures.[72]

Conn[73] describes these two aspects of self-transcendence as anchored in the two basic psychological needs of human beings: *autonomy* (to be a self) and *relationship* (to transcend the self). On the one hand, we experience ourselves as an ego, a first-person reality constituted by consciousness; on the other hand, the fundamental need to go out of oneself, to grow beyond oneself, and to transcend one's own self is inherent in every human being. Conn sees the latter as the source of everything in the human being, which is expressed in all instances of true, creative insight. This fundamental need for self-transcendence can express itself in different ways. In cases where it relates to overcoming individual isolation, and spontaneous action is directed not only toward oneself but also toward the welfare of others, Conn

---

[71] Nonaka and Toyama (2005: 421).

[72] Scharmer (2007: 180ff.).

[73] Conn (1998).

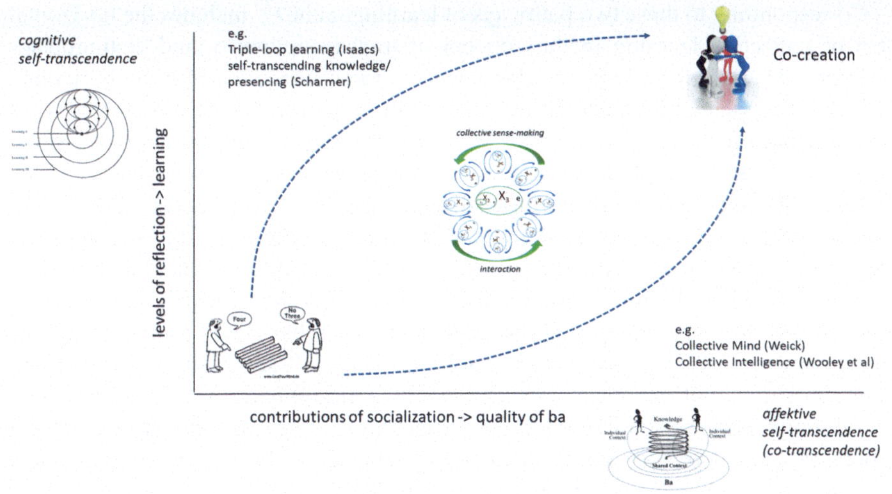

Fig. 6 Two pathways of learning toward higher knowledge and co-creation

speaks of *affective self-transcendence*. In cases where it refers to the unfolding of different levels of consciousness, Conn speaks of *cognitive self-transcendence*.[74]

The pathways of learning outlined above, *collective learning by co-transcendence* as well as *transdisciplinary learning by self-transcendence*, can thus be seen as two basic human pathways or modes of human learning. Fig. 6 illustrates them in a two-dimensional model of learning: The vertical axis refers to the *unfolding of levels of consciousness* and therefore maps the three different *levels of reflection* in learning processes L1–L3. There, *existential reflection* and *presencing*—as access to the *zone of non-resistance*—represent the third and highest form of learning and development of consciousness (*Scharmer's self-transcendence*). The horizontal axis refers to the *movement from the I to the we* and therefore maps the different manifestations of a *ba* dependent on the quality and density of the *contributions of socialization*. There, *ultimate basho or collective improvisation* represents the highest quality of a shared knowledge space where the individual consciousness, the self, is fully transcended and merged in the context of the greater whole (*Nonaka's co-transcendence*).

Development along one axis implicitly leads to some development along the other axis. Both pathways of learning lead to the same goal, which we have described above as *co-creation*: collective, transdisciplinary learning of a group, community, or organization, in which conscious individuals grow beyond themselves through a strong we. This means they can recognize future potentials with regard to a common concern and raise them into reality, so that new, meaningful possibilities for action emerge, also and especially in the face of existential crises and challenges.

[74] Conn (1998: 323ff.).

Corresponding to these two pathways of learning, Schüz[75] includes the horizontal path of collective learning in the concept of transdisciplinarity and distinguishes between a *horizontal and a vertical movement of transdisciplinarity*: the horizontal movement supplements gained knowledge with new perspectives from diverse ways of experience but remains on one level of reality. Science-society collaboration gives an example, as it complements scientific knowledge with the experiences of lay persons. The vertical movement of transdisciplinarity, however, reflects reasons and conditions of existing reality via transcendent forms of consciousness. In the example given, artistic and spiritual modes of experience would be integrated beyond scientific thinking. In this comprehensive sense, transdisciplinarity represents a process that triggers learning through encountering the other (horizontally) or encountering oneself and one's Higher Self (vertically). It describes the overcoming of one's own field of reference, be it a paradigm, a discipline, a theory, or one's own self through horizontal (affective) or vertical (cognitive) movement. Horizontal processes can allow new contexts (learning level 2) to be integrated into one's own thinking, thus expanding and adapting it. Vertical processes can question or even abandon the entire context of thinking. The highest perspective can be thought of as a non-perspective, the absolute, the all-one thinking in which all diversities are integrated (*ultimate basho, zone of non-resistance*). Transdisciplinary, organizational learning is thus a self-transcending process in horizontal and vertical aspects that goes beyond the learners and the organization itself, and that is accessible through three starting points:

Creating new differences in the learning process through multi-perspectivity as a horizontal starting point for new learning outcomes in existing organizational structures and cultures: recognizing that knowledge has subjective character and that inclusion of multiple experiences and perspectives is fruitful for one's own learning process promotes a certain "intellectual humility" and awakens the aspiration to include multiple perspectives individually or organizationally in search for solutions. In addition to horizontal expansion, opening to new perspectives can also provide an impetus for vertical movement to the (individual or organizational) learning level 2.

Reflecting patterns of interaction and their underlying attributions of meaning as a starting point for a vertical movement to organizational learning level 2: Realizing that organizational knowledge and action is based on shared attributions of meaning which again results in respective organizational structures and cultures suggests that these very structures and cultures should be questioned and adapted with regard to the joint concern as well as challenging this concern itself with regard to its power to bond people for a common knowledge space ba (we-intention). This questioning can also provide the impetus for a vertical movement toward learning level 3.

Existential reflection of individuals as a vertical starting point on learning level 3: Realizing that a reflection on the source from which people act and the exploration of their deepest concerns raises the whole learning process onto a new level suggests

[75] Schüz (2024: 5, 14).

searching for access to this source beyond the self. Individual insights gained in this process can pour in the basic assumptions of the organizational culture and thus embed the organizational identity and its actions in the context of the greater whole.

Supposing that all individuals have the full capacity for self-transcendence, i.e., they can reflect and continuously adapt not only the individual thought structures but also the interaction structures of the organization through 3rd level learning and existential reflection creates the image of an *organization with the capacity for collective self-transcendence*. This represents the perfect idea of a permanently learning organization, i.e., an *agile organization* as it is capable of continuously adapting its structures of thought, communication, and action to its environment without losing sight of its own meaning & purpose and its connection to a greater whole.[76]

5 Transdisciplinary Organizational Learning in Practice

To cope with an increasingly volatile world, organizations have been experimenting quite a while now with forms of learning and organizing that apply one or more of the above-mentioned approaches. Various examples can be found which anchor transdisciplinary elements in learning cultures, helping in their own way to address the crises of the organization mentioned at the beginning:

5.1 Multi-perspectivity and Collective Thinking

When different perspectives shall be integrated into a higher, better solution, all relevant and/or involved perspectives must first be present, heard, and exchanged. Various approaches already exist using specific conversation cultures and interaction structures to establish shared, fruitful, and creative thinking, e.g.

Spaces of practice for dialogical conversation skills: William Isaacs, co-founder of the Organizational Learning Center at MIT's Sloan School of Management, developed a trainable conversational culture within the Dialogue Project according to David Bohm, and used it successfully in organizational development for 30 years (www.dialogos.com/about/overview). It is based on practicing 4 basic dialogue competencies that mitigate the usual self-centered and reflexive reactions in communication and enable true dialogue: 1. listening and asking questions, being open to hearing something unexpected from other persons, recognizing and questioning one's own assumptions; 2. respecting divergent perspectives and conceding them to their owners; 3. letting go of our own opinion, stepping back and changing our focus or direction; and 4. speaking from an inner authority, giving up the need for

[76] Feldhusen (2014).

dominance. Practicing these competencies regularly in the team allows this dialogical attitude to flow into everyday organizational life, for example in usual meetings.

Circular discussion structure: Organizational frameworks such as Sociocracy (https://sociocracy30.org) or Holacracy (https://www.holacracy.org) use circular opinion forming with a predefined discussion structure for preparing certain decisions. Given a specific issue referring to a common goal, all perspectives involved are heard in turn and brought into interaction in a culture of listening. Contributions to the discussion are made in a circle. There is only one person to speak at a time in a kind of monologue. The other participants concentrate exclusively on listening. This ensures that every position is heard, but also that no one can dominate and give their perspective relatively more weight. The process runs in several rounds to ensure integration of the different contributions. Individual perspectives often broaden or change simply by perceiving an additional aspect from other contributions. This change then can be presented in the next round of talks. Usually after three rounds, a common perspective has emerged, which encompasses all individual aspects. A facilitator summarizes the final position, which is again voted on in the circle in a certain way, i.e., by "consent." This means a "no" is only accepted when the voter can give a justified, serious objection with respect to the common goal. In this (rather rare) case, an aspect seems to be missed in the preliminary rounds. This can be the subject of a new opinion-forming circle.

Workshop formats for raising collective intelligence:[77] Starting point is a common issue or question, which affects all participants. First, inputs from a wide range of perspectives are worked out individually, which are then shared and systematically condensed over several rounds of work with fine-tuned and reconciled interventions. This can, and often does integrate individual knowledge into true breakthrough solutions. An example in the field of creative teamwork is the Creative Rollercoaster by Udall (www.now-here.com, Udall 2014), an example of a cross-organizational large-scale approach is Syntegration by Malik (www.malik-management.com/de/malik-loesungen/malik-tools-and-methods/malik-supersyntegration-mss).

5.2 *Reflection of Interaction Patterns*

Practices for questioning and adapting interaction patterns can be installed at the organizational or team level or within meeting or certain decision-making processes. Examples for the ongoing reflection and, if necessary, adaptation of organizational structures can be found, for example, at the Gore company. There are no fixed positions ("no ranks no titles"), but each employee has the possibility to be involved in certain decisions, if he/she has valuable competences with respect to the goal or the decision to be made. For example, the receptionists can be consulted within

[77] Feldhusen (2021), Woolley et al. (2010, 2015).

recruitment processes based on their knowledge of human nature. At Gore, leadership is revocable and reversible. A leader is "chosen," i.e., results from everyday interaction and is determined by the trust of those who follow them.[78] Frameworks like sociocracy or holacracy (see above) have integrated this attitude in a systematic way. Responsibility for and distribution of tasks among staff members ("roles") can be restructured easily if necessary ("separate role from soul"). Roles are assigned by election (sociocracy) and are reviewed regularly with respect to the common goals (holacracy) at intervals of about 6 weeks in especially dedicated meetings. The agile framework Scrum provides a form of team reflection that can be easily integrated into every day organizational life: at fixed intervals, the team sits down for a "retrospective," which is exclusively about reflecting on the quality of the cooperation (such as communication attitudes and structures), not about the results of the cooperation in terms of content.

5.3 Reflecting on Underlying Sense-making

Most of the reflection and learning formats mentioned above refer to a common purpose. This brings up the question how to spark convergent sense-making to create a common, unifying purpose in a time where corporate goals are increasingly unclear and weak in bonding. The concept of purpose-driven organization delivers an approach as it consequently aligns the organization through focus on the corporate purpose which is defined by uncompromised customer and market orientation. A step further and deeper takes VIKOBAMA, a vision development process developed at the Vienna University of Economics and Business Administration. Outcome is meant to be a corporate vision that, in addition to market orientation, is also coherently linked to the personal life visions of individual employees (https://www.wu.ac.at/en/kbm/projects/vikobama). In a first step, each employee personally develops a deeply affirmed, forward-looking vision for him/herself, then, out of this strengthening experience, creates a vision for the company. The personal visions remain confidential, but the various company visions are shared and crystallized into a joint vision through systematic interventions. A corporate vision generated in this way represents a shared concern in the sense of Peter Senge's core disciplines for true learning organizations.[79] Regularly setting up such spaces—ba's—where different perspectives on the common concern can be reconciled again and again anew, keeps the common intention or corporate purpose alive. A good example is the consultancy company moocon (https://www.moo-con.com): twice a year, the whole organization meets for a two-day "campfire," called "I & We," where attitude and orientation of the organization is updated with and through the exchange of all employees. Examples and guidelines for an everyday level provide the concept of

[78] McKinsey (2008).

[79] Senge (1990: 251).

Art of Hosting (https://artofhosting.org/), a practitioners' network dedicated to meaningful conversation. For each and every meeting, it proposes to clarify the common goal of the meeting, so that participants can refer to it in case of differences or decisions.

5.4 Triple Loop-learning through Existential Reflection

In search for growth and innovation, trend researchers and innovation managers recognized human imagination and intuition as an economic knowledge resource already in the 1980s. Concepts on how to exploit this resource within the corporate context emerged at major institutions, such as MIT (Mintzberg's Emergent Strategy[80]), the London Business School (Hamel's Corporate Imagination[81]), or Berkeley (Teece's Sensing & seizing of possibilities[82]). However, learning from the future or triple-loop learning through existential reflection goes far beyond these concepts and cannot be taught as an academic concept or seminar content. A spontaneous, short-term access to the field beyond our thoughts, followed by a flash of inspiration, can be known by almost everyone. However, to consciously expand and cultivate this access for creating new possibilities from an inner transformation goes to the core of our self-image as human beings and requires a holistic approach to personal development with regular practice, mental time-outs, and a view of the big picture. Knowledge and innovation architects from The living core (www.thelivingcore.com) use precisely this in their work on innovation with companies. In a process that lasts several months they develop radical innovations that go far beyond existing thinking. It is also the approach of the Conscious Consulting Group (www.ccg-group.eu) to question certain basic assumptions so fundamentally that completely new realities can emerge. Their approach combines science and wisdom traditions for this purpose. The Upstalsboom hotel group (www.upstalsboom.de) provides a corporate example: Shortly after taking over the management from his father, owner Bodo Janssen was confronted with high fluctuation, high sick leave, and a decreasing number of applicants. A staff survey delivered devastating results for the management team. Shaken in his self-image, he withdrew to a monastery for six months for serious self-reflection and experienced a profound change in his understanding of leadership. Guided by the attitude that leadership should be a service to the system, his behavior changed in such a way that it also brought about change to the organization. Finally, he invited all employees to the monastery, which resulted in a company seminar on sustainable, value-led leadership, which is open to all new employees. In addition, common values such as trust, appreciation, openness, and responsibility were crystallized in a workshop and anchored in the mission

[80] Mintzberg (1985).

[81] Hamel and Prahalad (1991).

[82] Teece and Pisano (1994).

statement, which resulted in a deep bonding among the employees. While the hotel industry generally suffers from a shortage of staff, Upstalsboom is now receiving a flood of qualified applications, the sickness rate has dropped from 8% to 3% and both employees' satisfaction and earnings have increased (www.der-upstalsboom-weg.de/der-upstalsboom-weg/die-geschichte

Each of the examples given contributes in their own manner to addressing the crisis of the organization as outlined at the beginning. Each of it provides its own new answers to new challenges. Realizing that truly new solutions emerge in specific human interaction structures like co-creation highlights the value of humans and their social interaction compared to artificial intelligence. Shared processes of sense-making have the power to create goals and visions that can connect, align, and coordinate human activities. If these even engage with existential qualities, it can lead the organization to a proactive and constructive learning process that can contribute sustainable, fruitful solutions in the sense of a "*future-making practice*" at the overall societal and ecological level.[83]

6 Conclusion: Collective Learning on a New Level through Transdisciplinarity

An increase in data, faster evaluations, more complex analysis, or increased efficiency, i.e., learning 1, will obviously not deliver solutions to current challenges and multiple crises, if we persist in the same logic of thinking. As early as 2009, a distinguished group of scholars led by management thinker Gary Hamel called for essential *moonshots for management*, i.e., landings in territories outside our existing frameworks of thought and action, to survive in tomorrow's volatile world. Their suggestions include systems thinking, appreciating diversity, connecting to a higher meaning and purpose, and using human imagination as a source of knowledge to create a desired future.[84] A transdisciplinary view supports this call for growth as it expands the understanding of organizations and their capacity to learn into the mentioned direction. It leads to fundamental questions on what happens within and between people, in their structures of consciousness and interaction. Rather than a collection of employees and managers put together and functioning like a set of gears, organization can be seen as a living organism or collective mind that co-creates and shapes the world as it copes with its experiences and challenges. Sociologist of knowledge Nico Stehr puts it this way: knowledge is no longer just a means to unlock the secrets of the world, but itself represents a world in the process of being created.[85] In times of complex challenges, it is important for organizations to grow beyond themselves into a co-creative community that commits itself to a

[83] Peschl (2022).

[84] Hamel (2009).

[85] Stehr (2001).

new, higher perspective by incorporating a wide variety of perspectives and, by accessing the space beyond all known realities, creates new realities in favor for all participants. Types of knowledge that can recognize and unfold potential play a special role in this process. Subjective forms of knowledge such as human imagination, intuition, and vision must be made accessible in the sense of learning from the future and helped to become effective (according to Art. 11 of the Charter of Transdisciplinarity). Furthermore, for orchestrating and using distributed knowledge, organizations need collective learning at level 2, which questions existing organizational and value structures (cf. Art. 3 of the Charter of Transdisciplinarity). To be truly adaptive, what is currently addressed by concepts of agility, does not mean to exhaust existing paradigms with ever more efficient methods, but to be adaptive and willing to learn in a comprehensive sense: to continuously adapt processes and structures through ongoing internal negotiation. Above all, for facing existential crises and threats, organizations need to explore learning on level 3, which tackles questions about its higher meaning and purpose—far beyond the current trend of purpose-driven organizations—and embeds the identity of the organization in larger contexts of existence, encompassing humans as well as nature (according to Art. 5 + 8 of the Charter of Transdisciplinarity). Possibly, this can lead to overcoming the above-all-status of profit maximization (according to Art. 12 of the Charter of Transdisciplinarity). In their dynamic capabilities approach, Augier and Teece already outline an economic paradigm of co-specialized economies in which firms and markets co-evolve. They argue, rational optimization assumptions are insufficient for bringing non-tradable resources to their optimal place, thus profit-seeking instead of maximization is required.[86] And, companies need a higher perspective, as excellence can only arise from an orientation toward the well-being of the company, its employees, its customers, and other stakeholders, as well as society as a whole. Profit is not an end in itself, but the result of such comprehensive value creation.[87]

Finally, humility and a willingness to see ourselves as part of the problem are key to starting the process of higher learning. Starting with ourselves by understanding reality as our own construct allows us to realize that individual, as well as collective, desires to be absolutely right seems absurd; we only become wiser together through plurality[88] and by accessing sources beyond our cognition. In this way, we can begin to shape a future that is meaningful, fruitful, and sustainable for the system as a whole and for all those involved.

[86] Augier and Teece (2009: 412, 416).

[87] Nonaka et al. (2008: 3); Nonaka and Takeuchi (2021); see also concept of shared value according to Porter and Kramer (2011).

[88] Wüthrich (2022: 17ff.).

References

Argyris, C. (1993). *Knowledge for action*. Jossey-Bass.

Argyris, C., & Schön, D. A. (1978). *Organizational learning: A theory of action perspective*. Addison-Wesley Publishing Company.

Argyris, C., & Schön, D. A. (1996). *Organizational learning II. Theory, method, and practice*. Addison-Wesley Publishing Company.

Augier, M., & Teece, D. J. (2009). Dynamic capabilities and the role of managers in business strategy and economic performance. *Organization Science, 20*(2), 410–421.

Bateson, G. (1983). *Ökologie des Geistes. Anthropologische, psychologische, biologische und epistemologische Perspektiven*. Suhrkamp. (Original 1972: Steps to an ecology of mind. Collected essays in anthropology, psychiatry, evolution and epistemology, Chandler Publishing Company).

Chia, R. (2003). From knowledge-creation to the perfecting of action: Tao, Basho and pure experience as the ultimate ground of knowing. *Human Relations, 56*(8), 953–981.

Churchman, C. W. (1971). *The design of inquiring systems: Basic concepts of systems and organization*. Basic Books Inc.

Clark, A., & Chalmers, D. (1998). The extended mind. *Analysis, 58*, 7–19.

Conn, W. E. (1998). Self-transcendence, the true self, and self-love. *Pastoral Psychology, 46*(5), 323–332.

Erden, Z., et al. (2008). The quality of group tacit knowledge. *Journal of Strategic Information Systems, 17*(1), 4–18.

Feldhusen, B. (2014). *Organizing future: An integrated framework for the emergence of collective self-transcending knowledge*. Dissertation. Wirtschaftsuniversität Wien.

Feldhusen, B. (2018). Collective mind. Wie wir Organisationen täglich neu ins Leben rufen. Zeitschrift Führung + Organisation 06/2018, Schäffer Poeschel.

Feldhusen, B. (2021). *Kollektive Intelligenz und Psychologische Sicherheit: Haben wir Intelligenz im Gefühl? Organisationsberatung, Supervision, Coaching (OSC) 03/2021* (Vol. 28, p. 355). Springer Verlag.

Glasersfeld, E. V. (1984). An introduction to radical constructivism. In P. Watzlawick (Ed.), *The invented reality*. Norton.

Hamel, G. (2009). Moon shots for management. *Harvard Business Review, February*, 91–98.

Hamel, G., & Prahalad, C. K. (1991). Corporate imagination and expeditionary marketing. *Harvard Business Review, July-August*, 81–92.

Hutchins, E. (2011). Enculturating the supersized mind. *Philosophical Studies, 152*, 437–446.

Isaacs, W. N. (1993). Taking flight: Dialogue, collective thinking, and organizational learning. *Organizational Dynamics, 22*(2), 24–39.

Janis, I. (1972). *Victims of groupthink: A psychological study of foreign-policy decisions and fiascoes*. Houghton Mifflin.

Jordan, S., Messner, M., & Becker, A. (2009). Reflection and mindfulness in organizations: Rationales and possibilities for integration. *Management Learning, 40*(4), 465–473.

Kaiser, A., & Peschl, M. F. (2021). The role of self-transcending knowledge in Senge's understanding of learning organizations – Towards an interdisciplinary taxonomy of self-transcending knowledge. *The Learning Organization, 27*(6), 527–539.

Kieser, A., & Walgenbach, P. (2010). Organisation, 6., überarbeitete Auflage, Schäffer Poeschel Stuttgart.

Kirchhoff, M. D., & Newsome, W. (2012). Distributed cognitive agency in virtue epistemology. *Philosophical Explorations: An International Journal for the Philosophy of Mind and Action, 15*(2), 165–180.

Kolb, D. A. (1984). *Experiential learning: Experience as the source of learning and development*. Prentice-Hall, Inc..

Krippendorff, K. (1984). An epistemological foundation for communication. *Journal of Communication, 34*(3), 21–36.

McKinsey, (2008). Von der Amöbe lernen. Mc Kinsey Wissen 08/2008.
Mintzberg, H. (1985). Of strategies, deliberate and emergent. *Waters, James A, 6*(3), 257–272.
Nicolescu, B. (2008). In vitro and in vivo knowledge – Methodology of transdisciplinarity. In: Nicolescu, B. (Ed.), *Transdisciplinarity – Theory and practice*. Hampton Press, pp 1–21.
Nicolescu, B. (December, 2010). Methodology of transdisciplinarity – Levels of reality, logic of the included middle and complexity. *Transdisciplinary Journal of Engineering & Science, 1*, 17–32.
Nonaka, I., & Takeuchi, H. (1995). *The knowledge-creating company: How Japanese companies create the dynamics of innovation*. Oxford University Press.
Nonaka, I., & Takeuchi, H. (2021). Humanizing strategy. *Long Range Planning, 54*(4), 102070.
Nonaka, I., & Toyama, R. (2005). The theory of the knowledge-creating firm: Subjectivity, objectivity and synthesis. *Industrial and Corporate Change, 14*(3), 419–436.
Nonaka, I., Toyama, R., & Konno, N. (2000). SECI, Ba and leadership: A unified model of dynamic knowledge creation. *Long Range Planning, 33*, 5–34.
Nonaka, I., Toyama, R., & Hirata, T. (2008). Managing flow. In *A process theory of the knowledge-based firm*. Palgrave.
North, K. (1998). *Wissensorientierte Unternehmensführung*. Gabler Verlag.
Peschl, M. F. (2007). Triple-loop learning as foundation for profound change, individual cultivation, and radical innovation. Construction processes beyond scientific and rational knowledge. *Constructivist Foundations, 2*(2–3), 136–145.
Peschl, M. F. (2022). Learning from the future as a novel paradigm for integrating organizational learning and innovation. The Learning Organization 2022, pp 1–17.
Porter, M. E., & Kramer, M. R. (2011). Creating shared value. *Harvard Business Review, January-February*, 62–77.
Scharmer, C. O. (2001). Self-transcending knowledge: Sensing and organizing around emerging opportunities. *Journal of Knowledge Management, 5*(2), 137–150.
Scharmer, C. O. (2007). *Theory U. Leading from the future as it emerges*. SOL.
Schein, E. (2016). *Organizational culture and leadership* (5th ed., Jossey-Bass Business and Management Series). Wiley & Sons.
Schüz, M. (2024). Transdisciplinary management of polycrises: A philosophical approach. In M. Schüz (Ed.) *Transdisciplinary thinking and acting: Fundamentals and perspectives*, in this volume pp.3–32. Springer.
Seligman, M. E. P., et al. (2013). Navigating into the future or driven by the past. *Perspectives on Psychological Science, 8*(2), 119–141.
Senge, P. (1990). *The fifth discipline*. Doubleday Business.
Singer W (2016) *Neuronale Grundlagen des Bewusstseins*. Gastvortrag an der Universität für Angewandte Kunst Wien, 07.09.2016. https://www.dieangewandte.at/termine/wolf_singer_neuronale_grundlagen_des_bewusstseins_07-09-2016 (letzter Abruf 11.11.2022)
Stacey, R. (2000). The emergence of knowledge in organizations. *Emergence, 2*(4), 23–39.
Starr, A., & Torbert, B. (2005). Timely and transforming leadership inquiry and action: Toward triple-loop awareness. *Integral Review, 1*, 85–97.
Stehr, N. (1996). *Knowledge as a capacity for action. Science and technology redesign project*. Statistics Canada.
Stehr, N. (2001). A world made of knowledge. *Society, 39*(1), 89–92.
Teece, D. J. (2009). *Dynamic capabilities & strategic management: Organizing for innovation and growth*. Oxford University Press.
Teece, D. J., & Pisano, G. (1994). The dynamic capabilities of firms: An introduction. *Industrial and Corporate Change, 3*(3), 537–556.
Theiner, G., Allen, C., & Goldstone, R. L. (2010). Recognizing group cognition. *Cognitive Systems Research, 11*(4), 378–395.
Tosey, P., Visser, M., & Saunders, M. N. (2011). The origins and conceptualizations of "tripleloop" learning: A critical review. *Management Learning, 43*(3), 291–307.

Tsoukas, H. (2000). Knowledge as action, organizations as theory: Reflections on organizational knowledge. *Emergence, 2*(4), 104–112.

Tsoukas, H. (2005). New times, fresh challenges: Reflections on the past and the future of organization theory. In H. Tsoukas (Ed.), *Complex knowledge* (pp. 378–393). Oxford University Press.

Virtanen, I. (2013). In search for a theoretically firmer epistemological foundation for the relationship between tacit and explicit knowledge. *The Electronic Journal of Knowledge Management, 11*(2), 118–126.

Weick, K. E., & Roberts, K. H. (1993). Collective mind in organizations: Heedful interrelating on flight decks. *Administrative Science Quarterly, 38*, 357–381.

Weick, K. E., Sutcliffe, K. M., & Obstfeld, D. (2005). Organizing and the process of sensemaking. *Organization Science, 16*(4), 409–421.

Woolley, A. W., et al. (2010). Evidence for a collective intelligence factor in the performance of human groups. *Science, 330*(6004), 686–688.

Woolley, A. W., Aggarwal, I., & Malone, T. W. (2015). Collective intelligence and group performance. *Current Directions in Psychological Science, 24*(6), 420–424.

Wüthrich, H. A. (2022). *Manifest der intellektuellen Bescheidenheit. Problemlösung neu denken*. Versus Verlag Zürich. Vahlen Verlag München.

Dr. Birgit Feldhusen graduated in 1995 in business administration at Kiel University, Germany. Afterwards, she gained 15 years of experience in the international consumer goods business, where she held roles in auditing, controlling, marketing, and consulting. Joining academia and science again in 2010, she received a Ph.D. from the Vienna University of Economics and Business in 2014 for research on Emergence of Collective Self-transcending Knowledge. In 2015, she founded "organizing future," an institute offering talks, lessons, key notes, and other formats for new perspectives on organizing and leading the future. In 2018, she joined the Department for Management and Economics at the University for Continuing Education Krems, near Vienna, where she established the division "future organizations" and, in 2020, took over the Center for General Management and its operational responsibility for central MBA programs of the university's business school.

Transdisciplinarity through the arts

Annina Boogen and Fernando Obieta

1 Introduction

Discipline—the unity of science—comes from the Latin word *disciplina*: It refers to education and training, self-control and determination, knowledge in a field of study, and an orderly way of life, adherence to certain rules, fitting into the order of a community. Following this, an academic discipline also follows certain discipline specific methods, (unwritten) rules, and shows some ways of self-control such as peer-review processes. This guarantees the scientific rigour, the "quality" of scientific outputs.

These specific methods, (unwritten) rules, and ways of self-control are part of a *thought collective* held together by a *thinking style*.[1] Fleck does not speak of "disciplines" but of "thought collectives" instead. By this he means a community that exchanges ideas among its members. If this continues for a longer period of time, a specific style of thinking emerges for a particular thought collective. A "trained style of thinking" not only makes it possible to see certain things, it also excludes "seeing other things", since a thought collective exhibits existing—or even entrenched—patterns of perception.

[1] Fleck (1983).

A. Boogen (✉)
Center for Energy and the Environment, School of Management and Law, Zurich University of Applied Sciences (ZHAW), Winterthur, Switzerland
e-mail: annina.boogen@zhaw.ch

F. Obieta
School of Design, Film and Art, Lucerne University of Applied Sciences and Arts (HSLU), Lucerne-Emmenbrücke, Switzerland
e-mail: fernando.obieta@hslu.ch

© The Author(s), under exclusive license to Springer Nature Switzerland AG 2025

M. Schüz (ed.), *Transdisciplinary Thinking and Acting*, Transdisciplinary Management of Ecological and Social Crises,
https://doi.org/10.1007/978-3-031-70680-6_16

In the 1990s, an immensely increasing particularisation of the disciplines was claimed by Jürgen Mittelstraß: Borders of disciplines are increasingly threatening to become not only institutional borders, but also borders of knowledge.[2] This plays an important role for complex issues such as societal and ecological challenges that affect a wide variety of actors. Here disciplinary thinking only reaches a certain point, since it can only use its disciplinary lens with its methodological filters and its trained narrow style of thinking. Therefore, to tackle such complex issues transdisciplinary approaches are needed.[3] And this is because, a complex issue can only be dealt with using a more holistic lens—by "seeing other things"—in order to reach another level, thus researchers need to *transcend the disciplinary*.

> One cannot simply and immediately see something new and changed. First, the whole style of thinking has to change, the whole intellectual mood has to shake, the violence of directed thinking has to stop. A specific intellectual restlessness and a change in the moods of the thinking collective must arise, that only creates the possibility and the necessity to see something new and changed.[4]

But what means to work with a transdisciplinary approach? And what means to research through the arts? In this chapter, we lay out why it is a good idea to make use of artistic practices for acquiring a transdisciplinary stance. First, in a theoretical discussion, we discuss that artistic practices allow affective processes, and affective thinking may lead to individual and collective situatedness to be revealed and put up for negotiation. This allows a reflective process to happen. Following, we elaborate three examples that come from different angles and perspectives. Note, that these examples are not an exhaustive list of approaches for transdisciplinary work through the arts. Finally, we conclude with an appeal.

2 The arts as a Place of Reflection and Action on Social and Ecological Crises

The arts differ from the majority of other disciplines in that their practices are inherently tied to their authors and their products also remain tied to them in the perspective of the viewer. Works of art are always an expression of the perspective of their creators. This, of course, applies to any humanly created product, but in the arts this connection always remains in focus. These affective ways of working for

---

[2] Mittelstraß (2003).

[3] OECD (2020).

[4] Own translation from: "*Man kann etwas Neues und Abgeändertes nicht einfach und sofort sehen. Zuerst muss sich der ganze Denkstil verändern, muss die ganze intellektuelle Stimmung ins Wanken kommen, muss die Gewalt der gerichteten Denkbereitschaft aufhören. Es muss eine spezifische intellektuelle Unruhe und eine Wandlung der Stimmungen des Denkkollektivs entstehen, die erst die Möglichkeit und die Notwendigkeit dazu schafft, etwas Neues, Abgeändertes zu sehen*" (Fleck, 1983:78).

reflection, critique, speculation, and experimentation always confront the interpretations, associations and further affections of other perspectives and make them the central negotiation.

> Traditionally, the arts are the site of a preoccupation with affectations, the accidental and the incommensurable, so that in them 'a culture of action and a culture of work can refract and reflect'.[5]

Artists need to refract, disassemble, and/or at least partially understand (in varying degrees) the fields, disciplines, and/or practices that they're exploring to be able to be affected and commence with their own practices.[6] Thus, artistic practices often inherently employ transdisciplinary notions to achieve their results. This, in the context of transdisciplinarity—which, importantly, can also be carried out by non-artists—represents an affective approach to a given or chosen object of investigation, in order to reflect on one's own and collective thinking.

> This reflexivity [of art] takes place in particular by situating perception in perception, by turning the medial against itself and by unsettling implicit references in order to, in ever-new ways, bring that to light which is hidden by (the realm of) technē.

Central to this is that the reflexivity of the arts is not limited to the creators who carry out the artistic practices, but also extends to the viewers. This can be the visitors of an exhibition, but outside of art institutions it also concerns other contributors, colleagues, project partners, etc.: The disclosure of affective thoughts enables thinking beyond established disciplines and axioms.

> It is crucial here to remember that the point is not to go beyond those contradictions towards some kind of an inspired or transcendent vision, mysteriously able to discover a unified reality. [. . .] Any strategy of explaining away, of reducing some aspects of our experience to selected others, has to be resisted. In other words, the demand for coherence will authorize no simplification, no purification or selection of what would really matter. Everything we experiment must matter. As I said, friction must be maximised. Again, it is not a question of criticizing our specialized abstractions, or of dictating limitations. Rather of dramatizing them as achievements with a price [. . .] to produce both a restraint upon specialists and an enlargement of their imagination.[7]

Artistic practices pursue the tangibility of friction—for the creators themselves, others, or the audience of an exhibition. In this sense, transdisciplinarity through the arts and artistic practices does not aim to open up comprehensive new truths or

[5] Own translation from: *"Traditionell sind die Künste der Ort einer Beschäftigung mit Affektionen, dem Zufälligen und Inkommensurablen, sodass sich in ihnen 'eine Kultur der Tat und eine Kultur des Werkes brechen und reflektieren' kann"* (Busch, 2012: 16). Busch quotes: (Vogl, 2007: 24).

[6] Mersch (2015: 145). There are also artistic practices that focus on their own craft without incorporating other disciplines or topics. This article focuses on artistic practices which also incorporate a conceptual approach by (also) reflecting upon other topics and discourses—e.g. social and ecological crisis.

[7] Stengers (2008: 6).

realities, but to put individual, collective, and social positions up for negotiation, and not to rely on specialised ways of thinking of (different) disciplines, but to actively question and reflect them in other ways, with other spaces of thinking and possible conclusions. This allows each discipline and its practices to be understood as cultural practices, always linked to the "cultural capital"[8] of their practitioners and of the disciplines themselves—including the arts.

This approach allows not only to make one's own affective thoughts accessible—to others and to oneself—but also to reveal the connections with one's own disciplines, cultural capital, ideologies, virtues, and morals. This will never be an exhaustive or comprehensive revelation, but an expanded one in contrast to established practices. The point is that each practice is in relation to others and cannot be considered in isolation from them. Haraway proposed her concept of *situated knowledges* to think about the act and practice of positioning, to question the sanctioning of where something is true or not, and to consider the origin and surrounding positions, knowledges, and thinking, that led to such a sanction:

> I am arguing for politics and epistemologies of location, positioning, and situating, where partiality and not universality is the condition of being heard to make rational knowledge claims. These are claims on people's lives. I am arguing for the view from a body, always a complex, contradictory, structuring, and structured body, versus the view from above, from nowhere, from simplicity.[9]

When expressed through artistic practices, the focus remains on the embodied perspective. The question of rationality is left out; contradictions and complexities are exposed; affective thoughts become tangible, allowing them to be examined.

> Situated knowledges are about communities, not about isolated individuals. The only way to find a larger vision is to be somewhere in particular. The science question in feminism is about objectivity as positioned rationality. Its images are not the products of escape and transcendence of limits (the view from above) but the joining of partial views and halting voices into a collective subject position that promises a vision of the means of ongoing finite embodiment, of living within limits and contradictions—of views from somewhere.[10]

Summarising, artistic practices—in contrast to most disciplines that strive for a "view from above"—can open up a particularity and collectivity that enables a transcendence of disciplines. Habitus and mutual disciplinary affirmation are no longer necessary and are replaced by the recognition of situatedness. Artistic practices can therefore lead to a transdisciplinary stance that affectively addresses social and environmental crises.

---

[8] For cultural capital, see: Bourdieu (1987).

[9] Haraway (1988: 589).

[10] Ibid: 590.

3 Three Examples

The stance elaborated in the previous subchapter brings forth diverse results, perspectives, and approaches. In order to make three major streams accessible—1. investigation and communication; 2. participation and activism; 3. artistic research—three examples are discussed in detail below.

3.1 *Forensic Architecture*

Using architectural analysis to investigate human rights violations including violence committed by states, police forces, militaries, and corporations, Forensic Architecture (FA) works at the overlap of investigative journalism, research, and the arts.[11] FA was founded by Eyal Weizman in 2010 and is based at Goldsmiths, University of London, where Weizman is a Professor of Spatial and Visual Cultures. Weizman, originally an architect, researched Israeli architecture on the West Bank for his dissertation and realised how much he could learn about architecture and state power from satellite imagery. Nowadays he leads a team of over a hundred researchers, software developers, filmmakers, investigative journalists, artists, and lawyers. The project teams often use satellite images, photography, or surveillance videos as source material, but "*the group refers to our current time as 'post-photographic', an era when individual images are less telling than the relations between them*".[12] To find the in-between, they use methods including, for example, 3D modelling, audio analysis, fieldwork, photogrammetry, re-enactment, remote sensing, and situated testimony. FA's projects often are commissioned by various human rights organisations, thus the results of the investigations (including videos, reports, reconstructed images, and simulations) are presented in national and international courtrooms and parliamentary inquiries.

For example, an ongoing project, commissioned by Amnesty International, deals with the torture in Saydnaya prison, which is operated by the Syrian government. Until now, no human rights organisation or journalists was able to visit the prison or speak with those held there. Further, there are no images of Saydnaya in the public domain. Thus, FA researchers had to work with the memories of survivors of the prison. First, they used written interviews conducted by Amnesty International and satellite images of the building to construct a 3D model of the prison. Next, they conducted "situated testimonies": While using architectural and acoustic modelling (see Fig. 1), the researcher helped witnesses reconstruct the architecture of the prison and their experiences of detention and to correct their initial 3D model (see Figs. 2 and 3). Through this, witnesses became active participants in the investigation.[13]

[11] For more information on FA's working read here: https://forensic-architecture.org/about/agency.

[12] Watlington (2023).

[13] Read more about this specific project here: https://forensic-architecture.org/investigation/saydnaya.

Fig. 1 Torture in Saydnaya Prison: Saydnaya prison, as reconstructed by Forensic Architecture using architectural and acoustic modelling. Photo: Forensic Architecture (2016), re-print with permission of the image author

Fig. 2 Torture in Saydnaya Prison: Saydnaya prison is under a regime of silence; prisoners are not allowed to speak or scream when tortured. In these conditions, and since they are never allowed to leave their cells, prisoners develop an acute experience of sound and depend on their hearing in order to understand what happens outside their cell. Photo: Forensic Architecture (2016), re-print with permission of the image author

Fig. 3 Torture in Saydnaya Prison: A detainee works with Forensic Architecture researchers to recreate elements of the prison in April 2016. Photo: Forensic Architecture (2016), re-print with permission of the image author

The results of the projects are also presented in exhibitions in prestigious museums. FA's hope is that "…*publicity from museums will put pressure on the powers that be, and effect change*".[14] However, FA does not call its work art, nor distinguishes between art and investigation, but rather calls its practices *aesthetic investigations*.

In this context, aesthetic does not mean beautiful but is rather used in the sense of the Greek "aísthēsis", that is the study of sensory perception, perceiving with the senses such as feeling, hearing, and seeing. Thus, aesthetic knowledge is derived from the senses and particular situations and experiences.

> Aesthetic investigations have a double aim: they are at the same time investigations of the world and enquiries into the means of knowing it. This means that they seek accountability both for events and for the devices with which we perceive them. They deal with the production of evidence while questioning and interrogating the notion of evidence, and with it the cultures of knowledge production or truth claims that it relies upon. They engage in the presentation of facts while being aware of the way each presentation, indeed each media form, can distort the very facts they produce. They seek to establish claims to truth while criticising the institutions of power and knowledge with their monopoly over the mechanisms of truth production.[15]

Therefore, in the investigative practices by FA the boundaries between (sensory) research and activism become blurred. While the goal of activism is action to generate an effect, art is an expression that generates affect, which on a first glance

[14] Watlington (2023).

[15] Fuller and Weizman (2021: 15–16).

do not fit together, however they are complementary.[16] Activist art is grounded in the act of "doing"—it is not simply enough for artists to refer to political action, but that artists should be actively entangled into it. Therefore, the work by Forensic Architecture can be seen as a form of activist art but also transdisciplinary work through aesthetic practices in order to deal with social crises such as racist violence.

3.2 Anina Jendreyko

Anina Jendreyko uses theatre as a means to explore areas of conflict and crisis in both national and international contexts, especially in the Middle East. Born in Germany and raised in Switzerland, Jendreyko studied acting in Berlin, followed by engagements at various theatres. Then she left the theatre business, learned the Turkish language and worked on various projects. For example, she co-developed a literacy and post-literacy project for women. For several years she lived in Turkey, the Kurdish regions, and Greece where she worked in cultural centres and initiated theatre projects in schools and in refugee camps.[17] Thus, she uses education through theatre against oppression and for self-empowerment.

Since 2006, she has been living and working in Switzerland again as an actress, director, and as a lecturer for performance skills and transcultural work. In 2007, she initiated the transcultural theatre and education project "fremd?!" in Basel. For several years, an intercultural group of singers, dancers, and actors co-developed plays with school classes that dealt with the realities of their lives and staged them in various theatres in Basel and Chur. From this, the Volksbühne Basel was founded in 2012, which develops large theatre productions, and other cultural events such as readings, concerts, film screenings, and discussion events. A recent production by Jendreyko that premiered in 2020 at the Volksbühne Basel deals with the genocide of the Êzidi people and shows how new structures are built within resistance and reconstruction, which are decisively shaped by women. «Shengal—die Kraft der Frauen» is a documentary theatre play that reflects a contemporary historical testimony of an awakening that reaches far beyond the Shengal. The artistic work of interweaving the live music of the Êzidi musicians, the video and sound recordings shot in Shengal, as well as the text and play of the actors dissolves the objective distance of the spectators and creates a subjective connection between supposedly foreign worlds (see Figs. 4, 5, and 6).

Jendreyko's work is a form of collective working and thus a participatory artistic practice. The interest of the contemporary arts in participatory and collaborative processes has surged since the 1990s, after the fall of communism.[18] Participatory art engages the public in the creative process, letting people become co-authors, editors,

[16] Duncombe (2016).

[17] Jendreyko (2022).

[18] Bishop (2023: 3).

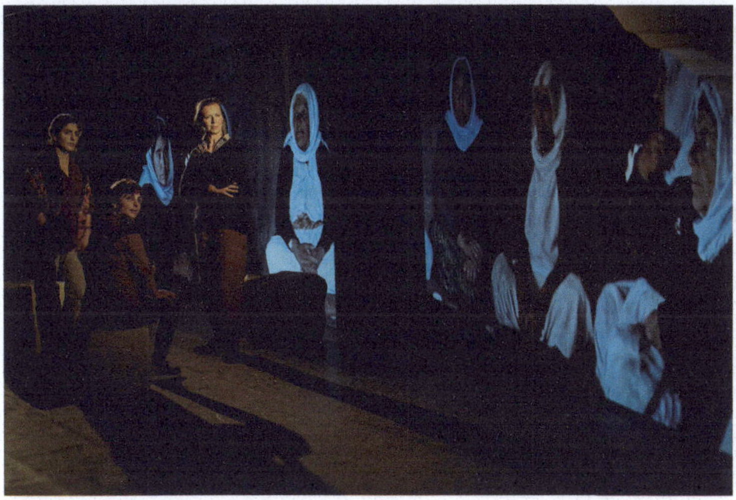

Fig. 4 From the production «Shengal—die Kraft der Frauen» at Volksbühne Basel. Photo: Matthias Wäckerlin (2020), re-print with permission of the image author

Fig. 5 From the production «Shengal—die Kraft der Frauen» at Volksbühne Basel. Photo: Matthias Wäckerlin (2020), re-print with permission of the image author

and observers of the work process. Another label for this practice is "post-studio" art, since the artist leaves her/his studio, in order to practise the artistic practice in other places, such as schools or public places while including non-artists in the process. Anina Jendreyko ways of working are thus inherently transdisciplinary processes

Fig. 6 From the production «Shengal—die Kraft der Frauen» at Volksbühne Basel. Photo: Matthias Wäckerlin (2020), re-print with permission of the image author

through artistic practices in order to deal with social crises such as issues in refugee camps.

3.3 *Hannes Rickli*

Hannes Rickli is a visual artist, thus his research focuses on the instrumental use of media. His practice of choice is the experiment. According to Rickli, the practice of experimentation as a method of material and conceptual work connects the practice of art and the practice of research.[19] This experimentation often includes research infrastructures such as scientific video material of fish behaviour analysis. Specifically, he is interested in the "waste" of scientific practices, the things that are not visible in scientific outcomes—the videos of the fish became data, thus the endless video material itself became "waste". But with technological progress, biologists' research practices have become more efficient and do not produce such "waste" anymore.

In order to further work with the research infrastructure and its materiality and ways of working Rickli had to produce his own material instead of collecting the "waste". For example, in the project "Computersignale" (computer signals, part I: 2012–2015, part II: 2017–2021) he and his team used the underwater station RemOs1 (Remote Observation System), which is located at the West Coast of Spitsbergen, Norway, as research infrastructures to observe (see Figs. 7 and 8).[20]

---

[19]Rickli (2015).

[20]For more information on the «Computersignale» project, read here https://computersignale.zhdk.ch/en/.

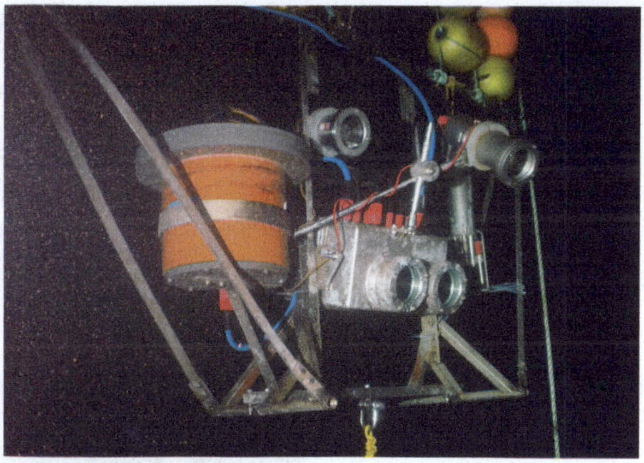

Fig. 7 RemOs1, Kongsfjorden, Ny Ålesund, Spitsbergen, 9.3.2013, 15: 30:12. Photo: Philipp Fischer (2013), re-print with permission of the image author

Fig. 8 RemOs1, Kongsfjorden, Ny Ålesund, Spitsbergen, 24.6.2013, 13: 58:20. Photo: Philipp Fischer (2013), re-print with permission of the image author

The RemOs1 station is used for a long-term survey of environmental changes in habitats of marine life. For this purpose, RemOs1 is lowered near the coast and takes images every 30 minutes to measure the development of flora and fauna of the flora and fauna. With the underwater station several questions were opened up for Rickli's artistic working as a visual artist: The station is underwater for a long period of time, and invisible for the human eyes for this time, how can this equipment then be observed by the artists? Further, with the technological progress and the data-driven disciplines of natural science the measuring devices became more and more digital black boxes, consequently the processes of scientific working became less visible and observable. This led to the question of the role of electricity and its availability in remote regions and the idea to wiretap the scientific equipment. To shift the scientific working processes of the underwater station into the realm of human perception, Rickli's PhD student Valentina Vuksic (an artist and computer scientist) inserted audio probes into the device of RemOs1 (see Fig. 9). Together with the images, that the station collects for the observation of marine life, the audio signals (power

Fig. 9 RemOs1, Circuit board electromagnetic sensor power consumption with preamplifier 12.3.2012. Photo: Hannes Rickli (2012), re-print with permission of the image author

supply, camera, on-board computer, webcam, and hydrophone in the outdoor area) are transferred through a fibre optic cable to a server. Like this, the artistic system collected material for several years.[21]

Between 2016 and 2020, Hannes Rickli was a fellow at Collegium Helveticum—a joint Institute of ETH Zurich, the University of Zurich, and the Zurich University of the Arts with the aim to provide a forum for dialogue between the humanities, social sciences, physical sciences, engineering, medical science, and the arts. During this time, he took an interest in digital infrastructure such as data centres in Switzerland, and the questions what digital data actually are, how they are created, and whether they have a body that can be touched or heard. However, data themselves are not directly perceptible, but they need infrastructure and electricity, which in turn leaves an ecological footprint in the physical world. The outcome of this fellowship period was an exhibition and a book co-edited with Monika Dommann and Max Stadler—both historians of science. The book "Data Centers—Edges of a Wired Nation" deals with making something invisible visible. It is a collage of texts and visual essays on supercomputers, the data bunkers, crypto valleys, and the social, political, and ecological impacts of such infrastructure. The objective of this co-produced endeavour was to make visible the large abundance of digital infrastructures in Switzerland as well as the institutions that are involved in the creation, maintenance, and regulation of these digital infrastructures and then to critically discuss them afterwards.[22]

Rickli's work is to be understood as artistic research. Defining artistic research is not straightforward: We start with what it's not: It's not research of or on art, nor it's art practices or their outcomes itself:

> The point, however, is to understand artistic practice and artistic research as kindred practices, to be sure, but not necessarily identical ones. Artistic research stands and falls

[21] Rickli and Vuksic (2015).

[22] For the full reference to the book, see Dommann et al. (2020).

> not on its connection to art but to aesthetics. The attribute of art is not essential to artistic research but rather the term aisthesis, sensual knowledge. It makes no claim to autonomy. But artistic research justifies itself in those places where it intermittently intervenes in scientific discourses as well as in everyday worlds, with the purpose of further developing these, transforming them or causing shifts in them and, in surprising and sometimes incomprehensible ways, driving them forward.[23]

Following this, artistic research includes an inherently transdisciplinary stance. In a nutshell, artistic research, that is not limited to the field of art, can draw on the specifically artistic-aesthetic practices, techniques, and procedures that have been developed in modernity, such as the focus on the means of representation and their productivity for our perception and understanding of reality. The methodological toolbox of artistic research contains the aesthetic practices that modern art has developed in its differentiation from science.[24] Therefore, Rickli's research practices are transdisciplinary work through artistic practices in order to deal with social and ecological crises such as ecological effects of data centres.

4 Conclusions: An Appeal

> Every established order tends to produce the naturalization of its own arbitrariness.[25]

The challenges that lie ahead require a different understanding of each other and ways of working that go beyond disciplines and interdisciplinarity. It is no longer enough to specialise deeper and deeper in one's own discipline until the discipline naturalises itself. Transdisciplinarity through the arts does not aim to replace specialisation, but to go beyond it. This can be motivated and conducted by participation, activism, artistic research, or other means that still need to be developed. What remains is a longing for further perspectives that are not constrained by axioms, to challenge assumptions, to explore the pressing social and ecological crises, and to inspire action for meaningful change in collaboration and beyond established methods and ways of thinking. The arts show an exemplary reflective stance towards transdisciplinarity that does not have to remain in its field and provides points of reference for any discipline. The arts show an exemplary reflective stance towards transdisciplinarity that provides points of reference for any discipline and thus should not only remain in its field. Transdisciplinarity through the arts can enable people from different disciplines to understand themselves as collectively situated and to find ever new ways through affective practices.

[23] Henke et al. (2020: 18).

[24] For an introduction to artistic research methods, see Badura et al. (2015).

[25] Bourdieu (1977: 164).

References

Badura, J., Dubach, S., Haarmann, A., Mersch, D., Rey, A., Schenker, C., & Toro Pérez, G. (2015). *Künstlerische Forschung—Ein Handbuch*. Diaphanes.

Bishop, C. (2023). *Artificial hells: Participatory art and the politics of spectatorship*. Verso Books.

Bourdieu, P. (1977). *Outline of a theory of practice* (Vol. 16). Cambridge University Press.

Bourdieu, P. (1987). *Distinction: A social critique of the judgement of taste*. Harvard University Press.

Busch, K. (2012). Passivität. Textem-Verlag.

Dommann, M., Rickli, H., & Stadler, M. (2020). *Data centers: Edges of a wired nation*. Lars Müller Publishers.

Duncombe, S. (2016). Does it work? The affect of activist art. *Social Research, 83*(1), 115–134.

Fleck, L. (1983). *Erfahrung und Tatsache: Gesammelte Aufsätze – Mit einer Einleitung herausgegeben von Lothar Schäfer und Thomas Schnelle*. Suhrkamp.

Fuller, M., & Weizman, E. (2021). *Investigative aesthetics: Conflicts and commons in the politics of truth*. Verso Books.

Haraway, D. (1988). Situated knowledges: The science question in feminism and the privilege of partial perspective. *Feminist Studies, 14*(3), 575–599.

Henke, S., Mersch, D., Strässle, T., Wiesel, J., & van der Meulen, N. (2020). *Manifest der Künstlerischen Forschung Eine Verteidigung gegen ihre Verfechter*. Diaphanes.

Jendreyko, A. (Director). (2022, October 21). Bildung gegen Unterdrückung – Theaterarbeit im Flüchtlingscamp. In *Schweizer Radio und Fernsehen (SRF)—Passage*. Acceded May 15, 2023, from https://www.srf.ch/audio/passage/bildung-gegen-unterdrueckung-theaterarbeit-im-fluechtlingscamp-w?id=12264070

Mersch, D. (2015). *Epistemologies of aesthetics*. Diaphanes.

Mittelstraß, J. (2003). *Transdisziplinarität-wissenschaftliche Zukunft und institutionelle Wirklichkeit*. UVK Universitätsverlag Konstanz.

OECD. (2020). *Addressing societal challenges using transdisciplinary research*. OECD Science, Technology and Industry Policy Papers No. 88. OECD.

Rickli, H. (2015). Experimentieren. In Badura, J., Dubach, S., Haarmann, A., Mersch, D., Rey, A., Schenker, C., Toro Pérez, G. (Hg.): Künstlerische Forschung—Ein Handbuch (pp. 135–138). Diaphanes.

Rickli, H., & Vuksic, V. (2015). *RemOs1. Beginn einer Datenarbeit in der Arktischen See*. Band 16 Der Schriftenreihe des Institute for Contemporary Art Research (IFCAR). Zürcher Hochschule der Künste (ZHdK).

Stengers, I. (2008). *Spekulativer Konstruktivismus*. Merve Verlag.

Vogl, J. (2007). *Über das Zaudern*. Diaphanes.

Watlington, E. (2023). When does artistic research become fake news? Forensic architecture keeps dodging the question. *Art in America*. Acceded May 15, 2023, from https://www.artnews.com/art-in-america/features/forensic-architecture-fake-news-1234661013/

Dr. Annina Boogen studied environmental sciences (BSc) and energy sciences (MSc) at ETH Zurich. In her doctoral thesis on energy economics and policy, she worked on the residential electricity demand and its efficient use. After finishing her PhD thesis in 2016, Annina's research focus shifted from more classical energy economics into the field of behavioural economics and individual behaviour change. In addition to her work as a postdoctoral researcher at ETH Zurich, she studied part-time in the MA Transdisciplinary Studies in the Arts at the Zurich University of the Arts (ZHdK) where she investigated the sensory perception of energy infrastructure in alpine landscapes using performative approaches and published a non-scientific synthesis report as a vinyl record. Currently she works at the Center for Energy and the Environment at the Zurich

University of Applied Sciences (ZHAW). In her recent work she is investigating the affective and emotional factors in individuals' energy-related behaviour and attitudes. She is also an artistic researcher working with sound, president of the Swiss Society for Acoustic Ecology (SSAE) and an associate researcher at the Uri Institute Cultures of the Alps. She is the author of several articles on her research areas. Contact: https://anninaboogen.ch/

Fernando Obieta is a conceptual media artist and artistic researcher. After studying Computer Science with a focus on media technologies, he studied Interaction Design and Transdisciplinarity in the Arts at the Zurich University of the Arts. His artistic work and research focus on the critique and reflection of (digital) media, artefacts, systems, services, and solutions and how they are contingent and situated. He works at the intersection of concept, design, art, and technology. He is currently doing his PhD in media theories at the University of Arts Linz funded by the Swiss National Science Foundation's (SNSF) Doc.CH program and is being hosted at the Lucerne School of Design, Film and Art. The dissertation "The Contingency of Design – a critique" is supervised by Prof. Dr. Gloria Meynen (Linz) and Prof. Dr. Rachel Mader (Lucerne). He is investigating how a feminist ethics and morality of machines and their production could be constituted. To this end, he is working on a series of posits under the title "machines congregating", in which he examines how different machines perceive people and how people observe others (or themselves) through machines. Contact: https://fernando-obieta.com/

Transdisciplinary Dimensions of NEW HEALTH: The New Paradigm of Medicine

Nanda Meister

1 Introduction

"The most effective medicine is the natural healing power that lies within each of us," is a statement that is attributed to Hippocrates.[1] How much of this guiding principle has flowed into classical medicine? How much attention is paid in these times of accelerated change to the healing power that lies within? Has there even been a burial of this healing power, as most people have lost access to their body and self? Does this phenomenon of disconnection from our self and our body even describe the greatest pandemic of this time?

From the moment we cut our finger, the wound begins to heal. We do not have to contribute anything to it in most cases. The entire organism is heading for wholeness at all times. Is this wholeness just physical integrity, or is the body basically meant to communicate with us, to give us clues to bring us back to our original wholeness? Do the physical symptoms bring us to our buried traumas, our primal pain, and ask us, not for suppression, but for resolution of the underlying conflict?

In times of great individual and collective upheaval, such as those brought about by the pandemic and the current world situation, the need for a quantum leap not only in consciousness but also in the healthcare system is omnipresent. The care situation in most nursing facilities and hospitals is increasingly worrying, financial resources are scarce, and optimal health care is not accessible to everyone without restrictions. Economic criteria are often paramount, leaving little room for profound doctor–patient interactions. The need for increasing autonomy and a say on the part

[1]Hippokrates von Kos (460 bis ca. 377 v. Chr.) griechischer Arzt, »Vater der Heilkunde«

N. Meister (✉)
NEW HEALTH Advocate, Co-Founder of The Sacred Shell - Ecosystem for New Health, New Leadership and Peace, Zurich, Switzerland
e-mail: nandarani@thesacredshell.com

© The Author(s), under exclusive license to Springer Nature Switzerland AG 2025

M. Schüz (ed.), *Transdisciplinary Thinking and Acting*, Transdisciplinary Management of Ecological and Social Crises,
https://doi.org/10.1007/978-3-031-70680-6_17

of those affected contrasts with the resignation of the masses, who often feel at the mercy and helpless in the face of the unfortunate circumstances. The individual has often lost his place and is replaced with a number. Despite the technical advances gained in the twentieth and twenty-first centuries, medicine seems to have lost what is essential, namely humanity. "For the crisis in human medicine there is only one medicine: the humane."[2] And it will always remain the humane that makes medicine real medicine.

2 The Fourth Revolution in Medicine: The Wave of Humanity?

As early as the 1980s, Arnold Relman spoke of the three revolutions in medicine in the New England Journal of Medicine.[3] The first corresponds to the "medicine of the feasible" in the postwar period, which was dedicated to medical advancement and specialization. This was followed by the second revolution in the 1970s, which involved cost containment and the introduction of flat rates per case and increased productivity. We are still confronted with these upheavals today. At a time when it must be possible to assess and treat an entire human being in all its complexity in just a few minutes. In the third revolution, the focus is now on assessment and accountability. This means that therapists, especially physicians, are confronted with economic efficiency, evidence, and quality assurance.[4] These are domains that did not primarily correspond to the intention to work in a healing profession. These three waves flow into each other and continue to move today. Parallel to the changing paradigm of the corporate world, it is about profitability and legality. As long as it is profitable and legal, it is a good deal. In this way, man becomes similar to a product.

The merits and achievements of classical Medicine are, without a doubt, impressive. NH aims to expand classical medicine to include the holistic, human dimension through transdisciplinary consciousness. Especially in a time when humanity and human interaction are becoming increasingly rare thanks to AI and digitalization.

Just as the corporate world is now craving more awareness, authenticity and purpose and conscious leadership is in demand, the world of medicine is changing. Could the coming wave be a revolution of humanity? In which a return to the essential principle of the therapist succeeds? The term therapist (old Greek θεραπευτής, therapeutés) originally means "the servant, the attendant, the caretaker, the guardian" and is used today for a provider of a healing profession or a healing procedure. Serving and being there, lingering in patient presence and active listening and communicating,[5] as the original principle of the therapist, is central and is an essential part of a successful therapeutic relationship.

---

[2] Uhlenbruck (1929).

[3] Relman (1980, 2009).

[4] Loss (2017).

[5] Neeraj (2003); Ha (2010).

The previous revolutions in medicine since World War II have created a distance between the patient (lat.: patiens, "suffering, enduring") and the practitioner, which leads to dissatisfaction on both sides. The increasing pressure and incongruence of initial values leading to the medical profession, in addition to other factors such as increased time demands and great responsibility, have made the physicians' profession a field with a higher suicide rate compared to the general population.[6] Emotional exhaustion and compassion fatigue are associated with career choice regret and lack of career prospects and lead to high burnout rates (approximately twice as likely compared to general population) among physicians.[7]

In all the years that I have worked at various clinics, the question has never been how the doctors are doing physically and mentally. How they were feeling? Do they need support? I know from my own experience what compassion fatigue feels like. Empathy for the patient and myself was not accessible any longer. And that lead me to awareness to take care of myself, to be self-compassionate. But for most, exhaustion is the rule, not the exception. Dealing with personal resources and the life–work balance that comes with them is not yet taught in medical school. Consequently, in this way, compassion in the encounter with the sick counterpart is lost, just as the relationship to oneself is lost. It is no wonder that the capacity to really engage with a person and his/her story is hardly present.

How should it be possible to remember the healing power first of all in oneself as a therapist as well as to subsequently train the affected person to get in touch with it? The *Declaration of Geneva*[8] 1948 by the World Medical association states: "I will take care of my own health, well-being and abilities to provide the highest level of treatment." It is part of our transdisciplinary responsibility to act accordingly.

3 Milestones in Medicine: From Victimhood to Self-responsibility

Medicine has always been in a state of flux. From the ancient humoral pathology ("Säftelehre," the teachings of the humors) to the cellular pathology of Virchow,[9] further to the human genetics with decoding of the complete DNA 2001[10] and beyond that to the epigenetics, which recognized how the cell environment affects our DNA and whose well-being depends essentially on mental and emotional factors.[11,12]

[6] Irigoyen-Otiñano (2022).

[7] Hodkinson (2022).

[8] Anonymous (2017): "World Medical Association Declaration of Geneva"

[9] Virchow (1871).

[10] Collins (1995).

[11] Lipton (2005).

[12] Gustafson (2017).

These recent findings have freed us from being victims of our genes. At the same time, they also bring responsibility, as we are much clearer about how the neurotransmitters and hormones produced by our thoughts can affect us negatively or positively. This is also described by the field of psychoneuro (endocrino) immunology, which clearly depicts the interactions that link the psyche, nervous system, endocrine system, and immune system.[13] These findings play an important role in the field of NH. In addition, for almost 20 years now, the field of socio-genomics has been added, which describes what influence social stress, such as the subjective perception of social isolation and loneliness, has on genes. This upcoming field seems significantly related to epigenetics in that it shows how associated emotions directly affect our DNA.

NEW HEALTH (NH) is a transdisciplinary approach to health maintenance and disease management. It incorporates the different perspectives and needs of affected persons and accompanying therapists and physicians as well as other involved persons of the social fabric, which are oriented toward the maintenance of health. It is based on the fact that we all may and must take responsibility for our individual health in order to promote the necessary change in the health care system and in ourselves. But it also goes far beyond that as it draws from diverse dimensions, some of which I discuss in this compilation. The transformation of the "sick-care-system" into a health-care system can only succeed if the socio-political conditions are in place, which will take time. This article strives to set impulses so that further steps can be taken toward the recovery, extension, and evolution of the medical system in general.

4 The Importance of Trauma-informed Care and the Four Core Dimensions of New Health: Including a Trauma-informed Perspective to Modern Medicine

In recent years, there have been regular scientific publications in prestigious journals showing the links between trauma, physical and mental health. However, there is still a lack of knowledge about this in the mainstream world of medicine. The mind-body construct is still largely ignored despite staggering evidence. Even in the mental-health field, knowledge about the connections between trauma and widely given diagnoses such as depression, anxiety, and panic disorders, is not taught.[14] Thus, most psychiatrists are trained to administer an adapted medication that works at the symptomatic neurotransmitter level causing a lot of undesired side effects. However, they do not get to the root cause of the disorder from a trauma-informed perspective.

---

[13]Pert (1999).

[14]Maté (2003).

NH, the next evolution of our health system, is a multidimensional system that strives for a trauma-informed, health-friendly, and conscious society. This concern inevitably involves all aspects of our being, as health is not just the absence of physical symptoms, but is a dynamic process essentially related to emotional well-being. The definition of Health by the WHO concludes: "*Health is a state of complete physical, mental and social well-being and not merely the absence of disease or infirmity.*"[15] I would like to add the aspect of resilience as "*complete*" overall well-being is a state that seems to be an ideal not easily achieved or kept. The ability to manage physical, mental, and emotional difficulty is relevant to stay equipoised during challenging times and can even be a consequence of difficult childhood circumstances. Furthermore, resilience is a skill that can be taught.

In the following section about the individual dimension of NH, you will find an overview of the *Archetypes of childhood trauma* by Nicole Le Pera to further explain the extended concept of trauma.

I can still remember well when I worked as a young doctor in internal medicine almost 20 years ago, which followed a special model that was called "Menschenmedizin" (human medicine) that shares the holistic approach of NH.[16] This pioneering model was incorporating art, expression-oriented psychotherapy, philosophy, medical ethics, and psychosomatics into the treatment of all patients. Every patient received the offer of additional accompaniment. This was revolutionary at the time—and probably still is, because it is practiced far too seldom nowadays. At that time I met a woman in her early 40 s who was admitted as an emergency patient with a first episode of atrial fibrillation, a significant irregularity of the pulse that exhausts the heart and can lead to further problems. What stayed with me from that encounter were her words, *"My whole life is out of rhythm."* Nothing could probably have described the condition of her heart better. From her story, it was clear that she was recently left by her partner and didn't know what to do next. Was her heart reacting to this? Many stories dealing with such contexts would confirm this. What can we learn from this in terms of treating and accompanying people?

5 Questioning the Common Dogma of Medicine

At the present time, the common dogma of medicine is untouched and still largely unchallenged. The evidence load of the above-mentioned correlations and the need for change will become even stronger in the coming years, and only then a real rethinking can take place. Despite all the developments and the so-called progress, it has not yet been possible to significantly reduce or prevent certain diseases. The likelihood of dying from cancer has decreased only marginally in recent decades,

---

[15] Schramme (2023).

[16] Hess and Hess-Cabalzar (2006).

and almost every second person still develops cancer one or more times in his or her life. Another big task are mental disorders that account for 13% of the global disease burden, and major depression alone is expected to be the largest contributor by 2030.[17] The economic effect is great, with mental disorders expected to cost nearly a third of the projected US$47 trillion incurred by all non-communicable diseases by 2030 predicted by the World Economic Forum 2011 in Geneva.[18]

If we take a critical look, we might well ask ourselves whether we are doing something wrong? Where is the potential for development here? Does the key even lie in recalling the words of Hippocrates? How is it possible to reconnect with the healing power that resides within us? Could this reconnection bring the change in our health and our health care system that we desire?

6 NEW HEALTH and its Four Core Dimensions: A Transdisciplinary Approach to Disease Prevention and Management

NH considers the human being a transdisciplinary being, who carries the physical, energetical, mental, emotional, social, and spiritual dimensions within and who is, without a doubt, affected by all these dimensions. If we aim to keep or restore health, we need to access all these different dimensions and cannot focus on just one of these, e.g. the physical dimension. Therefore, medicine, just like theology, as a single discipline, is a contradiction in terms. In Western medicine, everyone is trained to be a specialist and subspecialist and often loses sight of the whole person in all its dimensions. Let alone, it is taught to deal in a short time with the other dimensions of being human, which would be so precious to deal with the cause of the disease. In fact, this does not need much time, but only the right questions and a true interest of a healthy doctor.

This leads us to a challenge that can only be solved in a transdisciplinary context. In order to approach this challenge, I present four core dimensions of NH as concluded from my own experience and the experience of many experts in that increasingly recognized field of trauma-informed and -rooted medicine.

This article introduces the four core dimensions of NH and is far from including all further dimensions in their complexity.

It will provide insights into:

1. *The individual dimension of NH*
2. *The social dimension of NH*
3. *The educational dimension of NH*
4. *The spiritual dimension of NH*

---

[17] World Health Organization (2008).

[18] Bloom (2011).

The whole aspects of environmental influences on people, for example related to the microbiome, nutrition, orthomolecular medicine, physical activity, or the surrounding culture as well as collective trauma, are not further discussed here. Nevertheless, they are essential in the greater picture of NH. Other holistic systems of medicine, such as Traditional Chinese Medicine, Ayurveda, European Classical Naturopathy, and Indigenous Medicine, among others, also have a place in NH because of their holistic approach. Bringing the different schools together would not only enrich but also complement the fragmented view of the patient that exists in modern medicine. Unfortunately, modern medicine is still very hesitant in this regard.

7 The Individual Dimension of NEW HEALTH

"Healing Medicine from the inside out" is one of NH's guiding principles. This means that change begins in the microcosm of each person to significantly shape the macrocosm.

The prerequisite for being a qualified therapist in the field of NH—and here I naturally include the medical profession—is to be familiar with one's own life story and to be committed to a process of awareness. This means in particular a strong capacity for self-reflection, fearless self-inquiry,[19] and awareness of experienced traumas and how they affect our mind-body system. This experience, which is gained through personal growth processes, leads to greater understanding, respect, empathy, and compassion for one another. Qualities that are urgently needed in the accompaniment of people who in 80% of cases causally experience illness through inner stress.

As a doctor the intention is to provide healing. Nevertheless the term *healer* has unfortunately often been misinterpreted and is even despised among medical doctors. I wonder why they exclude themselves, aiming for the same goal, namely health. When I speak of a healer, I refer to someone who is able to remind others of their own inherent healing power and encourage its development. This person is also able to discern when further medical intervention is needed and understand what factors might be causing illness. Who also has the ability to guide a person to the root cause of the disease and accompany them through the resolution process. This needs high personal and professional competence.

A healer knows the path of healing by walking it himself. This means, he is committed to the processes of awareness and self-reflection and is able to competently navigate his own trauma and thereby facilitating his own healing with the support of others, if needed. As a result of his developed humanity, his communication is clear, compassionate and respectful, occasionally guiding while recognizing and integrating the needs and perspectives of the sufferer. In addition, there is a life–

[19] Maté and Maté (2022).

work balance, that allows the authenticity of recommended measures for stress reduction and prevention of illness to be genuinely embodied by the healer. As this Indian parable shows: a mother comes to Mahatma Gandhi and asks him to tell her son not to eat sugar. He replies: *"Come back in three months."* After three months the mother returns and Gandhi tells her child: *"Stop eating sugar."* When the mother asked why he did not tell this to her son earlier, he replied: *"I had to first stop eating sugar myself before telling your son to do the same."*

Not only is there an interdisciplinary team of experts (e.g., gastroenterologists, cardiologists, and surgeons) who think holistically, but the wisdom of how our mind is influencing our well-being (psychoneuroimmunology) and the development and processing of disease patterns is recognized and included.[20] Furthermore, the three core needs of every human being, according to Gabriela von Witzleben: autonomy, relationship, and safety, are integrated into the flow of information and accompaniment.[21] A good therapy decision is made, when all of these three core needs are met. Nowadays mainly safety is focused on, but autonomy and relationship are left behind. I have seen many patients that struggled, because they felt ignored in their need for autonomous decision-making.

8 The Extended Concept of Trauma

We are all traumatized. Whether it is a short-lived severe trauma, or chronic stressful situations over a longer period of time. Emotional neglect, pseudo-harmony and denial, not respected boundaries and much more, experienced in childhood and then repeatedly later, everyone, without exception. This is probably what life brings. However, it was not possible for everyone to emerge equally resilient from difficult situations. The term "trauma" is a widely misunderstood term. Most people understand it to mean exclusively, consequences of sexual abuse, physical violence, or war situations. In the well-known tests for questioning regarding child trauma, such as the ACE ("Adverse Childhood Experiences" of the Centers for Disease Control and Prevention), it is mainly about severe trauma, whereby physical, verbal, and sexual abuse experiences are asked in ten questions. A high score is associated with drug or alcohol abuse, suicide, and chronic illness.[22] This test is relevant because it clearly shows that childhood trauma has a formative influence on the further course of life.[23]

Now, it is the case that many people do not score highly and yet show all the signs of substantial trauma-response (fight, flight, freeze, fawn (people-pleasing)) in

[20] Pert (1999).

[21] Von Witzleben (2019).

[22] Felitti (1998).

[23] LePera (2021:96ff.).

everyday life.[24] It is often the case that one cannot even remember events from childhood, or has only photographically stored memories. Many children who were, e.g., emotionally neglected often still live as adults split off, dissociated from themselves and have built up fantasy worlds.[25] The access to their own feelings is often missing and the knowledge of their own needs and truth is practically non-existent. Our psyche protects itself through various defense mechanisms, that leads to unawareness of the deep imprints and scars that each one carries. A certain amount of *trauma* belongs to life. How we are able to grow through it creates our resilience. Still, traumatizing behavior is normalized in society and leads to a lot of imbalance on the physical and mental level.[26]

In the following I would like to mention the most common archetypes of childhood trauma that Dr. Nicole Le Pera has presented in her book *"How to do the work: recognize your patterns, heal from your past, create your Self."*[27] These are essentially related to the individual dimension of NH. However, it is just not enough to look at the classical traumas as they are asked in the ACE. According to neurologist Robert Scaer, a trauma is any life event experienced in a state of *relative helplessness.*[28]

Archetypes of the most common childhood trauma according to Nicole Le Pera[29]:

- Parents denying your reality; leads us to not trust our own judgment and we look to others to shape our reality.
- Parents not seeing you, or hearing you; leads to survival mode and emotional disconnection from the parental figure, self-expression is not seen and it is hard to trust our inclinations and follow intuitive needs.
- Parents who lead a proxy existence through their child, or want to shape it; incredible pressure to succeed leads to self-loss and indecision as well as procrastination; respectively, obsessive striving for success, addictions, and in extreme cases suicide
- Parents who cannot set boundaries; children internalize that not respecting boundaries is "closeness" or even "love" and this influences their relationship patterns later in life. On the contrary, they keep everything to themselves and are inaccessible to closeness.
- Parents for whom appearances are overly important; forms basis for belief that being liked depends on outward appearance. Parents not in control of their emotions; feels lack of emotional resilience and disdain for feelings per se, this in turn leads to suppression of difficult emotions.

[24] LePera (2021:93).

[25] LePera (2021).

[26] Maté and Maté (2022).

[27] LePera (2021).

[28] Scaer (2014: 83/253)

[29] LePera (2021:103–124)

I conclude that all these above mentioned most common archetypical traumas create stressful life situations leading to a stressful environment in the body that can be described as a psychoneuroimmunological imbalance that causse disease if not resiliently managed.

We as human society have adapted these patterns and behaviors as normal.[30] As a consequence follows a disturbed emotion regulation that has a direct effect on our immune system and leads to chronic diseases, including cancer. Implementing emotional intelligence and resilience through conscious education of all ages and professions is central to NH to prevent disease and maintain health. It is not just about being trauma-integrating or trauma-informed, but trauma-rooted, where the cause of most chronic diseases has its roots in the mental-emotional system.[31] Consequently, healing approaches must also be linked to the resolution of this emotional imbalance. Thus, *not only the eyes are the gateway to the soul*,[32] as it is attributed to Hildegard von Bingen a Benedictine nun, German mystic, and polymath in the Middle Ages, but I would suggest adding to it: *the body is the gateway to the soul.* Through various methods that connect the body and the expression of unlived emotions, the vital force (also: India: Prana, China: Chi) can flow freely again and the immune system can be supported in its ability to regulate. Such holistic systematic healing methods are found in ancient Chinese medicine or Ayurveda. Newer approaches that integrate ancient wisdom are, for example, offered by the Root Cause Institute founded by Paul Leendertse,[33] Neo Emotional Release that is merging psychosomatics, Bodywork, and integral trauma founded by David Manning,[34] as well as well-known Somatic experiencing by Peter A Levine,[35] just to name a few.

I can recall that during my work in oncology, I interviewed many patients who had cancer and asked them, what they themselves believed to be the reason for their illness. To be clear, this question has nothing to do with guilt, because at no point can you act differently than you do. However, the discussions were so informative that it quickly became clear to me that there is a connection with the development of the disease. For example, it was about having to leave a job due to a bullying situation, but lacking the courage to do so and fearing the loss of ones identity if not performing, ultimately being driven to become ill as a result. Various levels play a role here, such as existential fears, excessive identification with the work/activity, a feeling of inferiority, and many more factors, which often have their cause in childhood. Since they are part of the cause of the problem, these aspects must be included in a transdisciplinary treatment. A further conspicuousness could be observed in women with breast cancer, who sacrifice themselves in relationships,

[30] Maté and Maté (2022).

[31] Maté (2003).

[32] Von Bingen ca.(1151).

[33] Leendertse (2020).

[34] Manning, https://emotionalreleases.com

[35] Levine (1998).

do not know their primary needs and often could not develop a sense of boundaries and self-worth in childhood. To give just a few examples.

During my 20 years of practice in the medical field I experienced that the individual opinion of what caused the illness, the inner chronic stress, was completely irrelevant in conventional treatment and often not taken seriously by the medical professionals. If it came up at all. One main reason is that these connections are insufficiently known in the area of the university, classical medicine and thus also not taught. According to my therapeutic experiences most of imbalance in health state are indications for deeply rooted causes. Bringing the latter to conscious awareness they lead one to the starting point for healing—the willingness for change and inner transformation. People who report spontaneous healings have often made radical changes. First in the area of nutrition/supplements, as this is the area that can be most easily addressed in self-competence.[36] However, more important components were of taking responsibility for one's own healing process and following one's intuition, admitting and releasing suppressed emotions and allowing positive emotions, allowing social support, overall having strong reasons for living, and deepening spiritual connection.[37] Likewise, higher levels of meaning are clearly associated with better physical health, as well as with behavioral factors that decrease the probability of negative health outcomes or increase that of positive health outcomes.[38]

The prerequisite for being a competent and autonomous affected individual is the awareness that healing comes from within, responsibility for health must be taken, and no one can make you "whole," because wholeness already exists at any point in time. Illness is merely a way of recognizing this.

Another quote attributed to Hildegard von Bingen says, "Every disease is curable—but not every patient."[39] In NH, people learn to deal maturely with the physical imbalance we call disease. If this conscious transdisciplinary approach to disease is not yet in place, then the empathetic guidance of a therapist encourages this. The therapist-affected individual encounter takes place at eye level, respecting the qualifications of each (therapist as well as affected person), and decision-making about necessary steps are not rushed, so that the autonomy and trust of the affected person are preserved at all times.

Another important virtue a therapist should provide is openness to confront his or her patient's history and the emotional pain often associated with it. Radical honesty about one's own life instead of looking away brings the necessary change, for both, patient and professional. Unfortunately, most patients are only willing to do this form of healing work when they are in need. And so far, doctors show the same behavior of not confronting themselves with the warning signs and language of their bodies that could lead to health even before serious illness needs to occur.

[36] Turner (2014:21ff.).

[37] Turner (2014).

[38] Roepke (2014).

[39] Von Bingen ca.(1151).

In our fragmented modern world, experts of different disciplines are engaged to heal different problems of a patient. The doctor focuses on physical problems. Psychological problems are treated by psychotherapists, while if someone has spiritual problems he seeks out a priest or a pastor. But if we now, as companions of individuals who go through a transformation process called disease, ourselves remember our fellow humanity, we see the human being as a “transdisciplinary being” with the whole history that has allowed illness to arise. Thus, this approach of deeply understanding life and its very individual stories can lead to the fact that people can overcome illness, instead of merely treating symptoms in chronic dependence on medication for long times even up to decades. The individual dimension of NH leads to more consciousness on the part of sufferers and therapists from which a change on a societal level can emerge that goes beyond the hospital or health center and can enter every household.

9 The Social Dimension of NEW HEALTH

“All actual life is encounter” says the Jewish religious philosopher Martin Buber.[40] We are relational, social beings. The need to give love and to receive love is anchored in us. Our life is about being in relationship: with our counterpart, the world around us, ourselves, and the larger context. Therefore, the social environment in which we move is highly relevant and formative. Just as it is relevant in which milieu the cell in the body resides.[41] If we are confronted with social isolation and loneliness, we have approximately twice the risk of dying earlier.[42] Researchers and practitioners should consider loneliness, social isolation, and their interaction whenever possible. Especially at a time marked by the collective trauma of the Covid-19 pandemic, it is more important than ever to focus on social isolation and loneliness.[43]

Relationships can be beneficial to health, but they can also be maximally stressful as we realize the correlation with a traumatic surrounding. They have a direct correlation, e.g., with cardiovascular disease such as high blood pressure or heart attack, with mental health issues, autoimmune disease, cancer, and many more.[44] Because in relationship the pain of past trauma gets triggered and as the psyche needs to protect its owner by keeping unawareness, the body ultimately reacts physically to the chronic stress when the emotional load becomes unbearable. A rediscovery of trust in humanity is a prerequisite for healing and needs to be supplied by therapists.

[40]Buber (1984).

[41]Lipton (2005).

[42]Long et al. (2023); Barnes et al. (2022); Timothy et al. (2022); Wang et al. (2023).

[43]Nagesh et al. (2022).

[44]Holt-Lunstad (2011).

On the individual level the Personal determinants of health (PDOH) are considered the individual-level resources that can help to support positive health outcomes; the key PDOH include resilience, optimism, purpose in life, and social connectedness.[45]

NH aims to move from social isolation to social interaction, thereby creating a sense of resilience, optimism, and purpose through social connectedness. The realization that we are all "in the same boat" has a comforting effect and at the same time it leads to more fellow humanity and compassion.

In my practice, I noticed this circumstance in particular when treating women with breast cancer. Many of them explicitly wish to accompany other women with the same fate on their way. On their journey through cancer they experienced so much posttraumatic growth and inner transformation that felt like a *personal awakening to real life* to them. They became experts in that particular field that they wish to share their personal treasures and support others. In this way, those originally affected become companions on the transformative path of healing. This can be either in a professional setting or in the field of volunteering. It is a matter of offering vessels where this is possible. A team of women I assisted on their healing journey has founded esha-heal.com[46] that supports women physically, mentally, emotionally, and spiritually through retreats and a variety of therapy offers. Nothing is more powerful than a person who can share his or her wealth of experience with authenticity and who knows the concerns and questions of those affected. This is also the background of many renowned teachers of Health and Wholeness of our time. They report from their personal wealth of experience. This leads to giving further meaning to their own difficult experience by ultimately being able to serve someone else through personal realization. Definitely, people who have gone through an intense healing experience and have dealt with the underlying theme are ideal to find a task in the field of health care, prevention, and education.

The individual is the smallest unit in this collective system of the trauma-informed community. This means that we create self-awareness, self-managing, and self-leadership on an individual level and bring this awareness and clarity into the entire collective system. This generates a constructive dynamic for understand ing, growth, and adaptation. As a collaborative process, the community facilitates improvement for the greater good of all. For this to happen it is essential to adapt the social environment in which we live as human beings who have all experienced major or minor traumas. As an example a clear compilation of this can be found in the Sanctuary Model by Dr. Sandra L. Bloom.[47] There, The Seven Commitments of a trauma-aware Sanctuary Community are presented, which include the following aspects: Nonviolence (Safety), Emotional Intelligence, Inquiry and Social Learning, Democracy, Open Communication, Social Responsibility, Growth and Change. All facets adding to the social dimension of NH. Unfortunately, this environment is not

[45] MacLeod et al. (2020).

[46] https://esha-heal.ch/

[47] Esaki et al. (2013).

yet present in most hospitals, municipalities, and other institutions. Neither is the knowledge about the importance of creating a healthy environment taught.

Therefore, NH strives to bring people together at community level, so that the social influences create closeness in which safety and healing can be provided. This creates safe places that incorporate the above aspects to create strong reasons for a successful life.[48] Nowadays, social interaction is often provided via apps and online, which is a contemporary, reasonable approach. However, the physical interpersonal interaction is significantly more relevant, as the different individual nervous systems react to each other and thus a sense of coherence, real closeness, and safety can develop.[49] To cite the doctor, theologian, and philosopher Albert Schweitzer (1875–1965):

> Medicine is not only a science, but also the art of letting our own individuality interact with the individuality of the patient.[50]

10 The Educational Dimension of New Health

The educational dimension of NH includes all levels of society, but focuses as a first step on the establishment of *conscious education* in schools and health care systems. ***According to my opinion these*** important fields should be included to establish *conscious education* in the form of courses for: health care professionals, social workers, pastoral care, professionals of social insurance institutions, children and youth protection authorities, teachers, parents, children, youth workers, public health institutions, volunteers, survivors, and many more. The wisdom gained from these fields will also influence the corporate world and other organizations in the future. Individual and collective health is at the core of all future social development.

The *conscious education* in NH the way I see it will put great emphasis on trauma training (recognition, prevention, transition, resolution). However, it also includes emotional and mental health training and re-establishes contact with one's own body wisdom, and thus with one's own inner healing power.

Ideally, the discussion of emotional, and thus also physical health, begins with parents before they conceive a child. It aims to create healthy internal and external circumstances in families that will allow children to grow into mature adults. Relevant is the training of teachers, social workers, and professional staff in schools and all social institutions, regardless of social class. Up to now, trauma training has been offered to teachers within the framework of studies, which mainly deal with children from lower social strata.[51] The assumption is that complexly traumatized, "difficult" children occur more often in lower social strata. However, we can assume

[48] Turner (2014).

[49] Porges (2009).

[50] Unknown source.

[51] Goodrum (2022).

that, if we look at the extended concept of trauma, in the higher social strata it is usually only better concealed.

The current crisis in mental health care is the consequence of a lack of emotional care, and this permeates all social classes. In the future, it will certainly not get easier for young adults in this regard, as they are at the mercy of enormous social pressure in a world shaped by social media and virtual realities. In most cases, both parents are also working and prefer to pay expensive institutions to take care of their children. Thus, the emotional lack that they themselves experienced in a different way in their generation is passed on. Therefore, an essential aspect, which also has an educational component, as mentioned above, is communal care. As the quote attributed to African cultures says, *"To raise a child, it takes a whole village."* Thus, education and upbringing are not the sole responsibility of the parents or the school. The social context and the social support in which an individual grows up is essential and can definitely have a preventive effect.

In *conscious education* it is not only about establishing theories, but within the framework of personal awareness processes, to experience the inner, necessary work for one's own transformation and then to convey it. As listed, the processing of inner traumas has a longer-term preventive effect and brings people closer to themselves and their purpose of existence.

This form of *trauma-inclusive education* will particularly be relevant for health care professionals—especially physicians from diverse disciplines—and deserves its place in the academic context. It can lead to a broader understanding of the interrelationships between body/mind/trauma/disease.

Finally, we don't just need to know how to treat people who experienced trauma, we need to improve our understanding between trauma and disease. How to prevent the consequences of experienced trauma by cleansing the emotional space inside our body? There are manifold approaches today, that are dedicated to exploring and healing this emotional space and they are a crucial part of the *conscious education* in NH. Our thoughts and our feelings and their effect on our body are inextricably linked. By letting go the identifications in our mind of who we think we are and what we think is best for us, we slowly move to the core of the holistic approach, that is *the Self*, literally.

If we move on that path of emotional and conscious reconnection, it means that we are able to identify what is us and what is just the story surrounding and influencing us. If we identify with being "the cancer patient," "the heartsick," or "the depressed" we stay in a victimhood mentality that deprives us of our sense of self-responsibility. Without that resilience it is very difficult to heal. As long as we do not change perspective and move to that level of consciousness, it will be difficult to primarily prevent disease, because we are still identifying with the underlying story. The educative dimension of NH is huge and multi-layered and needs to be established in the years to come. These are but a few aspects. *Conscious education* ultimately teaches how to move from control to surrender while being together in this human experience. It starts where trauma has not happened yet—and it moves beyond trauma.

11 The Spiritual Dimension of NEW HEALTH

If we look at the whole transdisciplinary spectrum in relation to being human, in which health without question plays an essential role, we cannot avoid including other fields such as theology, philosophy, psychology, physiology, metaphysics, quantum physics, and spirituality. For the great questions of humanity: *"Who am I?"* and *"What am I here for?"* are and will always be drivers to remind us of our original consciousness. Thus, illness can also be experienced each time as a crisis, a turning point (lat.: crisis, turning point) to experience personal transformation. In working with people with cancer in particular, these questions are essential, as the urgency of the disease gives one the opportunity to confront one's real being beyond the facade. In the accompaniment, especially of women with breast cancer, from a certain point on, it was always very much about the topic of their calling. After the question of the *why* has lost significance, the *what for,* the question for meaning and purpose in life, becomes much more relevant.

This brings us to the three levels of human existence, which can be summarized as follows:

The *first level*: one is oriented to one's own will, to desires that one tries to fulfill, according to which one directs one's life. For example: Desire for high socio-economic status. To achieve this, one works energetically throughout one's life and submits to compromises and circumstances that do not serve one in essence, but which correspond to the image one has of happiness. The goal after retirement is to realize all the dreams that one has denied oneself in life. All too often I have seen that exactly at this time illness sets in, because the meaningfulness in life was never really given space, and in the absence of work a vacuum is created that one does not know how to fill. This is probably the level at which the largest proportion of people find themselves.

The *second level*: this is the level where one pursues one's calling for this life and experiences great satisfaction in fulfilling the purpose of one's current existence. This leads to peace of mind and thus inner contentment and balance, which automatically has a positive effect on physical well-being. An important PDOH (Personal determinant of health) is fulfilled, namely "purpose" is lived.[52] There, too, it can happen that after the identification about the activity is gone, a vacuum arises, but much less, because the intention "to have lived a fulfilled life" has been fulfilled. On this level the condition is given to move to the third level. For the mind is calm and the possibility to incorporate contemplation, meditation, and meaning-oriented action into one's life as an essential component is given.

Nevertheless, for many people it is often sufficient to have done in life what they had dreamed of and where their talents and creativity could be used. Consequently, there was no need to move beyond the created comfort zone of the second level. Almost no one I have accompanied in illness over the years has not asked the question, what is his/her life's mission? And often one of the answers is obvious:

[52] MacLeod (2020).

part of their life's mission is to have that very moment experience that brings them closer to their authentic self. In fact, it's not about searching so far, but acknowledging what is right there. Instead of living as a human doing, become a human being to realize what it means to be fully alive. With this quality it is possible to get closer to the moment and to experience a state of inner contentment and peace. This conscious awareness then flows into all areas of life and becomes a great resource in dealing with the physical imbalance called disease.

Now there are also the transitional phases, especially on the way from experienced illness to perceived health. The times when everything feels foggy, unclear, disoriented, helpless, and dark. Dealing with transitions is a tool that no one teaches. It's just either this way, or that way, sick or healthy. But the nuances in between are unknown constants and what it takes to deal with them is simply assumed. It feels like going through a long tunnel, a client once described it—and you don't know when the end is in sight. During this time, you wish for change. There is this certainty that change is necessary. Only it is not yet clear how this can take place. In this time, it makes sense to move to the *third level of* existence.

The *third level*: There, the question "What am I doing for eternity?" is in the foreground. It is about personal search for meaning, the concept of time is extended beyond this life. The examination of spiritual questions comes to the fore. People increasingly devote themselves to activities that have to do with inner development. For some, this resource is found in faith and prayer. It is well known that faith is associated with a better quality of life.[53]

Still others are just beginning to become interested in the field of consciousness and are opening up to the big world questions: *"What is the meaning of life?"*, *"Who am I really?"*, and *"Where do I come from?"* A kind of "awakening" takes place, which is often accompanied by a change of interests and a lack of understanding on the part of those around them.

If it is a time of transition in the realm of the *second level* and the person does not want to return to the *first level* under any circumstances—which is very often the case, because this level has led to self-denial and illness—it makes sense to move to the third level until the *second level* emerges clearly from it. In order to connect with the inner wisdom or the original consciousness, there are various practices, and it is highly individual what is beneficial for personal development. It is free from dogmas or external standards. It is about getting in touch with inspiration and gaining insight into the big picture and one's own essence.

Unfortunately, there is no room for this in the current time window given for recovery and reintegration processes as these processes often require space and time. Would it not make sense to rather provide this time of convalescence sufficiently (lat. convalescere: to become healthy, to recover, to become strong) to become strong and heal from the core? So that after a few weeks or months it does not come again to the next disease failure and in the course to chronic diseases?

[53] Rediger (2020:339ff); Cunningham (2002).

My hypothesis is that trauma is basically nothing other than disappointed love—disillusioned love. The unfulfilled desire for the exchange of love that lies dormant in each of us. When we are born we need safety, nourishment, and love. The disillusionment we experience if this cannot be provided to a sufficient extent is the point where trauma starts in this lifetime and insecurity starts manifesting. Ultimately, it is a starting point for the quest to reach out for the highest good, love, that will only find fulfillment in an unconditional space.

As long as we identify with the stories and associated emotions we experience as human beings, it is difficult to penetrate to the core of our being. Then the time comes on the healing path when it is a matter of letting the stories, like clouds in the sky, move on ... and to move beyond trauma. Is illness there to accelerate these processes? To put us in touch with our original healing power, of which we ourselves are part? I think this is the case. Only when we grasp the human being in its full extension, transdisciplinarily; penetrate its personal, transgenerational, cultural, and spiritual history with understanding and compassion, can we recognize where deep within the conflict festers that has led to chronic stress and thus to illness.

> Just as the earth cannot be changed unless the consciousness is changed, our inner ecosystem and consequently our bodies cannot change. "Earth cannot be changed for the better unless the consciousness of individuals is changed."[54]

This is part of the spiritual dimension of NH and serves as an important foundation for walking pathways out of the disease.

12 Conclusion and Outlook: Moving Beyond Trauma

Could New Health, which advocates for individual and collective health care, be the New Paradigm of medicine? In this article, I have described the four core transdisciplinary dimensions of NH and illustrated how expanded knowledge of the connections between the origins of disease can lead to greater physical and psychological balance, meaningfulness, and ultimately, ourselves. It breaks ground in offering a new approach to health and prevention via emotional health, emotional intelligence, and spiritual wisdom. For this, it is essential to expand the concept of trauma. Because trauma is always only *what* it does to the person, not *how* it looks from the outside. It also argues for a movement away from the unconscious practitioner to the self-reflective, conscious therapist, from the lost patient to more self-responsibility and patient competence.[55]

The next revolutionary wave in medicine is ideally characterized by lived humanity that leads to compassion. To get to this point, NH proposes *conscious education* primarily in health care and schools. This also includes the creation of a

---

[54] Küng (1998:31).

[55] Bopp et al. (2005).

social framework for the recovery of individuals, which can have various forms. In order for this upheaval in consciousness to succeed, the healing of our "inside" is needed. That is why the revolution begins in each one of us. Only in this way can medicine heal and be healed from the inside out.

The spiritual dimension of NH is a building block for recognizing the transformational power present in the illness and to actually be able to experience these crises as turning points toward the good. In this way, it is possible for everyone to recognize that, apart from one's own experienced history, there are other experiential spaces of human existence. NH stands for a living, conscious, transdisciplinary medicine, in which being with humanity has a place again and which carries us beyond trauma.

References

Anonymous. (2017). World Medical Association Declaration of Geneva. *African Health Sciences, 17*(4), 1203. Available in https://pubmed.ncbi.nlm.nih.gov/29937893

Barnes, T. L., et al. (2022). Loneliness, social isolation, and all-cause mortality in a large sample of older adults. *Journal of Aging and Health, 34*(6–8), 883–892. https://doi.org/10.1177/08982643221074857

Bloom, D. E., et al. (2011). *The global economic burden of non-communicable diseases*. World Economic Forum. Harvard School of Public Health. Available in https://www3.weforum.org/docs/WEF_Harvard_HE_GlobalEconomicBurdenNonCommunicableDiseases_2011.pdf

Bopp, A., et al. (2005). *Was kann ich selbst für mich tun? – Patientenkompetenz in der modernen Medizin*. Rüffer & Rub.

Buber, M. (1984). *Das Dialogische Prinzip*. Lambert-Schneider.

Collins, F. S., et al. (1995). The human genome project. *Alcohol Health and Research World, 19*(3), 190–195. Available in https://pubmed.ncbi.nlm.nih.gov/31798046

Cunningham, A. J. (2002). *Bringing spirituality into your healing journey*. Key Porter Books.

Esaki, N., et al. (2013). The sanctuary model: Theoretical framework. *Families in Society, 94*(2), 87–95. https://doi.org/10.1606/1044-3894.4287

Felitti, V. J., et al. (1998). Relationship of childhood abuse and household dysfunction to many of the leading causes of death in adults. *American Journal of Preventive Medicine, 14*(4), 245–258. Available in: https://pubmed.ncbi.nlm.nih.gov/9635069

Goodrum, N. M., et al. (2022). Family-based prevention of child traumatic stress. *Pediatric Clinics of North America, 69*(4), 633–644. Available in: https://pubmed.ncbi.nlm.nih.gov/3593

Gustafson, C., et al. (2017). The jump from cell culture to consciousness. *Integrative Medicine (Encinitas), 16*(6), 44–50. Available in: https://www.ncbi.nlm.nih.gov/pmc/articles/PMC6438088

Ha, J. F., et al. (2010). Doctor-patient communication: A review. *Ochsner Journal, 10*(1), 38–43. Available in: https://pubmed.ncbi.nlm.nih.gov/21603354

Hess, C., & Hess-Cabalzar, A. (2006). *Menschenmedizin. Für eine kluge Heilkunst*. Suhrkamp.

Hodkinson, A., et al. (2022). Associations of physician burnout with career engagement and quality of patient care: Systematic review and meta-analysis. *BMJ, 378*, e070442. Available in: https://www.bmj.com/content/378/bmj-2022-070442

Holt-Lunstad, J., et al. (2011). Understanding the connection between spiritual well-being and physical health: an examination of ambulatory blood pressure, inflammation, blood lipids and fasting glucose. *Journal of Behavioral Medicine, 34*, 477–488. https://doi.org/10.1007/s10865-011-9343-7

Irigoyen-Otiñano, M., et al. (2022). Suicide among physicians: Major risk for women physicians. *Psychiatry Research, 310*, 114441. Available in: https://www.sciencedirect.com/science/article/pii/S0165178122000555
Küng, H. (1998). *Declaration of the parliament of world religions*. SCM Press.
Leendertse, P. (2020). *The root cause of cancer*. Wheel of Life.
LePera, N. (2021). *Heile. Dich. Selbst*. Harper Wave.
Levine, A. P., et al. (1998). *Trauma-Heilung: Das Erwachen des tigers; unsere Fähigkeit, traumatische Erfahrungen zu transformieren*. Synthesis.
Lipton, B. H. (2005). *The biology of belief: Unleashing the power of consciousness, matter and miracles*. Mountain of Love/Elite Books.
Long, R. M., et al. (2023). Loneliness, social isolation, and living alone associations with mortality risk in individuals living with cardiovascular disease: a systematic review, meta-analysis, and meta-regression. *Psychosomatic Medicine, 85*(1), 8–17. Available in: https://pubmed.ncbi.nlm.nih.gov/36441
Loss, J. (2017). *Wandel in der Medizin*. Walter de Gruyter.
MacLeod, S., et al. (2020). Defining the personal determinants of health for older adults. *Innovation in Aging, 4*(Suppl 1), 924. Available in: https://www.ncbi.nlm.nih.gov/pmc/articles/PMC7741943
Manning, D.. (n.d.) No books available yet. https://emotionalreleases.com.
Maté, G., & Maté, D. (2022). *The myth of normal: Trauma, illness, & healing in a toxic culture*. Avery.
Maté, G. (2003). *When the body says no: The cost of hidden stress* (1st ed.). A.A. Knopf.
Nagesh, P., et al. (2022). The physical and mental health consequences of social isolation and loneliness in the context of COVID-19. *Current Opinion in Psychiatry, 35*(5), 305–310(6). https://doi.org/10.1097/YCO.0000000000000806
Neeraj, K. A. (2003). Interacting with cancer patients: the significance of physicians' communication behavior. *Social Science & Medicine, 57*(5), 791–806. https://doi.org/10.1016/S0277-9536(02)00449-5
Pert, C. B. (1999). *Molecules of emotion: Why you feel the way you feel*. Touchstone.
Porges, S. W. (2009). The polyvagal theory: New insights into adaptive reactions of the autonomic nervous system. *Cleveland Clinic Journal of Medicine, 76*(2), S86–S90. Available in: https://www.ncbi.nlm.nih.gov/pmc/articles/PMC3108032
Rediger, J. (2020). *Cured: The power of our immune system and the mind-body connection* (p. 339). Penguin UK.
Relman, A. (1980). The new medical-industrial complex. *The New England Journal of Medicine, 303*(17), 963–970. https://doi.org/10.1056/NEJM198010233031703
Relman, A. (2009). Assessment and accountability. *Journal of Health Services Research & Policy, 14*(4), 249–250. https://pubmed.ncbi.nlm.nih.gov/19770124
Roepke, A. M., et al. (2014). Meaning and health: A systematic review. *Applied Research Quality Life*, 1055–1079. https://doi.org/10.1007/s11482-013-9288-9
Scaer, R. (2014). *Das Traumaspektrum. Verborgene Wunden und die Kraft der Resilienz*. G.P. Probst.
Schramme, T. (2023). Health as complete well-being: The WHO definition and beyond. *Public Health Ethics, 16*, 210. https://doi.org/10.1093/phe/phad017
Timothy, L., et al. (2022). Loneliness, social isolation, and all-cause mortality in a large sample of older adults. *Journal of Aging and Health, 34*(6–8), 883–892. https://doi.org/10.1177/08982643221074857
Turner, K. A. (2014). *Radical remission: Surviving cancer against all odds*. HarperOne.
Uhlenbruck G. (1929). Quote by German immunobiologist and aphorist.
Virchow, R. (1871). *Die Cellularpathologie in ihrer Begründung auf physiologische und pathologische Gewebelehre*. Hirschwald.
von Bingen, Hildegard (1151–1158). (n.d.-a). https://www.aphorismen.de/zitat/16940.
von Bingen, Hildegard (1151–1158). (n.d.-b). https://www.aphorismen.de/zitat/16932.

Von Witzleben, G. (2019). *Das triadische Prinzip, Minimalinvasive Psychologie mit Bauch, Herz und Kopf.* Carl-Auer Verlag.

Wang, F., et al. (2023). A systematic review and meta-analysis of 90 cohort studies of social isolation, loneliness and mortality. *Nature Human Behaviour, 7*, 1307. https://doi.org/10.1038/s41562-023-01617-6

World Health Organization. (2008). *The global burden of disease: 2004 update.* World Health Organization. Available in: https://apps.who.int/iris/handle/10665/43942

Nanda Meister (MD) studied Medicine at the University of Zurich, Switzerland and specialized to be a doctor of Medical Oncology and Internal Medicine. Since the third year at University it was clear to her, that she will move into the field of working with cancer patients as the opportunity for personal growth through physical and emotional crisis is omnipresent in that domain. To experience what medical professionals confront in their everyday life and to understand more what it means to be a cancer patient, she went through a hospital career in general oncology with her main emphasis on palliative care and breast cancer. Continuing her journey as a leading doctor at Center for Integrative Oncology in Zurich, she realized the need to dive deeper to the root cause of the disease that stayed untouched by conservative medical approaches and still seemed to be so utterly relevant. That was when she decided to found Transformative Oncology that strives to resolve the underlying trauma associated with the development of cancer. Her further education included different approaches, including the Simonton Method, Process- and embodiment focused Psychology, Systemic Therapy and Constellation work with the Triadic Principle, Phytotherapy, and more. But what she sees as the foremost ingredient to support people in healing is her own personal healing journey, moving through trauma and then beyond, again and again. Her primary intention is to create a healthcare system with a transdisciplinary approach, where the individual, social, educational, and spiritual dimensions are at the core. Therefore, she shares NEW HEALTH with the world to prevent disease and co-create a healthy, trauma-aware environment, where people experience purpose and meaning. For that purpose she gathers conscious leaders in the fields of Health and Wholeness to collaborate and to rise in consciousness. Since her early twenties she travels to India regularly for her own spiritual development and offers pilgrimages to different sacred sites in the world to nourish the inner, eternal dimension.

The Seven Gifts of a Vegan and Vegetarian Nutrition: A Transdisciplinary Reflection

Christian Halper

1 The Transdisciplinary Aspects

The question of proper nutrition is an application of transdisciplinary thought and action. For appropriate nourishment touches disciplines such as health maintenance (medicine), well-being (psychology), taste (aesthetics), getting along well with each other (ethics), good relationship with nature (ecology), inner peace (stress management; religion), and mental growth (spirituality). All these disciplines are addressed by "vegan and/or vegetarian nutrition," which connects the disciplines and even reaches beyond them. It is thus a typical problem-solving field of transdisciplinarity.

The following remarks now treat the mentioned disciplinary aspects of vegetarian nutrition as its "the seven gifts" to mankind, without forcing them on it. Each person is free to accept them or not, and to check to what extent he or she serves the greater whole with it or not. The seven gifts are based on scientific studies, personal experience of the author, who has been dealing with this for many years, both theoretically and practically, as a gastronomic entrepreneur,[1] as well as obvious conjectures and hypotheses. Accordingly, the following considerations are substantiated or put up for discussion as worth considering, depending on the available sources.

[1] The author is founder and owner of two vegetarian restaurants, called "Tian"—one in Vienna and one in Munich—each of them got one Michelin star due to their excellent food.

C. Halper (✉)
Entrepreneur, Vienna, Austria
e-mail: halper@licht7.at

© The Author(s), under exclusive license to Springer Nature Switzerland AG 2025

M. Schüz (ed.), *Transdisciplinary Thinking and Acting*, Transdisciplinary Management of Ecological and Social Crises,
https://doi.org/10.1007/978-3-031-70680-6_18

2 The Seven Gifts

A vegetarian and vegan nutrition provides humanity seven gifts. These gifts are: (1) Nutrition and Taste, (2) Well-being, (3) Health, (4) Empathy and Love, (5) Global Healing, (6) Inner Peace, and (7) Spiritual Growth. These presents influence the entire world. They are interconnected with each other—just like waves that expand when you throw a stone into water.

2.1 #1: Nutrition and Taste

The first gift of a vegetarian cuisine is that it nourishes and tastes well. There is a myriad of combinations of ingredients, which are maybe unknown or have not been tested yet, which include, for example, morels with leek and vanilla, vegan "Weisswurst" made from mushrooms, etc. This vast variety of different tastes in vegetarian products is found in land as well as in water plants. Thus, it is easily possible to create an extraordinary vegetarian and vegan Gourmet cuisine. This is based on the big variety of ingredients, such as, for example, flaxseed oils, lentil sprouts, Tempeh made from soy beans and oyster mushrooms for example. These and many other plant-based ingredients can lead to an extraordinary taste experience. These ingredients become essential food items as they give you energy and power and make you feel more awake and present. Furthermore, spices and oils serve as a variety of taste. Naturally, these taste experiences are very individual and not everyone will like everything that has been invented as combination or even dislike certain ingredients or have an allergy. However, as individual as the taste of each person is, as individual are the possibilities to mix and match the variety of ingredients available and create extraordinary vegetarian and vegan dishes. Given the high numbers of available products, almost infinite combinations are possible and can be experimented with. To put this in context, humans know about 250,000 plant species of which around 30,000 are possible to be consumed and 7000 of which have been cultivated for human consumption.[2] As a comparison, there are around 360 water-based animals and plants as well as 35 types of domesticated land animals for human consumption. Thus, we have 94.5% land plants, 5% water animals and plants and finally 0.5% land animals. To further spread the variety, these individual plants have sometimes over 100,000 different varieties, such as rice plants.[3]

These numbers alone show the variety that a plant-based diet offers in terms of different tastes as compared to meat consumption. Take, for example, the variety of taste thinking of "Tomato-Oregano-Vanilla-Lentils-Coffee" as compared to "Beef-Pork-Lamb-Goat-Chicken." The spectrum of taste in the latter is much more

[2]Bundesanstalt für Landwirtschaft und Ernährung (n.d.)

[3]Think Rice (2020).

constrained as compared to the mentioned plants. Thus, it follows that a vegetarian gourmet cuisine is much easier to create rather than a meat-based.

We see a clear trend toward the vegetarian diet. In 2022 in Germany, about 1.58 million people claimed to not consume animal products, which was an increase of 170,000 persons (or 10.75%) from 2021 and even more so since 2015 which saw an increase 730,000 people within 7 years. This shows that this is a trend that has been ongoing for years now and will likely continue to grow in the future as well.[4] This is especially due to the rising number of vegetarians and vegans in the younger generation as they are more likely to relinquish meat (and meat products) which has also been shown in various statistical analyses. These show that around 21.8% of the vegetarian population in Germany is in the age group 20–29 which constitutes the largest group of vegans and vegetarians. In the older age groups (40 years and older), the average number is around 12.35%.[5] Furthermore, it could be shown that since 1990 the meat consumption faces a continuous downwards trend.[6]

2.2 #2: Promotes Well-being

One important demand of a healthy nutrition is that I want to feel well after eating and that it does me a favor. Food should bring you energy rather than draining it. I want to feel awake and clear in my mind. My personal experience is that I need less sleep when eating vegetarian or vegan as compared to a meat diet. Personally, this ill-being after eating meat was the reason why I started to find alternatives to my intake over 20 years ago already. What I found however was that "only vegan" does not necessarily constitute a good diet and I do not have this well-being automatically. Looking for French fries with Ketchup it is vegan, but I do not feel light and energized after finishing my plate. This point is undoubtedly complex as food alone is not an isolated reason to feel good or bad. Other factors such as stress (workload, deadlines, etc.) or the overall health status (infections, chronic diseases, sleepless nights, etc.) also play an important role in ones well-being. Although, the diet also has an important role in the mentioned confounding factors.[7] In a study it was found that veganism and vegetarian lifestyle is not only associated with the choice of food but also has an holistic approach toward life and its challenges including psychological stress and a positive attitude toward life itself which makes people happier and more balanced. However, as mentioned earlier with the "fries and Ketchup" example, there is an established connection with junk food and ill-being, which holds true for both meat consumption and vegetarian/vegan junk food choices.[8]

---

[4] IfD Allensbach (2022a); cf. statista (2022).

[5] IfD Allensbach (2022b).

[6] Thünen Institut (2022).

[7] Krizanova and Guardiola (2020).

[8] Gallagher et al. (2021), Gehring et al. (2021).

What I found was that A) food must be as close to nature as possible, including features such as: ecological, whole foods, rich in crudités, seasonal, regional, and fresh. B) The diet should be versatile and diverse rather than eating heavily processed food items.[9] As example, if you would only eat ecologically grown baby carrots, this would be not only one-sided but probably also bad for your overall health in the long term. And C) you should however also be aware of your own type and preferences. Not everyone can tolerate raw or spicy foods the same way. You have to observe and feel your own body rather than adhering to strict rules to find out what does you good and what not. This also includes finding the right balance regarding the amount and volume of food you eat. It also has been shown that satiety only starts after around 10–20 minutes after food intake, thus eating speed is also of relevance when it comes to eating (and potentially adverse overeating).[10] If you are overweight, it should also be noted that the feeling of satiety might be shifted back and it takes longer for you to feel full.[11] D) Finally, a healthy lifestyle beyond food intake also consist of drinking sufficient amount of water, exercise, and breathing which is not limited to one type of person but includes all human beings regardless of gender or age.[12]

You can think of the program as the ancient philosophical teachings of Empedocles (who lived in the middle of the fifth century BCE) who is claimed to be the inventor of the four elements.[13] In this case, these elements include nutrition (Earth), drinking water (Water), movement (Fire), and breathing (Air). This metaphorical comparison can serve you as an easy reminder of what is necessary for a healthy life and well-being without losing yourself in complicated scientifical explanations on what effects these rather simple rules have on our day to day lives.

Based on my personal experience, I found that when I eat a vegetarian/vegan diet I am awake for an extra hour each day as I do not need as much as sleep as before. Summing up an hour daily more time for myself adds up to 15 days (or 3 weeks) of extra holiday each year. This has already been proven scientifically as the quality of sleep is positively correlated with a vegetarian diet.[14]

Most people lack the vitamin B12 irrespective of the diet form and thus it should be supplemented.[15] Overall, it is suggested to slowly change your behavior and your diet rather than an approach where you change everything at once.

This present of well-being also includes one further point. It is recommended to avoid fish, meat, and dairy products the day before running a marathon as your body needs the right nutrients for such an extraordinary athletic performance. Runners and endurance athletes are aware of this since many years already and it has been shown

---

[9] Gallagher et al. (2021), Gehring et al. (2021).

[10] Liu et al. (2000).

[11] Matsuda et al. (1999).

[12] Ayotte Jr and Corcoran (2018), Fadda et al. (2012), Pross et al. (2013).

[13] Kingsley (2020).

[14] St-Onge et al. (2018), Kohyama (2021).

[15] Obeid (2019).

in to the long history of more than 100 years of running modern day marathons that avoiding these animal products benefits everyone attending such events.[16]

Not only for endurance athletes, but also for strength athletes, a vegetarian diet has been shown to be beneficial. As Patrik Baboumian, the strongest man in Germany, says: "The strongest animals are plant-eaters. Gorillas, buffalos, elephants and me"). The movie "Gamechangers" also shows multiple examples on how a meatless diet helps athletes to achieve high performance in their fields which has also been proven in many scientific studies already.[17]

Furthermore, thinking outside of athleticism that includes high performance that you achieve with your body, the nutrition also plays a vital role in sports that require attention and thinking. Viswanathan Anand who was world champion in playing chess from 2007–2013 claims that his success came due to a vegetarian diet.[18] Other famous, exceptional people such as Albert Einstein and Thomas Alva Edison also ate a vegetarian diet.[19]

2.3 #3: Being Healthy

One effect of a vegetarian diet is a healthy life. It has been shown that the prevalence of cancers such as breast cancer or heart attacks can be reduced when adopting a vegetarian diet that serves as protective mechanism in these illnesses.[20] As already mentioned earlier, this also includes a healthy lifestyle overall which is also found commonly in people adopting a vegetarian diet.[21] This might be new to some people, but already in ancient times, personas like Plutarch already found a correlation between eating and health. In his advice about keeping well, it is written that "Especially to be feared are indigestions arising from meats; for they are depressing at the outset, and a pernicious residue from them remains behind. It is best to accustom the body not to require meat in addition to other food. For the earth yields in abundance many things not only for nourishment but also for comfort and enjoyment, some of which it grants to our use just as they are with no trouble on our part, while others we may make savory by all sorts of combination and preparation."[22]

When looking at our bodies, the question that arises is "Which fuel is it built for?". In case of nutrition this means: "For which diet is the body built for?".

---

16 Burke (2007), Tiller et al. (2019).

17 Pohl et al. (2021).

18 Happy Cow (n.d.); cf. Szot et al. (2022).

19 Krüger (2018), International Vegetarian Union (n.d.); cf. Dreißig (2022), Einstein Webseite (n.d.), Heim & Jochmann (2000), Ludwig (2015).

20 Dinu (2017), Jabri et al. (2021).

21 Olfert (2022).

22 Babbitt (1928).

In the following, I metaphorically use the comparison of a human body with a car which was produced and handed over to its first owner. The car producer provide to the buyer: (a) manufacturer information. If you do not have that, then you can do a (b) technical analysis or (c) learn from experience.

Manufacturer Information

Looking at religion, what does the word of God mean throughout the spiritual traditions of Christianity, Buddhism, and Hinduism when it comes to the usage of the human body?

Judaism/Christianity/Islam: "And God said, 'Behold, I have given you every plant yielding seed that is on the face of all the earth, and every tree with seed in its fruit. You shall have them for food.' (Genesis 1:29).[23] All three religions reference the fifth commandment "Thou shalt not kill" which can be said not only about humans, but animals as well.

Buddhism: "Whatever there may be with breath of life, whether they be frail or very strong, without exception, be they long or short, or middle-sized, or be big or small, (. . .) Let no one bring about another's ruin and not despise in any way or place; let them not wish each other any ill from provocation or from enmity" (Metta-Sutta 1.18).[24] Buddhism teaches that all beings in the world–irrespective of their size—"without exception" have the right to live and that no one should harm the other irrespective of their being. This is a central demand in Buddhism and many Buddhists (but not all) practice a life of vegetarianism or veganism. Next to the better known Buddhism, there is also Jainism which practices not even eating the roots of plants as this would kill the plant and not allow regrowth.[25]

Hinduism: In Hinduism, a central principle is "Ahimsa" which means non-violence and is a central part of the worldview. This does not only apply to humans, but all beings. In India, up to 39% of the population lives vegetarian which is the highest number of vegetarians in the world.[26] Even when cutting plants for human consumption, it is seen as important to keep the damage as little as possible, acknowledging that complete freedom of violence is not possible, but more of a "more or less" principle.

[23] Bible Gateway (2016).

[24] Mills (n.d.)

[25] Shah (2004).

[26] Sahgal et al. (2021).

Technical Analysis (Physical Characteristics of the Body)

There are multiple indicators that humans are built for a vegetarian diet. This includes flat cuspids, the gastro-intestinal tract, the shape of the hands as well as the endurance when it comes to physical activity. On the other hand, flesh-eaters have sharp cuspids (fangs) and flat, broad molars.[27]

The factor regarding the length of the digestive system differs between plant- and meat eaters.[28] The latter retain their food in the digestive tract longer than the former. The fiber-rich plants are excreted more quickly, while the rotting meat remains in the body longer. And, as a Stanford study has shown, the vegan twin siblings have an overall better cardiovascular health than their meat-eating counterparts.[29]

Additionally, we can see a difference in the hands/paws of herbivores and carnivores. The latter have claws to catch, hold, and kill their prey, while herbivores like the human cannot use their fingers to do that, just imagine trying to grab and hold a fish out of flowing water with your hands, which are not designed to do that. Rather than that, the dexterity of the complex hand allows the human to crack open nuts, pull roots (vegetables), and grab fruit from trees rather than operating on meat with their bare hands.

Learn from Experience

There is a correlation between spendings for health and life expectancy. In Japan, people do not spend much on their health, yet they are getting older compared to the people in the USA who spend a lot on their health yet die younger. For comparison, in Japan, the per capita spending on health is 4666 USD, while the population of the US spends 12,914 USD being almost three times more, while the average life expectancy of Japan is 84.62 years and, in the USA, it is 77.28 years[30] (see Fig. 1). There are multiple reasons that play into this discrepancy, but my personal assumption is that this is directly correlated with their nutrition. In Japan, there is the island of Okinawa that is also called "Island of the 100 year olds." People there traditionally live eating mainly fruit and vegetables, as Craig Willcox (Professor of public health and gerontology at Okinawa International University) says: "The sweet potatoes, bitter melon, carotenoid-rich marine foods like seaweeds, green leafy vegetables, and fruit in the diet are anti-aging as they reduce inflammation and oxidative stress."[31]

[27] Mu et al. (2021).

[28] Withers et al. (2016), Karasov and Martínez del Rio (2007).

[29] Beasley et al. (2015); regarding the Stanford Study refer to Moskal (2023).

[30] Peterson-KFF (2023), Data Commons (2023a), , 2023b).

[31] Goss (2020).

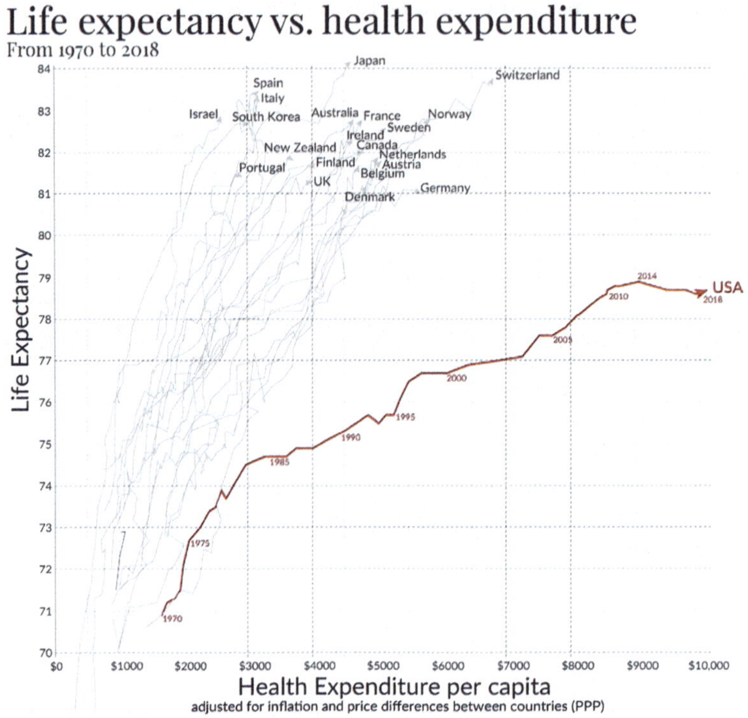

Fig. 1 Life expectancy vs. Health Expenditure per capita of different countries. Note the difference of almost 6 years comparing Japan and the USA despite spending more than twice as much for individual health. Source: Roser (2020), cf. Our World in Data, OECDstat (2017): https://ourworldindata.org/us-life-expectancy-low

Japanese People live with the saying "nuchi gusui" which translates to "let food be your medicine" and "hara hachi bu" which means they stop eating when they feel around 80% full.[32]

The influence of our food on our health has been extensively studied, for example by Dr. med. Ludwig Manfred Jacob who published the book "Dr. Jacobs Weg des genussvollen Verzichts" in which he analyzed 1400 studies to get an idea of what can be considered healthy and came to the conclusion that a lot of vegetables, herbs, spices, berries, most fruit, 2 liters of water daily, herbal and green tea as well as a moderate amount of complex carbohydrates with low/medium glycemic index such as wholegrains, legumes, and nuts.[33] This has also been displayed in the film "Forks over Knives."[34]

[32] Goss (2020), Willcox (2002).

[33] Jacob (2018).

[34] Klein (2023).

Another study (meta-analysis) found a correlation between the intake of animal (and dairy) fat linked to breast cancer risk in women which was not found with the intake of plant-based fats.[35]

Furthermore, there is an established link between meat consumption and heart attack risk. Interestingly, this does not only hold true for processed meats (18% increase in risk) but also for unprocessed meat (9% increase).[36] Even more so, eating a plant-based diet even shows a reduction of the risk of suffering from coronary heart disease.[37]

This evidence shows that those who nurture themselves according to our personal constitution, being vegan or vegetarian, live healthier and have a better life overall. My personal opinion is that humans are natural vegetarians as this fits our constitution, but this seems to have been forgotten. There is enough protein, iron, and calcium available in plants. The often cited lack of Vitamin B-12 comes from extensively washing the vegetables or fruits. Looking at gorillas who live a vegan diet, no lack of Vitamin-B12 can be detected as they eat their food unwashed.[38] In fact, it has been shown that apes in captivity are also at risk of being Vitamin B12 deficient.[39]

2.4 #4. Empathy and Love

The first three presents are more about the "myself," meaning: my food and taste, my well-being, my health. The fourth present however relates to the "you," the relationship to other people and living beings, the social component of life.

When you think of the resonance principle from physics, you think of an oscillating tuning fork, which resonates to others and the oscillation is being transported to them. When we think about this physical principle, we can also utilize this for our nutrition. There is also the saying "What goes around, comes around," we experience ourselves as we experience others. Or think about "like attracts like" and the golden rule "What you do not wish done to yourself, do not do to others" or, in other words, "treat others as you want them to treat you." So, ask yourself the question: Do you want to feel love, empathy, happiness, and love? Or do you want to feel pain, violence, and fear? Out decisions lead to our related feelings being sent out. If we sow violence, we will reap violence and thus if we sow love, we will reap love. These simple principles have been preached (and lived) since thousands of years and are still repeated to these days. All these principles do not only hold true for human interactions, but to all living beings.

---

[35] Khodarahmi and Azadbakht (2014).

[36] Papier et al. (2023).

[37] Quek et al. (2021), Esselstyn (2014).

[38] Klaper (n.d.)

[39] Oxnard et al. (1970).

And when we kill and devour animals, we also ingest their stress hormones such as adrenalin that they produce during transport, handling and their death.[40] This could potentially lead to adverse effects on the quality of meat and in further consequence on our personal health.[41] While there is no direct link of the released stress hormones directly influencing us established yet, if we keep our nutrition to a plant-based diet, then we are on the safe (and healthy) side by avoiding these reactions entirely.

2.5 #5 Global Healing

We are currently facing a loss of biodiversity including extinction of up to one million plants and animals caused by humans.[42] This leads in further consequence to a disturbance of the natural order and balance and the natural resilience of nature. The way we eat (and produce our food) acts as one of the main parts in this development.[43] Meat consumption is directly linked to deforestation as these areas are being used to grow food for livestock and it is assumed that 12–18% of global greenhouse gasses are attributed to the livestock industry.[44] To quantify the loss of forests (especially in the Amazon regions and the rainforests worldwide), it is assumed that around 10 million hectares are lost due to deforestation and the main usage of this land is to plant crops for feeding animals planned for consumption.[45] Naturally, the question is why it is necessary to use that much land for producing food (i.e., "second-hand" food). This question can be answered by the inefficiency of calories when they come from animal products. Humans use only around 59% of the available calories in the world (accounting for both plants and meat).[46] The rest is being consumed by animals, which is inefficient. So it takes 25 calories of grains to produce 1 calorie of beef, 15 for pork, and 9 for chicken.[47] Logically, it follows that the direct consumption of the calories is a more efficient way of nutrition. Apart from this obvious reason, also the global greenhouse gas emissions differ greatly between different types of food. It has been shown that meat has a 250 times higher emission rate as compared to legumes.[48]

Looking at another habitat of consumed animals, i.e. fish it is well-known that the oceans are being overfished. Around half of the earths' oceans are in a state of having

---

[40] Carrasco-García et al. (2020).

[41] Korneliussen (2012).

[42] Tollefson (2019), Naeem et al. (2016).

[43] Tollefson (2019)

[44] González et al. (2020).

[45] Ritchie and Roser (2021).

[46] Cassidy et al. (n.d.)

[47] Yale Center for Business and the Environment (2016).

[48] Tilman and Clark (2014).

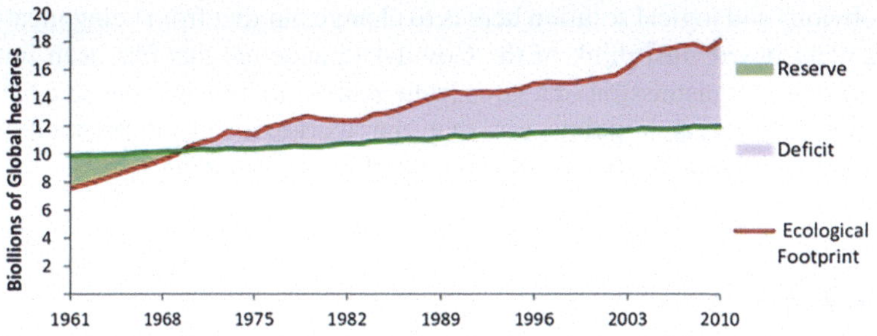

Fig. 2 Ecological footprint and loss of biodiversity since 1971. Source: Lazarus et al. (2015)

exceeded critical thresholds.[49] Experts even warn about a seafood collapse by 2050 due to overfishing, pollution, and global warming.[50] Globally, only 1.6% of the total calories consumed are coming from fish. The question everyone should ask therefore is: "Is it worth overfishing and destroying ecosystems for this small percentage?"[51] While fish from agriculture is seen as an alternative, here also high energy expenditure comes into play to produce sufficient amounts of fish for consumption. Looking at the data, farmed salmon needs 26–48 Megajoules of energy to produce one kg. As comparison, beef needs between 43 to 64 Megajoules, thus there is nearly no benefit in terms of energy savings when consuming farmed fish.[52]

From all this it follows, that the ecological footprint of the human species is too high. Around the year 1971 was the turning point, where the biological capacity of the earth was overtaken by the pollution humans cause and did not decline ever since (see Fig. 2).[53] This simply means: Humans consume more resources than the earth can produce and thus we have an ecological deficit.[54]

Finally, the use of antibiotics in farm animals is overly high. It is estimated that 4 times more antibiotics are used for farm animals than for humans. The reasoning behind this is clear since keeping the animals alive with medication allows for consumption rather than having to throw them away. This however leads to a multitude of negative side-effects. In Germany, data suggest that the pharma industry supplies more than twice as much antibiotics for animal use compared to human use, while only 1% of these are being given to domestic animals.[55] In the USA, this is even more pronounced with numbers going up to four times as much.[56]

[49] Link and Watson (2019).

[50] Worm et al. (2006).

[51] Tacon and Metian (2018)

[52] Henriksson et al. (2010).

[53] Lazarus et al. (2015).

[54] Poppe (2013).

[55] Wallmann et al. (2012).

[56] Budde (2015).

The obvious and logical solution here is to change our diet from being meat-based to being plant-based. Just think of the Covid-19 pandemic that has been linked to being a zoonosis (transmission of a virus from animals to humans) due to a bat being eaten.[57] Current research points out that the worldwide meat production (and consumption) acts as a catalyst toward potential future pandemics.[58]

2.6 #6 Inner Peace

A plant-based diet does not only improve our physical constitution, but also our emotional well-being and balance as pointed out earlier already.[59] Thus, as we find our inner peace, we will also find peace with the people around us.

Overall, inner peace is based on three main components, which are sufficient satiety, the right breathing, and the homeostasis between acids and bases.[60] What is the value of inner peace? The three major components are bliss, wisdom, and world peace. If humanity ever achieves world peace, we would stop killing each other. But does this also hold true for animals? Globally, around 180–200 million animals are being slaughtered each day.[61] To compare this, in both world wars, about 60 million people were killed in total, which means that each day more than three times as much animals are killed as humans were killed in the 11 years that war was raging in Europe in the early to mid-twentieth century. It is easy to optimize your own inner peace and find your balance through a vegetarian and vegan diet. To achieve this, many old and new ways have already been implemented and are lived by hundreds and thousands of people worldwide already.

We know about the Ayurveda medicine for example. Here, when it comes to food there is the principle of "Sattva," which means having spiritual purity and balance. A Sattva diet places emphasis on seasonal foods. It includes fruits, ripe vegetables, nuts, oils, seeds, legumes, whole grains, and proteins that are not meat-based. Dairy is allowed, but only if the cow is fed and milked properly.[62]

Also in traditional Chinese medicine (TCM), such principles are well-known. Here, food items are balanced between warming/heating and cooling as well as drying and damping, corresponding to the principle of Ying and Yang. Warming foods include onion, tomatoes, and coffee while examples for cooling foods are sweet fruit like banana, leafy vegetables, and bitter herbs while drying foods are nuts, beans, and legumes and damping foods are gluten-containing foods and soy

[57] Burki (2020).
[58] Dhont et al. (2021).
[59] Dinu (2017), Olfert (2022).
[60] Hopkins et al. (2022).
[61] Food and Agriculture Association of the United Nations (2022).
[62] Turner (2013).

products. The principle that the diet follows is to find a balance between these aspects, e.g. decrease heat by eating more cooling foods.[63]

Finally, also western medicine has a balancing principle, namely the balance of acids and alkaline that we consume. Grouping foods into alkaline, e.g. herbs, vegetables, mushrooms and acidic, e.g. meat, fish, fried foods, cheese gives an impression on what effects these foods have on our body. Generally, the more alkaline the diet, the better we feel. This was already shown in research as well.[64] As a rule of thumb: The less processed a food item is, the less acidic it is. For example, fried food items are being considered highly acidic.[65]

2.7 #7 Spiritual Development

Our personal reality is just part of the entire reality. Interconnected with this is the age-old question who we actually are. We are more than we believe to be. Our own progress and the way to self-awareness comes through our decisions that we take every day and this either slows or accelerates our progress.

We argue that a vegan and vegetarian diet is ideal for your own spiritual journey of development and that we can both develop our mental but also our spiritual side, as in a famous quote attributed to Mahatma Gandhi it is stated that "The greatness of a nation and its moral progress can be judged by the way its animals are treated."[66]

The spiritual path is encouraged by many philosophies and religions, notable are Hinduism, Buddhism, and Jainism and as a matter of fact the approaches they are following are rooted in becoming happy, peaceful, and in touch with nature, e.g. via the concept of Karma in Hinduism or that good behavior brings you to Nirvana in Buddhism.[67]

3 Conclusion

The adventure is not only individual, but also a collective one. You start to realize these gifts when you go the first step on your own. This, however, requires that we have to realize the dangers and problems of our own diet and work on finding solutions to change it. From this awakening, we come to a transformation, which starts within ourselves and then inspires others to follow us and hopefully ends by changing humanity overall.

---

[63] College of Naturopathic Medicine (2023).

[64] Osuna-Padilla et al. (2019).

[65] Rehberg (2023).

[66] PeTA (2022).

[67] Merriam-Webster (2023), Dictionary (2023).

So the first step you have to undertake is that you need to encourage yourself to question your own habits in order to improve them. Try vegetarian/vegan alternatives to what you normally eat. How do you feel? How does it taste? In the long run, you will adapt and change the composition of your diet. Whatever you change, there are three possible outcomes. You either like it more, you dislike it, or there is no difference for you. Find out what you can change to feel better and continue so. Don't forget about the golden rule "In everything, do to others what you would have them do to you"(Mat. 7, 12).[68] And if you want to make a change to your surroundings, be loving and kind and kindly help others finding the path to a peaceful life through a vegetarian and vegan diet.

References

Ayotte, D., Jr., & Corcoran, M. P. (2018). Individualized hydration plans improve performance outcomes for collegiate athletes engaging in in-season training. *Journal of the International Society of Sports Nutrition, 15*(1), 27.

Babbitt, F. C. (Ed.). (1928). *De tuenda sanitate praecepta. Plutarch Moralia*. Harvard University Press. Retrieved February 23, 2023, from http://www.perseus.tufts.edu/hopper/text?doc=Perseus%3Atext%3A2008.01.0177%3Asection%3D9

Beasley, D. E., et al. (2015). The evolution of stomach acidity and its relevance to the human microbiome. *PLoS One, 10*(7), e0134116.

Bible Gateway. (2016). Genesis 1:29, English Standard Version. Retrieved February 23, 2023, from https://www.biblegateway.com/passage/?search=Genesis%201%3A29&version=ESV

Bible Gateway. (2019). Matthew 7:12, English Standard Version. Retrieved March 20, 2023, from https://www.biblegateway.com/quicksearch/?quicksearch=In+everything%2C+do+to+others+what+you+would+have+them+do+to+you&version=NIV

Budde, J. (2015). Mensch und Tier - Welche Rolle der Antibiotika-Einsatz im Stall spielt, Tagesspiegel. Retrieved March 20, 2023, from https://www.tagesspiegel.de/wissen/mensch-und-tier-welche-rolle-der-antibiotika-einsatz-im-stall-spielt-3629716.html

Bundesanstalt für Landwirtschaft und Ernährung. (n.d.): Nationales Fachprogramm pflanzengenetische Ressourcen. Retrieved February 20, 2023, from https://www.ble.de/DE/Themen/Landwirtschaft/Biologische-Vielfalt/Nationales-Fachprogramm-Pflanzen/nationales-fachprogramm-pflanzen_node.html

Burke, L. M. (2007). Nutrition strategies for the marathon: Fuel for training and racing. *Sports Medicine, 37*(4–5), 344–347.

Burki, T. (2020). The origin of SARS-CoV-2. *The Lancet Infectious Diseases, 20*(9), 1018–1019.

Carrasco-García, A. A., et al. (2020). Effect of stress during slaughter on carcass characteristics and meat quality in tropical beef cattle. *Asian-Australasian Journal of Animal Sciences, 33*(10), 1656–1665. Retrieved February 24, 2023, from https://www.ncbi.nlm.nih.gov/pmc/articles/PMC7463084/

Cassidy, E. et al. (n.d.) Environment reports: Food matters. Retrieved March 17, 2023, from http://www.environmentreports.com/change-your-diet-change-our-destiny/

College of Naturopathic Medicine. (2023). The energy of foods in Chinese medicine. Retrieved March 17, 2023, from https://www.naturopathy-uk.com/news/news-cnm-blog/blog/2020/07/16/the-energy-of-foods-in-chinese-medicine/

[68] Bible Gateway (2019).

Data Commons. (2023a). Ranking nach Lebenserwartung. (Alle Länder in Nordamerika). Retrieved February 24, 2023, from https://datacommons.org/ranking/LifeExpectancy_Person/Country/northamerica?h=country%2FUSA&hl=de

Data Commons. (2023b). Ranking nach Lebenserwartung. (Alle Länder in Asien). Retrieved February 24, 2023, from https://datacommons.org/ranking/LifeExpectancy_Person/Country/asia?h=country%2FJPN&hl=de

Dhont, K., Piazza, J., & Hodson, G. (2021). The role of meat appetite in willfully disregarding factory farming as a pandemic catalyst risk. *Appetite, 164*, 105279.

Dictionary. (2023). Nirvana. Retrieved March 20, 2023, from https://www.dictionary.com/browse/nirvana

Dinu, M. (2017). Vegetarian, vegan diets and multiple health outcomes: A systematic review with meta-analysis of observational studies. *Critical Reviews in Food Science and Nutrition, 57*(17), 3640–3649.

Dreißig, K. (2022). Faktencheck: War Hitler Vegetarier? Retrieved February 21, 2023, from https://vegpool.de/magazin/hitler-vegetarier.html

Einstein Webseite. (n.d.): Einsteins "Lieblingsspeisen". Retrieved February 21, 2023, from https://einstein-website.de/lieblingsspeisen/

Esselstyn, C. (2014). *Essen gegen Herzinfarkt: Das revolutionäre Ernährungskonzept*. TRIAS.

Fadda, R., et al. (2012). Effects of drinking supplementary water at school on cognitive performance in children. *Appetite, 59*(3), 730–737.

Food and Agriculture Association of the United Nations. (2022). Food and agriculture data. Retrieved March 17, 2023, from https://www.fao.org/faostat/en/#home

Gallagher, C. T., Hanley, P., & Lane, K. E. (2021). Pattern analysis of vegan eating reveals healthy and unhealthy patterns within the vegan diet. *Public Health Nutrition, 25*(5), 1–11.

Gehring, J., et al. (2021). Consumption of ultra-processed foods by Pesco-vegetarians, vegetarians, and vegans: Associations with duration and age at diet initiation. *The Journal of Nutrition, 151*(1), 120–131.

González, N., et al. (2020). Meat consumption: Which are the current global risks? A review of recent (2010-2020) evidences. *Food Research International, 137*, 109341.

Goss, R. (2020). This island unlocked the secret to long life—and knows how to get through tough times, National Geographic. Retrieved February 24, 2023, from https://www.nationalgeographic.com/travel/article/uncover-the-secrets-of-longevity-in-this-japanese-village

Happy Cow. (n.d.). Viswanathan Anand is a vegetarian World Chess Champion. Retrieved February 23, 2023, from https://www.happycow.net/vegtopics/famous/viswanathan_anand

Heim, H., & Jochmann, W. (2000). *Monologe im Führerhauptquartier 1941–1944*. Orbis Verlag.

Henriksson, P. et al. (2010). Global seafood alliance. Energy efficiency of aquaculture. Retrieved March 17, 2023, from https://www.globalseafood.org/advocate/energy-efficiency-aquaculture/

Hopkins, E., Sanvictores, T., & Sharma, S. (2022): Physiology, acid base balance. *StatPearls*. Retrieved February 24, 2023, from https://www.ncbi.nlm.nih.gov/books/NBK507807/

IfD Allensbach. (2022a). Personen in Deutschland, die sich selbst als Veganer einordnen oder als Leute, die weitgehend auf tierische Produkte verzichten, in den Jahren 2015 bis 2022. *Statista*. Retrieved February 23, 2023, from https://de.statista.com/statistik/daten/studie/445155/umfrage/umfrage-in-deutschland-zur-anzahl-der-veganer/

IfD Allensbach. (2022b). Vegetarier in Deutschland nach Alter im Vergleich mit der Bevölkerung im Jahr 2022. *Statista*. Retrieved February 23, 2023, from https://de.statista.com/statistik/daten/studie/745049/umfrage/vegetarier-in-deutschland-nach-alter/

International Vegetarian Union. (n.d.). Thomas Alva Edison (1847–1937). Retrieved February 23, 2023, from https://ivu.org/history/northam20a/edison.html

Jabri, A., et al. (2021). Meta-analysis of effect of vegetarian diet on ischemic heart disease and all-cause mortality. *American Journal of Preventive Cardiology, 7*, 100182.

Jacob, L. (2018). *Dr. Jacobs Weg des genussvollen Verzichts*. nutricaMEDIa.

Karasov, W., & Martínez del Rio, C. (2007). *Physiological ecology: How animals process energy, nutrients, and toxins*. Princeton University Press.

Khodarahmi, M., & Azadbakht, L. (2014). The association between different kinds of fat intake and breast cancer risk in women. *International Journal of Preventive Medicine, 5*(1), 6–15.
Kingsley, K. (2020). Stanford encyclopedia of philosophy. *Empedocles*. Retrieved March 17, 2023, from https://plato.stanford.edu/entries/empedocles/
Klaper, M. (n.d.) Vitamin B12 basics. Retrieved February 24, 2023, from https://www.doctorklaper.com/b12
Klein, K. (2023). Forks over knives. Forks over knives documentary: watch online for free! Retrieved March 17, 2023, from https://www.forksoverknives.com/the-film/
Kohyama, J. (2021). Which is more important for health: Sleep quantity or sleep quality? *Children (Basel), 8*(7), 542.
Korneliussen, I. (2012). Is meat from stressed animals unhealthy? Retrieved February 24, 2023, from https://sciencenorway.no/agriculture%2D%2Dfisheries-animal-welfare-food-safety/is-meat-from-stressed-animals-unhealthy/1426527
Krizanova, J., & Guardiola, J. (2020). Happy but vegetarian? Understanding the relationship of vegetarian subjective well-being from the nature-connectedness perspective of university students. *Applied Research in Quality of Life, 16*, 2221–2249.
Krüger, M. (2018). Fünf historische Persönlichkeiten, von denen Sie vielleicht nicht gewusst haben, dass sie Vegetarier waren. Retrieved February 23, 2023, from https://www.dw.com/de/f%C3%BCnf-historische-pers%C3%B6nlichkeiten-von-denen-sie-vielleicht-nicht-gewusst-haben-dass-sie-vegetarier-waren/a-46154625
Lazarus, E., et al. (2015). Biodiversity loss and the ecological footprint of trade. *Diversity, 7*(2), 170–191. Retrieved March 23, 2023, from https://www.mdpi.com/1424-2818/7/2/170
Link, J. S., & Watson, R. A. (2019). Global ecosystem overfishing: Clear delineation within real limits to production. *Science Advances, 5*(6), eaav0474.
Liu, Y., et al. (2000). The temporal response of the brain after eating revealed by functional MRI. *Nature, 405*(6790), 1058–1062.
Ludwig, E. (2015). Hitler war Nichtraucher! (Aber kein Vegetarier.). Retrieved February 21, 2023, from https://www.vegan.at/inhalt/hitler-war-nichtraucher-aber-kein-vegetarier
Matsuda, M., et al. (1999). Altered hypothalamic function in response to glucose ingestion in obese humans. *Diabetes, 48*(9), 1801–1806.
Merriam-Webster. (2023). Merriam-Webster. Karma. Retrieved March 20, 2023, from https://www.merriam-webster.com/dictionary/karma#h1
Mills, L. (n.d.) Loving-kindness. Retrieved February 23, 2023, from https://suttacentral.net/snp1.8/en/mills?reference=none&highlight=false
Moskal, E. (2023). *Twin research indicates that a vegan diet improves cardiovascular health*. Stanford Medicine News Center. https://med.stanford.edu/news/all-news/2023/11/twin-diet-vegan-cardiovascular.html
Mu, Y., et al. (2021). Molecular evolution of tooth-related genes provides new insights into dietary adaptations of mammals. *Journal of Molecular Evolution, 89*(7), 458–471.
Naeem, S., et al. (2016). Biodiversity and human well-being: An essential link for sustainable development. *Proceedings of the Royal Society B: Biological Sciences, 283*(1844), 20162091.
Obeid, R. (2019). Vitamin B12 intake from animal foods, biomarkers, and health aspects. *Frontiers in Nutrition, 6*, 93.
Olfert, M. D. (2022). Life of a vegetarian college student: Health, lifestyle, and environmental perceptions. *Journal of American College Health, 70*(1), 232–239.
Osuna-Padilla, I. A., et al. (2019). Dietary Acid Load: Mechanisms and evidence of its health repercussions. *Nefrologia (Engl Ed), 39*(4), 343–354.
Our World in Data, OECDstat. (2017). Life expectancy vs. health expenditure, 1970 to 2015. Retrieved April 18, 2023, from https://ourworldindata.org/grapher/life-expectancy-vs-health-expenditure
Oxnard, C. E., Smith, W. T., & Torres, I. (1970). Vitamin B12 deficiency in captive monkeys and its effect on the nervous system and the blood. *Laboratory Animals, 4*(1), 1–12.

Papier, K., et al. (2023). Meat consumption and risk of ischemic heart disease: A systematic review and meta-analysis. *Critical Reviews in Food Science and Nutrition, 63*(3), 426–437.
PeTA. (2022). PETA honors Gandhi's lifelong commitment to animal liberation. Retrieved March 20, 2023, from https://www.peta.org/features/gandhi/
Peterson-KFF. (2023). How does health spending in the U.S. compare to other countries? Retrieved February 24, 2023, from https://www.healthsystemtracker.org/chart-collection/health-spending-u-s-compare-countries/#Health%20consumption%20expenditures%20per%20capita,%20U.S.%20dollars,%20PPP%20adjusted,%202021%20or%20nearest%20year
Pohl, A., et al. (2021). The impact of vegan and vegetarian diets on physical performance and molecular signaling in skeletal muscle. *Nutrients, 13*(11), 3884.
Poppe, E. (2013). Der Rebound-Effekt. Herausforderung für die Umweltpolitik. Retrieved March 17, 2023, from https://www.researchgate.net/publication/317259481_Der_Rebound-Effekt_Herausforderung_fur_die_Umweltpolitik
Pross, N., et al. (2013). Influence of progressive fluid restriction on mood and physiological markers of dehydration in women. *The British Journal of Nutrition, 109*(2), 313–321.
Quek, J., et al. (2021). The association of plant-based diet with cardiovascular disease and mortality: A meta-analysis and systematic review of prospect cohort studies. *Frontiers in Cardiovascular Medicine, 8*, 756810.
Rehberg, C. (2023) Saure und basische Lebensmittel – Die Tabelle. Zentrum der Gesundheit. Retrieved March 17, 2023, from https://www.zentrum-der-gesundheit.de/ernaehrung/ernaehrungsformen/basenueberschuessige-ernaehrung/saure-und-basische-lebensmittel
Ritchie, H., & Roser, M. (2021). Deforestation and forest loss. Retrieved March 17, 2023, from https://ourworldindata.org/deforestation
Roser, M. (2020). Why is life expectancy in the US lower than in other rich countries?. Published online at OurWorldInData.org. Retrieved June 17, 2023, from https://ourworldindata.org/us-life-expectancy-low
Sahgal, N., et al. (2021). *Religion in India: Tolerance and segregation*. Pew Research Center.
Shah, N. (2004). *Jainism the world of conquerors*. Motilal Banarsidass Publishers.
statista. (2022). Vegetarier in Deutschland. Retrieved February 20, 2023, from https://de.statista.com/statistik/studie/id/48117/dokument/vegetarier-in-deutschland/?locale=de
St-Onge, M. P., Crawford, A., & Aggarwal, B. (2018). Plant-based diets: Reducing cardiovascular risk by improving sleep quality? *Current Sleep Medicine Reports, 4*(1), 74–78.
Szot, M., et al. (2022). Can nutrients and dietary supplements potentially improve cognitive performance also in esports? *Healthcare (Basel), 10*(2), 186.
Tacon, A., & Metian, M. (2018, April 30). Global seafood alliance. Food matters: A comparative analysis of fish, income and food supply. Retrieved March 17, 2023, from https://www.globalseafood.org/advocate/food-matters-comparative-analysis-fish-income-food-supply/
Think Rice. (2020). U.S. rice varieties. Retrieved February 23, 2023, from https://www.usarice.com/thinkrice/discover-us-rice/us-rice-varieties
Thünen Institut. (2022). *Versorgung mit Fleisch in Deutschland*. Statistisches Bundesamt.
Tiller, N. B., et al. (2019). International society of sports nutrition position stand: Nutritional considerations for single-stage ultra-marathon training and racing. *Journal of the International Society of Sports Nutrition, 16*(1), 50.
Tilman, D., & Clark, M. (2014). Global diets link environmental sustainability and human health. *Nature, 515*(7528), 518–522.
Tollefson, J. (2019). Humans are driving one million species to extinction. *Nature, 569*, 171. Retrieved February 24, 2023, from https://www.nature.com/articles/d41586-019-01448-4
Turner, P. (2013). *FOOD YOGA – Nourishing body, mind & soul*. Food for Life Global.
Wallmann, J., Preuss, J., & Bender, A. (2012). *Erfassung von Antibiotika-Abgabemengen gemäß DIMDI-AMV*. Bundesamt für Verbraucherschutz und Lebensmittelsicherheit.
Willcox, B. (2002). *The Okinawa Program: How the world's longest-lived people achieve ever-lasting health--and how you can too*. Harmony.

Withers, P., et al. (2016). *Ecological and environmental physiology of mammals*. Oxford University Press.

Worm, B., et al. (2006). Impacts of biodiversity loss on ocean ecosystem services. *Science, 314*(5800), 787–790.

Yale Center for Business and the Environment. (2016). Disrupting meat. Retrieved March 17, 2023, from https://cbey.yale.edu/our-stories/disrupting-meat

Christian Halper, **Dipl. Ing. (DI)** is an Austrian entrepreneur and former co-owner of the SuperFund Group. He graduated from the Vienna University of Technology and earned a degree in communications engineering. After graduating, Halper co-founded the SuperFund Group, a renowned investment firm specializing in alternative investment strategies, together with Christian Baha. Parallel to his career in the financial industry, Halper also devoted himself to other entrepreneurial interests. He founded the vegetarian restaurant "Tian," which gained international recognition with its exquisite cuisine. Tian Restaurant has been awarded a Michelin star and has become a popular destination for foodies. Christian Halper is the owner of the "Strandhotel am Weissensee" that combines the beautiful and peaceful nature of this unique Austrian mountain lake with the enjoyments of vegetarian cuisine and a Spa directly at the crystal clear lake. He invested also in other vegetarian Concepts like "Divya's" in New York, that is on a mission to help people find joy, nourishment, and balance through food. Halper was also involved in the founding of the company "Healy," which develops innovative products in the field of holistic well-being. Healy offers personalized solutions designed to increase people's well-being and energy levels. Halper's dedication to Healy shows his eclectic interests and striving for progress in various fields. In addition, Halper is involved in various social and charitable projects. Christian Halper is an outstanding personality in the Austrian economy and has made a significant contribution with his entrepreneurial successes and his social commitment. His diverse interests and pursuit of excellence make him an inspiring figure in the business world and beyond.

A Transdisciplinary Approach to Peace-building: Towards a Narrative Leading to Inclusion and Justice through the Lens of Traditional Wisdom

Patrick Nickisch

We have to defeat tyranny in the realm of thought, and create a will for world peace
Shriman Sarvapalli Radhakrishnan (Sarvapalli Radhakrishnan in 1944 Back to Godhead VOLUME 1—Parts I, II, III & IV edited and founded by Abhay Charan Den).

1 Introduction

What does the domains of theology and science of religion have to offer to humanity at the present day threshold? Humanity seemed to have enshrined a question in its collective memory: Was religion not the reason for so many wars? Causalities of war are complex and so is religion. When both of them come together an explosive mix all too often occurred. While analysing a conflict one needs to consider various populations involved living under conditions of war and conflict for decades or even centuries. Religious rhetoric or any type of ideology will be in customary usage for a prolonged period of time. As such religion is influencing the story society is telling itself to make sense of the world and justify what we do. However, the teachings of religions and using its language for some other end have to be clearly distinguished. Radicalization of any kind and extremist approaches are problems in and of themselves. Religion and/or other ideological agendas can serve as cover up of political and economic interests misleading with simplified conclusions.

P. Nickisch (✉)
Department of International Expansion, Kilchberg, Switzerland

Hindu University of the America, Orlando, FL, USA
e-mail: patrick@thesacredshell.com

© The Author(s), under exclusive license to Springer Nature Switzerland AG 2025

M. Schüz (ed.), *Transdisciplinary Thinking and Acting*, Transdisciplinary Management of Ecological and Social Crises,
https://doi.org/10.1007/978-3-031-70680-6_19

> While religious intolerance and extremism are a frequent source of conflict, religion is more often the convenient scapegoat for underlying political and economic tensions.[1]

Oversimplification and an idealized approach when discussing peace and religion need to be avoided. Where is the journey as one humanity leading us? How can a theology of peace offer an inclusive, just, and transdisciplinary approach to peace-building? Grounded in a historical perspective the awareness of the presence leads to a long-term view to peace-building. The promises of religion and spiritual traditions on peace resemble their fulfilment for what it can invoke in the heart of the human being. As the Parliament of World Religions declared in 1992 in Chicago:

> Earth cannot be changed for the better unless the consciousness of individuals is changed. We pledge to work for such transformation in individual and collective consciousness, for the awakening of our spiritual powers through reflection, meditation, prayer, or positive thinking, for a conversion of the heart.[2]

Such declarations by scholars of religion, theologians, and religious leaders express the yearning to be of service to society by following the teachings of their traditions with sincerity and a humble heart. Theology as a discipline to systemically study religious teachings by a community of scholars in a particular sociocultural context is transdisciplinary by default offering a wealth of experience. The peace-building global community has been looking to religion to prevent war and maintain peaceful relations. Organized religions can be loud and constructive moral voices in society. As such they can provide trusted leadership in war torn areas.

> Historically and currently, we all realize that religious differences have often been a cause or a pretext for war. Less well known is the fact that the actions of many religious persons and communities point in another direction. They demonstrate that religion can be a potent force in encouraging the peaceful resolution of conflict.[3]

We will look at transdisciplinary approaches to peace-building through the lenses of traditional wisdom. Religion is a vital factor for peacebuilding and can become a major ally for preserving peace inside and across the nations.

I am proposing to embark on a collective journey to find new forms for traditional knowledge and modern science to bear on contemporary challenges. A transdisciplinary approach to peace-building strives to harness all resources and activate relevant stakeholders to progress to lasting peace. To arrive at a civilization ensuring human rights and happiness all are invited to cooperate like a crew on a ship. Each Earth crew member has unique talents and gifts to contribute. Traditional knowledge of the wisdom traditions, religions, and indigenous communities can offer guidance to a civilization in such great transition times we are living in.

We are first taking a historical perspective to look into the roles of religion how it had changed in Europe at the emergence of modernity and the Age of Reason. The central voice of such traditions and its role in shaping narratives and culture will be examined drawing conclusions for our modern context. We will draw from examples

[1] Vendley (2005: 90).

[2] Küng (1998: 31).

[3] Johnston and Sampson (1994: VII).

of religious leaders and grassroots organization across sectors working on a culture of dialogue to foster cooperation. I am arguing for a cultural sensitive approach to advance transdisciplinary forms of peacebuilding beyond the dominant Western cultural narrative with its colonial legacies and still lingering effects of its toxic past. New narrative building in support for inclusive multi-stakeholder peace-building initiatives will be shown contrasting it with old narratives of exclusion and monopolizing power.

2 In the Beginning was the Word: The Power of the Voices of Religion

The central role of religion in the lives of the people, in governance and academia started to decline in Europe over the centuries leading to modernity. Formally, theology was the main discipline in universities and what resembled European style universities across the world. Knowledge in Middle Age Europe was mostly gathered in centres of learning like small enclaves of Christian monks preserving the remains of the Roman era. Natural sciences, medicine, political science, economy, and other such subjects were deployed and guided by the law of religion with its moral guidelines and spiritual orientation. Knowledge about God, law, and the teachings of the saints were seen as the foundation for all other subjects. Similar in other cultures learned priest societies of ancient India would know mundane subject matters as scientists would know them today but also perform worship in temples and ceremonies for birth, marriage, and death. In ancient times either an institutionalized priesthood got established deriving its power from autocratic systems or more charismatic prophets[4] often considered saintly or endowed with an intrinsic ability to connect to the Divine attracted a following. They were and are the wisdom teachers, elders, and shamans and would act as guides, mentors, and teachers. They still play a central role in the overall direction and orientation of society. Such intellectual classes taught and served the public as guides and teachers in a wide area of knowledge and its specializations. Many disciplines were working together intertwined by a theological understanding offering ethical and spiritual orientation as well as practical guidance on day to day life issues, maintaining law and order.[5]

Greek philosophers and scholars of the ancient worlds, of Islam and Christianity as well as modern ethicists all shared a central concern: How to live a healthy, good, and just life? Law is and was written in the religious textbooks since the beginning of written records. The International Law Commission (ILC) and other Bodies of Law use customary law as found in religion and other traditions as a source for the codification of International law.[6] It took a longer time to come to this point as since

[4] Weber, Max (1922).

[5] Scott in Hatzopoulos and Fabio (2003: 32).

[6] https://www.icrc.org/en/doc/resources/documents/misc/customary-law-q-and-a-150805.htm

the confessional wars in Europe religion was cast out of the domain of international relations and law.

As some scholars have argued, for the modern nation state to appear, religion had to disappear from sight and into the private domain, at least in the European context.

> Religion was the object that needed to vanish for modern international politics to come into being.[7]

Hence, since the *Peace of Westphalia* established in Germany (1648), a narrative took shape to form the basis of modern governance and international law. Religion was banned from politics, at least officially. Since then, religion and its associated theology has been marginalized in Europe and the Western world. The continent was divided into those following the papacy of the Roman Catholic Church and all others who protested such rule, taxation, and perceived moral corruption as pointed out by Martin Luther spoken in the Assembly of worms in Germany in 1521. He famously remarked:

> I cannot and will not recant anything, for to go against conscience is neither right nor safe. Here I stand, I can do no other, so help me God. Amen.[8]

Even though this war was political, still historians called it a confessional war as the faultlines became denominational. It was not the intention of Martin Luther to polarize Europe or create a new denomination of Christianity. He was seeking to reform what he perceived to be a corrupted version of following the pure teachings of Jesus Christ. The voice of conscience of this famous German Catholic monk emerged at a time of political unrest and led to the 30 years long war which decimated one third of the German population. The spirit of the time was to become independent of the political and spiritual rulership of the Roman Catholic Church's papacy. As rulers choose to either side or part with the pope, entire populations changed their confession: *Cuius regio, eius religio* was the compromise to ensure peace.[9] Religion was no more a matter of personal belief, but depended on the political orientation of the ruler and where one was living and growing up. Such areas of prolonged conflict, even today are often based on long grown sentiments of desiring more autonomy in decision-making, freedom of thought and speech—which centuries later developed as Human Rights. Rising protests in many nations testify to the growing willingness to shift the narrative and look for viable alternatives. To move beyond mere polarization itself is a unique challenge.

In our globally intertwined world dividing lines of friends and enemies across state borders or religions are increasingly blurry. More important than ever is the moral voice of religious scholars, priests, and leaders as the voices of religion have been as important back then as it can be today, for good or bad.

---

[7] Hatzopoulos and Fabio (2003: 1).

[8] https://www.luther.de/en/worms.html

[9] https://www.juraforum.de/lexikon/cuius-regio-eius-religio

> It is the duty of religions and of their leaders above all to foster in the people of our time a renewed sense of the urgency of building peace. They must weave a priority for peace into the mindset of the global community.[10]

One outstanding example for such leadership and voice has been Pope Francis in his *Laudato Si* published in eight languages in 2015:

> LAUDATO SI', mi' Signore—"Praise be to you, my Lord". In the words of this beautiful canticle, Saint Francis of Assisi reminds us that our common home is like a sister with whom we share our life and a beautiful mother who opens her arms to embrace us. "Praise be to you, my Lord, through our Sister, Mother Earth, who sustains and governs us, and who produces various fruit with coloured flowers and herbs". This sister now cries out to us because of the harm we have inflicted on her by our irresponsible use and abuse of the goods with which God has endowed her. We have come to see ourselves as her lords and masters, entitled to plunder her at will. The violence present in our hearts, wounded by sin, is also reflected in the symptoms of sickness evident in the soil, in the water, in the air and in all forms of life. This is why the earth herself, burdened and laid waste, is among the most abandoned and maltreated of our poor; she "groans in travail" (Rom 8:22).[11]

This statement came at a time where world leaders were in the process of formalizing and ratifying policies to combat climate change paving the way for what could become a regenerative and just economy. While religious leaders could spark violence as in the confessional wars, intentionally or unintentionally, in our modern context they became more known to inspire their followers to preserve our environment and foster peace among the nations. Platforms and conferences have attracted world leaders to discuss the issues of our time and to inspire solutions and the "Laudato Si" had a significant influence to draw more attention to climate change. Organized religion and spiritually inspired initiatives on local and global scale have been part of these stakeholder groups for many years and inspired many campaigns ever since. This can be seen by countless faith based initiatives on Climate change like the Act Alliance[12] and Interfaith efforts worldwide.

> Contrary to what Karl Marx predicted (and contrary to what some Western commentators seem to suppose), religion is not withering away. It remains the context, or at least a part of the context, in which the majority of people alive on this planet today live their lives or from which they derive important inspiration and judgment for their lives.[13]

Religion as a vessel for knowledge and values is able to build healthy communities. The voices of communities, religious, and secular had become stronger in the

[10] Curaming (2002: 8–9).

[11] Statement on how to deal with "Mother Earth" by Pope Francis (2015: 3). https://www.vatican.va/content/francesco/en/encyclicals/documents/papa-francesco_20150524_enciclica-laudato-si.html

[12] https://actalliance.org

[13] Bowker (1986: 416).

process of gaining more civil liberties since the twentieth century. Since the seventies of the 20th century, *Religions for Peace* (RFP)[14] established national councils in many nations of the world with emphasis on religious leaders and high level peacebuilding interventions. The international community also launched the *Alliance for Civilizations* as a response to Huntington's *Clash of Civilizations*[15] in the early nineties.

> World politics is entering a new phase, and intellectuals have not hesitated to proliferate visions of what it will be-the end of history, the return of traditional rivalries between nation states, and the decline of the nation state from the conflicting pulls of tribalism and globalism, among others. Each of these visions catches aspects of the emerging reality.[16]

The responses to such narratives influencing the course of the world inspired both high-level and grassroots-oriented peacebuilding networks to work together for the realization of the three main aims of the United Nations: peace, sustainable development, and human rights.

The UNO made a request to the religious leader the Rt. Rev. and former Bishop William E. Swing who became the Founder President of the URI.[17] to unite the religions to support their work. A few years later the *United Religions Initiative*[18] (URI) was formally created in the year two thousand in New York City after years of consultations hosted in the San Francisco Bay Area. The structure for a decentralized movement guided by a democratically elected global council with national and regional leadership was developed together in consultation with the credit card company VISA and the founders of URI. VISA had some organizational expertise URI was able to tap into in developing one of the first decentralized autonomous organizations (DAO), another brain child of the San Francisco Bay area.[19] The basic structure are grassroots *Cooperation Circles* (CC) which are self-governed, autonomous associations with participatory decision-making of individuals and movements. It adopted this model as an organization and created a strong proof of point attesting the merit of the CCs. The *Decentralized Autonomous Organization* (DAO) model is built on similar principles but is more closely associated with the Blockchain based developments of recent years. These technologies present a remarkable opportunity for the peacebuilding community due to its transparent governance potential and distribution of funds, among many other applications. Such inclusive elements were preceded by the Nine Stakeholder groups as defined by the international community of the *United Nations Organization* (UNO) system since the fundamental *Earth Summit* in Rio Di Janeiro 1992. Since then, these groups

[14] https://www.rfp.org

[15] Considered one of the most influential essays published 1992 by the Council of Foreign Relations CFR defining much of the debate on peace and conflict.

[16] Huntington (1996).

[17] Founder President of the United Religions Initiative (URI)

[18] https://www.uri.org

[19] https://www.investopedia.com/tech/what-dao/#:~:text=A%20DAO%20is%20a%20decentralized,DAO%27 s%20operations%20is%20publicly%20disclosed.

contribute to global policy with their expertise on high-level policy negotiations and implementation.

Alliances were created, small and big, to follow the recommendations of world leaders urging all to take meaningful action. One example is the *Global Interfaith WASH Alliance*[20] (GIWA) founded in New York City 2013 during the Millennium Development[21] Review Summit, in the wake of Rio + 20[22] in 2012. Leaders of science, religion, UNO, governments, economy and more came together to work on Water, Sanitation, and Hygiene issues worldwide. In India, it has been most successful due to the strong role of religious leadership and the voice of its leaders in that country. Religions, spiritual institutions, *New Religious Movements* (NRMs), and indigenous peoples and first nations have been sitting alongside governments and business leaders, scientists, and numerous civil society organizations (NGOs) advocating for a human rights based approach to sustainable development.

> The United Nations and international development organizations like UNICEF are beginning to seek out religious networks for their ability to reach vast numbers of people and their capacity to affect change. More and more experts acknowledge the potential of religious communities as partners to address common concerns such as armed conflict, human rights violations, and poverty.[23]

As part of such efforts the return of religion to international relations as a force for good matured even further at the dawn of the New Millennium. Momentum increased in 2015 when the International community was achieving the monumental policy frameworks of the *Sustainable Development Goals*[24] (SDG), *Disaster Risk Reduction* (DRR) in Sendai,[25] Japan, followed by the Paris Agreement[26] on Climate Change. To manifest the "*World We want*" and to "*Leave no one Behind*" has been two of the main slogans of post 2015 UN Sustainable Development Agenda.[27] Governments have the responsibility to protect and therefore they are held accountable for what is happening on the territories they govern. A government is comprised of the people of a nation reflecting as much the people as a government does represent the decisions of a nation and its course. A United Peoples movement of Gandhian proportions would be required as the late Maurice Strong was remarking calling for a large movement to preserve the environment and ensure a sustainable future for humankind.[28] United Peoples, alongside a United Nations, leading from the Grassroots up to high-level leadership are mobilizing. Voices of religions, like

[20] https://washalliance.org

[21] https://www.un.org/millenniumgoals/

[22] https://sustainabledevelopment.un.org/rio20

[23] Vendley (2005: 99).

[24] https://sdgs.un.org/goals

[25] https://www.undrr.org/implementing-sendai-framework

[26] https://unfccc.int/process-and-meetings/the-paris-agreement/the-paris-agreement

[27] https://sustainabledevelopment.un.org/post2015

[28] https://www.mauricestrong.net/index.php?option=com_content&view=article&id=219:necreate-greater-public-interest-and-awareness-in-rio-20&catid=11&Itemid=21

the Pope's, have sparked unprecedented movements towards love and unity. They have informed, inspired, and guided narratives, which were powerful enough to have started changing the course of our modern civilization.

3 Shifting Global Narratives: From Separation to Transdisciplinary Inclusion

The dominating narrative of our times is clearly the Western worldview which is also known as the political West. It grew out of the Greco Roman worldview in which early Christianity spread and became a worldwide influence since the colonial age. What is being told by a religion or worldview informs all sectors of society by shaping its civilization and inner orientation. The Christian Civilization as it evolved out of Europe sanctified its global conquest to spread Christianity. This was leading to the Colonial Age and a Eurocentric global economy with its structural injustice reaching to the present. Learning from such narratives we can transition to building new ones, which are based on a culture of peace with a *transdisciplinary* approach to *peacebuilding*.

> The debates concerning the relation of religion, its study, and colonialism and its aftermath have increasingly engaged many scholars around the globe. In the aftermath of 9/11 these debates have taken on a new intensity and character. The significance of religions in the global context has become ever more apparent beyond the confines of the university, and the necessity of understanding how religions intersect and interact with other social, political, and economic realities has moved to the fore of public consciousness.[29]

The impact of the Christian narrative shaped much of the political and economic agendas of the West. It is therefore a good case to study for how a transdisciplinary approach can take shape in worlds of politics, academia, and beyond. The Western superiority complex has been partly evolving out of Christianity presented as the best and only real religion and its attempt to subdue the world. Jesus taught the way to God by his own example and he himself spoke about himself coming to fulfil the law and Jewish prophecies. "*Do not think that I have come to abolish Law or the Prophets; I have not come to abolish them but to fulfill them*" (Book of Matthew 5: 17).

In Jewish traditions, we can find two fundamental principles. There is the way to follow and there is the law to be guided by. As the Christian tradition left the mother tradition of the Jewish faith, this developed into an exclusive claim that Jesus is the only way to God, going so far as to seek to monopolize the belief in One God. Christianity, Islam, and Judaism are seen as the monotheistic traditions making it appear as if belief in "*the One God*" is the exclusive hallmark of the Abrahamic faiths. Subsequently, knowledge of the Personality of Godhead, known in the Americas, Africa and Asia were subdued and all traditions not following "the only

[29] Davaney and Laderman (2005: 980).

way" had to convert or perish as the conquests continued. Despite such atrocities, thousands of ancient lineages beyond the Abrahamic traditions are still active today and emerging more in recent decades. Alliances between military conquest and religious conversion campaigns followed by economic exploitation are not uncommon even up to today unfortunately. Especially in North, Central, and South America, European nations conquered these areas and forcibly converted them to Christianity or committed genocidal policies which still continue in some of the postcolonial nations in the Americas. The West was expanding their markets and forcing access to resources, which are some of the legacies which frequently disrupt into conflict even up to day.

> The West in effect is using international institutions, military power and economic resources to run the world in ways that will maintain Western predominance, protect Western interests and promote Western political and economic values. That at least is the way in which non-Westerners see the new world, and there is a significant element of truth in their view.[30]

For Peace studies and practices it is therefore critical to examine the role of the toxic past and find ways to access tools for healing and reconciliation such as trauma-informed therapies. Also, post Nazi Germany called this process of reconciling with its gruesome past "Vergangenheitsbewaeltigung", something all western nations ought to seriously consider to do for themselves rather than still maintaining a nostalgic sence of a glorious past of conquest and global dominance. Instead, a spirit of inclusivity and open-mindedness is required to build narratives which move beyond artificial divides like sectarianism, tribalism, and nationalism. A transdisciplinary perspective would intrinsically understand that narrow-minded exclusivity leads to conflict and bigotry.

Exclusive and monopolizing narratives unfold due to a variety of reasons. The perception of the need to protect ones' followers towards "the other" is deeply enshrined. Political conquests by Christian, Islamic, and Buddhist armies in the name of One World Religion followed. Their global colonial empires started to cover all the continents of the world by the nineteenth century.

> Religion due to its existential character can fall into a pattern of securitization that turns into an ideology, which is defined by using religious language and the like for an end that is not otherworldly but this worldly oriented. In other words, it is misusing religious language and symbol for its own ends".[31]

Due to this baggage many people of our contemporary society still blame religion and the Christian West in particular for having caused so much conflict and violence in the world. Consequently, many modern people lost their faith in religion and turned to secularist worldviews. Such modern forms resemble religion, promising salvation in their own ways with the hope to create some kind of heaven on Earth the way they imagine it. Such move was seen as a *"justified protest in the name of human freedom against Christendom's heteronymous God of power"*.[32]

---

[30] Huntington (1996).

[31] Hasenclever, quoted in: Hatzopoulos and Fabio (2003: 169).

[32] Revd Williams (1988: 78).

With the help of reason, science, and economic development, humanity would finally find peace and happiness. Religions and their associated civilizations were disregarded as backwards by modern Europeans and the emerging Western modern world.

The Eurocentric world view would lead the world to unprecedented prosperity, happiness, and peace—historically though mostly confined to the West. The bitter truth is that severe Human Rights violations happen within all nations. A rejection of certain applications of human rights law is often justified when they go against ones' own interests. Now, an inclusive global society requires a narrative which dismantles systemic injustices and broaden the dialogue to solve the issues of our day and time. However, one essential principle contributing to global peace emerged from the West which sees all human beings as equals based on individual basic human rights. Now, how to give space to the many voices in the world which look for a direction forward out of the quagmire of modern progress? Justice and inclusion are closely related. They point us to consider other world views sitting on the dialogue table as a fundamental prerequisite to foster a climate of peace. The intellectual climate of a global society strives to look beyond mere East and West and North and South divides.

A globally oriented web of interconnections in economic and cultural fields is already beyond the confines of national boundaries. Global companies, special economic zones, international organizations and trade agreements, the United Nations system, religions, science, art, and universities are some of the forms and platforms global societies interact within their own specialization and focus areas. For the ones who seek to see and to hear *the other* with clarity are able to recognize *each other* as partners for mutual security and prosperity. In this way innovative forms of knowledge acquisition, sharing, and decision-making can develop. Models based on such approach help our various civilizations to adapt, cross-fertilize, and build its resilience to survive the test of time.

> Postmodernity strives for a new basic consensus of integrative human convictions in a new world constellation towards which democratic pluralistic society is inexorably directed if it is to survive.[33]

In this way solutions across sectors can grow under mutual consent and with transparent decision-making. Such inclusive processes can ensure that the variety of voices are being heard. It is not a mere matter of luxury or idealism but of sharp realism for truth and justice to survive. Fact is, the dynamic of our times is "*to form a cosmopolitan public sphere that encourages civil communication between diverse communities on the local, national, and transnational level*".[34] From the grassroots all the way to high level forms of governance need to interact both organically and systematically. In this way such a public sphere will be resilient, adaptable, and innovative to address many issues, such as climate consequences. There are many

[33] Küng (1998: 22).

[34] Doe (2003: 97).

challenges in such a diverse mix of cultures, world views, and beliefs and therefore dialogue is essential. In this way, we cannot only find common ground, but also have access to the accumulative wisdom of the civilizations and their traditions to create a sustainable way of living.

Such culture of dialogue has been widely developed in the interfaith movement and contributed to a climate of openness, respect, and willingness to learn from one another. Within the context of a religious tradition Islamic banking evolved, and ethically oriented businesses aligned with an ecological way of living became a global phenomenon. People of faith share the gifts of their traditions and their narratives to understand and shape the world and our place in it. The environment and how to govern our affairs sustainably are some of the shared concerns of most faith traditions. Religions count as one of the largest investment sectors. People of conscience and open minds are part of the new participatory and just economic and governance models. Governance and policies can increasingly be informed by sacred knowledge due to its wealth of experience and therefore its merit for decision-making.

> They must learn from each other, and explore each other's deepest resources for faithful adaptation to the context of our global village.[35]

The integrity and truthfulness of such initiatives also require the transparency of an open engagement and informed public debate. The return of the sacred in policies and business practice and its inclusive character can be seen as a transdisciplinary movement.

To isolate the society from decision-making affecting all people has created a gap of power. As this gap seems unsurmountable, frustration grew and protests leading to social unrest. However, by looking at the old and new narratives, sharing and learning from each other, a conscious global community has gradually formed. Looking at globalization, the Interfaith movements, religious organizations, global professional communities, and grassroots world wide web of interconnections, it becomes increasingly apparent that "*the movement is toward a truly dialogical community, in which each member lives and is itself through dialogue with others*".[36] As dialogue has such a central role in transdisciplinary approaches to peacebuilding, a variety of methods evolved and are required. Global convening platforms of dialogue such as the United Nations are endowed with protocols and an etiquette making all participants feel safe to express themselves to promote cooperation. What is urgently required is a mobilization of all sectors of society to discuss and engage with the issues considering a broad area of solutions. Complex and simple systems of creating strong relationships have widely developed by specialized disciplines.

Conflict resolution is a particularly good example of such a discipline which is transdisciplinary due to its all-encompassing character. One system was developed

---

[35] Forward (2002: 2).

[36] Knitter (2002: 10).

where each party contributes to the issue at hand, drawing from its strengths rather than its weaknesses. This is known as *Appreciative Inquiry* (AI) and serves as an effective tool in the peace process. This system has been early adapted and developed by the Mennonite scholar and conflict resolution expert John Paul Lederach for URI to apply it systemically. As the United Religions Initiative's founding Executive Director Emeritus, the Rev. Canon Charles P. Gibbs explained:

> Grounded in a belief that calling a group to focus on a problem creates a practice of deficit, thinking and ties people to what has failed in the past, AI asks people to recall and share the best of their past experience.[37]

AI prompts us to be solution oriented appreciating each others' resources rather than to compete. This requires a fundamental move away from a narrative which sees scarcity and lack. Here, the result all too often is that we fight. However, as all increasingly become aware that we are sitting on an abundant table nobody needs to fight. Instead, we can share abundantly with each other from the grassroots up. AI requires us to look deeply into the local level[38] which Lederach emphasized and which resulted in a big shift in the field of conflict resolution and peace studies.

> I believe that the nature and characteristics of contemporary conflict suggest the need for a set of concepts and approaches that go beyond traditional statist diplomacy. Building peace in today's conflicts calls for a long-term commitment to establishing an infrastructure across the levels of a society, an infrastructure that empowers the resources for reconciliation from within that society and maximizes the contribution from outside.[39]

Each is called due to their own merit to gather as free people to meet and work on practical solutions.

The transition is for a return to a more balanced way of living with firm feet on the ground and in harmony with nature. In some of the ancient holy lands of indigenous and ancient civilizations across the globe the biodiversity is highly dense. The combination of local spiritual leaders and tribes, science, the economy and policy can combine to protect these last remaining bio diversity hotspots—with the creation of Eco-sanctuaries. The Island Nations of the Seychelles and other Big Ocean States have started to created vast ocean sanctuaries[40] around a blue economy.[41] The hope is that the same will increase with the nations who still have vast rainforests and interfaith efforts have combined here with indigenous tribes and secular policy makers.[42] Remaining species on land, water, and air can start to spread and thrive from there again. This local level of governance contrasts with the established trickle-down theory of the realist school within the discipline of international

---

[37] Gibbs (2002: 117).

[38] Lederach (1995).

[39] Lederach (1997: XVI).

[40] https://www.bluenaturealliance.org/seychelles, https://sanctuaries.noaa.gov/pacific-remote-islands/

[41] https://natureseychelles.org/what-we-do/blue-economy

[42] https://www.interfaithrainforest.org

relations with its emphasis on high-level national governmental governance with little regard of other stakeholders in the process of decision-making.

In this theory, decisions at the top with its tendencies to protect vested interests, all too often influenced by bribes, the so-called donations, and nepotism eventually and automatically effect the lower levels, rather negatively. The latter is still the dominant narrative which sees no wrong in hoarding enormous amounts of wealth with not much trickling down to the general population whatsoever. The economic divide between rich and poor could grow in a context which favours the increase of wealth amongst the wealthy. For a very tiny few to hold the vast amount of wealth is an indicator how the present day narrative supports such dysfunctional and ultimately harmful behavioural patterns. Loss of biodiversity, poor living conditions for billions of people, suffering from hunger, lack of water, lack of meaningful employment, proper housing, etc., are willingly taken for granted by a narrative which has been woven to support such exploitation of nature and humanity. Such dominating and exclusivist behaviours would be contrary to a self-regulative society endowed with an inner sense of justice and fairness.

> What becomes paramount, perhaps, is access to as wide a range of means of reconciliation as possible, which includes both the word and the deed.[43]

One can experiment with a variety of forms of wealth and resources within peace-oriented narratives. All kinds of knowledge and resources are brought to the table to be considered outside of harsh ideological divided and cultural biases.

The emergence of such a society is possible by a dialogical culture of peace[44] which was called forth by the UN General Assembly with policy and public campaigns such as the *International Decade for a Culture of Peace and Non-Violence for the Children of the World (2001–2010).*[45] Among others, it includes an *"economy of fair exchange"* not only with all involved stakeholders but also with nature. Such a paradigm shift in business practices is demanded by Mathias Schüz. It must replace the robber economy that had been practiced worldwide for centuries. As Schüz has pointed out, this was already called for by Aristotle and later Adam Smith.[46] In such a culture, elements which disturb the social order by selfish actions to gain profit over the suffering of others could not thrive for too long until responsible members and organs would intervene. The global situation is in need of being counterbalanced by a long-term perspective and grassroots involvement that civil society can offer to their governments. People increasingly demand a more democratic and horizontal system that represents shared governance on local, national, and international levels. As the old narratives of war, exploitation, and conflict continue to fade, new narratives of peace and fairness emerge and gain momentum. Therefore, as a matter of pragmatic survival, not sentimental religious

[43] Gopin (2002: 3).

[44] https://www.un.org/en/ga/62/plenary/peaceculture/bkg.shtml

[45] https://daccess-ods.un.org/tmp/9753335.11829376.html

[46] Schüz (2021: 40ff, 45 ff.)

nostalgia, governance systems of any sort need to rely on the wisdom humanity has been acquiring and relying on for thousands of years. Such emerging global civilization would honour the shared wisdom and heritage of humankind to integrate the advancements and progress made during the secular era—moving into an ecologically based globally diverse civilisation.

4 Conclusion

We discussed how Religion was cast out of politics eventually and the central role of its priesthood and associated networks started to fade in the state, academia, and the general culture of Europe. Examining the power of religious voices on economic and political developments I wanted to point towards valuable lessons for the presently emerging narratives. I argued to activate resources of the religious domain by recognizing its resourcefulness and leadership from the grassroots up to high-level leadership. Some resources to peace were presented as tools, which make them relevant to act as a universal force for good. I pointed out that the vast majority of humanity is guided by some form of religion and/or spirituality having a role for sustainable development, climate justice, peace, and human rights. Science, education, economy, and religion apply methodologies and hold wisdom and knowledge, which combined can form the keys to solve our present day challenges.

We are on a journey where the old narratives are making place for the new to emerge informed by a dialogical and inclusive process. We are moving out of silos to bridge the gaps between sectors where transdisciplinary approaches to peacebuilding take a fresh look sparking innovation. Such culture of peace leads to impactful change driven by empowered citizens claiming their sovereignty based on human rights.

Such coming of age would be of integrative sciences and inter-spiritual religious approaches grown out of efforts of concerned global citizens. They would come from all walks of life, being drawn to a vision for a world that honours diversity. One basic premise is taught in the Mahabharata of ancient India in the section of the Bhagavad Gita. One is alerted there to have a fixed orientation with our intelligence *(bhuddi)* on what unites us all in expanded consciousness as *…without there is no possibility of peace. And how can there be any happiness without peace?"*.[47] This quest is also a spiritual calling and inner experience, which can find many forms of contemporary expressions. It has the ethics, values, and spiritual wisdom of humanity alive in all subjects creating the foundation for a global civilization. Such an awakened humanity is what this world is calling for, and by the grace and wisdom of those pioneers who have lived this, and showed the way, each one of us is called to follow in their footsteps.

[47] Bhaktivedanta Swami (1972: Chap. 2 Verse 6).

References

Bhaktivedanta, Swami Prabhupada, A. C. (1972, 1983). *Bhagavad-gita as it is*. Bhaktivedanta Book Trust.

Bowker, J. W. (1986). The burning fuse: The unacceptable face of religion. *Zygon, 21*(4D), 415–438.

Curaming, L. M. (2002). Religion: A motivating factor in the pursuit of peace. *Dialogue & Alliance, 16*(1) Produced by Paragon House, St. Paul, U.S.A.

Davaney, S. G., & Laderman, G. (2005). Introduction: Contesting religion and religions contested: The study of religion in a global context. *Journal of the American Academy of Religion., 73*, 979–985.

Doe, S. S. (2003). Faith based community organizations for increasing global social capital. *Dialogue & Alliance, 17*(2) 2003/2004.

Forward, M. (2002). Homes for the human spirit: Religions, the United Nations, and the global ethic. *Dialogue & Alliance, 16*(1) Produced by Paragon House, St. Paul.

Gibbs, C. (2002). The united religions initiative at work. In R. Smock (Ed.), *Interfaith dialogue and peace-building*. Endowment of the United States Institute of Peace.

Gopin, M. (2002). The use of the word and its limits: A critical evaluation of religious dialogue and peacemaking. In R. Smock (Ed.), *Interfaith dialogue and peace-building*. Endowment of the United States Institute of Peace.

Hatzopoulos, P., & Fabio, P. (2003). *Religion in international relations: The return from exile*. Palgrave Macmillan.

Huntington, S. P. (1996). *Clash of civilizations and the remaking of world order*. Simon & Schuster.

Johnston, D., & Sampson, C. (1994). *Religion, the missing dimension of statecraft*. Oxford University Press.

Knitter, P. F. (2002). *Theologies of religions*. Orbis Books.

Küng, H. (1998). *A global ethic for global politics and economics*. Oxford University Press. Printed in US.

Lederach, J. P. (1995). *Preparing for peace: Conflict transformation across cultures*. Syracuse University Press.

Lederach, J. P. (1997). *Building peace: Sustainable reconciliation in divided societies*. United States Institute of Peace Press.

Pope Francis. (2015). Laudato Si. Encyclical Letter of the Holy Father Francis on care for our common home.

Revd Williams, T. S. M. (1988). Book review of "*Enlightenment and alienation: An essay towards an Trinitarian theology*", Colin E. Cunton, Marshall, Morgan and Scott, Basingstoke (1985) in Modern Theology 5 no 1, 1988.

Schüz, M. (2021). *Angewandte Unternehmensethik – Grundlagen für Studium und Praxis*. Pearson. (2nd actualized and extended edition).

Vendley, W. F. (2005). The power of inter-religious cooperation to transform conflict. *Cross Currents, 55*(1), 90–99.

Weber, M. (1922). Wirtschaft und Gesellschaft. Grundriss der Sozialökonomik [Economy and Society: Outline of Social Economics]. J. C. B. Mohr (Paul Siebeck).

Patrick Nickisch grew up in Cologne, Germany and currently lives in Switzerland. He started travelling as a pilgrim after completing school and entered open religious studies by the University of Lampeter in the Bhaktivedanta College seminary in 2006. He got initiated in the Chaitanya-Vaishnava Tradition as Purushottama Krishna Das in 2012.

In his trans-confessional ministry he provides services of pastoral care, teaches meditation, conducts inner trance journey and guides past life regression. His integral approach is informed deeply by his traditions' ancient knowledge and practice as well as contemporary cutting edge research of modern sciences. He is counselling members of the global community of diverse background and cultures as a guide and mentor being a certified life coach as well. He continued formal education with a Master in the world's religions in 2014 with Trinity Saint David. His main research since has been the role of religions for peace and security. In the second year of his seminary he got introduced to Religions For Peace attending the founding of its Belgium Chapter attending as a Youth Representative of his tradition. Since then he got connected with, supported and consulted a number of global initiatives promoting peace, environmental protection, and justice based on spiritual values and indigenous wisdom.

His volunteer commitment to good causes led to him to organizing programmes and taking on managerial duties especially in the United States of America, Europe, and India. In 2012, he became the Main Representative of the United Religions Initiative (URI), based in San Francisco, to the United Nations Organization in Geneva, Switzerland. He got involved in water policy working with UNICEF and the Kingdom of the Netherlands and faith based initiatives in 2013 culminating in the Ocean Conference in New York in 2017. He gained his primary insights into policy and practice attending UN conferences worldwide from 2011 till the present as an UN ambassador for civil societies. In this capacity he is participating in Human rights council sessions in Geneva, New York Headquarters high level political dialogue sessions concerning the Sustainable Development Goals, Climate Change conferences and contributing to other global assemblies on issues such as biodiversity, peace, disarmament, youth, women and the cause of indigenous peoples.

Fueled from Within: How Meditation Forges the Inner Tools Required for Change

Meditation's Contribution to Solving Current Social, Economic, and Ecological Crises

Sacinandana Swami and Maika Steinborn

1 Introduction

Climate change, financial crises, violence, racism, sexism, exploitation, corruption, divided nations, poverty, hunger, mental illness, and other diseases... the list of challenges we face is long. To manage these, humanity requires solutions on a variety of levels and from different perspectives. Not only that, these solutions need to step into a cooperative dialogue with each other. Transdisciplinary thinking calls and challenges us to expand our thinking to overcome the narrow borders of sectoral problem-solving. As the manifest of transdisciplinarity declares: "The transdisciplinary vision is resolutely open, as it steps beyond the domain of the hard sciences and calls for its dialogue and reconciliation with the humanities, social sciences, as well as art, literature, poetry and spiritual experience."[1]

Meditation offers an invaluable contribution to transdisciplinary thinking because it moves the practitioner away from rigidly and narrowly identifying with segmented perspectives and disciplines. Meditation detangles the fusion of awareness with the thinking process and individual thoughts. This broadens the practitioner's view point to include a greater range of interconnected agents, factors, and topics. Unhooked from the entanglement with individual thoughts and preoccupations, the practitioner is enabled to notice more of what is happening in their environment and to connect

[1] Freitas et al. (1994).

S. Swami (✉)
Meditation Teacher, Neuruppin, Märkisch Linden, Germany
e-mail: info@sacinandanaswami.com

M. Steinborn
Psychologist and Meditation Teacher, Berlin, Germany
e-mail: info@maikasteinborn.com

© The Author(s), under exclusive license to Springer Nature Switzerland AG 2025

M. Schüz (ed.), *Transdisciplinary Thinking and Acting*, Transdisciplinary Management of Ecological and Social Crises,
https://doi.org/10.1007/978-3-031-70680-6_20

the dots between areas that previously seemed unrelated. Transdisciplinary thinking and problem-solving relies on the cognitive flexibility and broadmindedness that meditation instills.

In this chapter, we present the contribution meditation has to offer to a transdisciplinary management of social and economic crises. We summarize empirical evidence from the vast research on the benefits of meditation—starting with its benefits on health, and continuing on to human values beyond health like calming the nervous system, personal growth, and spiritual experience. A discussion of meditation's positive effects on empathy, prosocial behavior, societal and ecological challenges concludes this chapter.

Throughout this chapter, we oscillate back and forth between presenting the empiric evidence and discussing how these effects are possibly achieved through meditation, relying both on contemporary Western and traditional theories and models of meditation.

2 A Healing Elixir: Meditation & Health

Every year the World Health Organization (WHO) publishes data on the burden of disease. A measure called *disability-adjusted life years* (DALYs) expresses the loss of healthy years in life due to premature mortality or disability caused by a certain disease. With some regional differences, the latest estimate shows that amongst the leading causes for lost DALYs are many diseases that can be prevented, improved, or cured through psychological and behavior change, like cardiovascular diseases, diabetes, depressive disorders, anxiety disorders, drug use disorders, and self-harm.[2] As every reader has probably experienced for themselves, changing the way we relate to our inner world, our habits and behavior is easier said than done. However, with effective tools and support, it is possible. This is where meditation can be of help and is becoming more and more popular as an integral part of many health prevention and intervention approaches.

2.1 Meditation as a Therapeutic Tool in Facing the Mental Health Crisis

The mental health crisis includes several challenges, predominantly a lack of mental health awareness and limited access to mental health services which creates a lot of individual suffering. In the EU every year more than one third of the population suffers from one or more psychological disorders, most prominently anxiety disorders, insomnia, and depression.[3] From an economic perspective, every year mental illness creates tremendous costs. For example, Gustavsson et al. found that in Europe

[2] WHO (2020).

[3] Wittchen et al. (2011).

neuropsychiatric disorders cost 798 billion Euro every year, the greatest contributor being psychological disorders.[4] The prevention and treatment of mental illness is as relevant as ever.

In the past decades, studies demonstrating the therapeutic effects of meditation have become so numerous that many meta-analyses (studies summarizing individual studies) and meta-syntheses (studies summarizing meta-analyses) have emerged.[5] For example, meditation has been shown to help with ADHD,[6] anxiety,[7] depression,[8] PTSD,[9] and stress.[10] A variety of third-wave psychotherapeutic approaches has and continues to emerge which include meditation as an inherent technique in their collection of interventions. Examples are acceptance and commitment therapy,[11] dialectical behavior therapy,[12] mindfulness-based cognitive therapy,[13] and stress reduction,[14] as well as compassion-focused therapy.[15]

Why does meditation have these beneficial effects on mental health? How does meditation add to the effectiveness of preventative and therapeutic interventions? The reason meditation is so effective therapeutically could be because it potentially reduces some of the main risk factors to developing mental illness transdiagnostically. The concept of risk factors as vulnerabilities for the development of psychological disorders is based on vulnerability-stress models that postulate, in short, that when a person experiences acute or cumulative stress or trauma, and they have other specific vulnerabilities, it is likely that they will develop a mental illness.[16] Amongst the most disruptive risk factors are experiential avoidance and isolation.

Experiential Avoidance

Avoidance of experience plays a key role in the development of mental illness.[17] When avoidance of experience is used as a coping mechanism for unpleasant and challenging feelings, a person retreats from the signals coming from their body, mind and emotions, and in that way loses contact to their inner world. This contributes to the typical experience of overwhelming numbness in depression. Additionally, avoiding experience in the sense of avoiding challenging opportunities

[4] Gustavsson et al. (2011).
[5] Sedlmeier (2022).
[6] Poissant et al. (2020).
[7] Singh and Gorey (2018).
[8] Goldberg et al. (2019).
[9] Gallegos et al. (2017).
[10] de Abreu Costa et al. (2019).
[11] Hayes (2005).
[12] Van Dijk (2012).
[13] Teasdale et al. (2014).
[14] Kabat-Zinn (2013).
[15] Gilbert (2009).
[16] Wittchen et al. (2020).
[17] Gros (2021); Harris (2019).

or situations that are anxiety provoking deprives people from the chance to expand their confidence and sense of ease in these situations. Instead, their anxiety connected to the situation grows. This is one of the main developmental mechanisms in anxiety disorders.

Learning to open up to experience and stop avoiding is one of the main goals in many therapeutic approaches (for example, in cognitive behavior therapy,[18] acceptance and commitment therapy,[19] psychoanalytic therapies, and all experientially based therapies like accelerated experiential-dynamic therapy[20] and internal family systems therapy[21]). Therapists teach their patients how to be present with their emotions and bodily sensations, how to notice, name, accept, understand them, and how to relate to them with compassion. This restores the patient's inner compass, dissolves numbness and anxiety, and makes way for health and courage.

Across most meditation techniques, practitioners learn focusing on their experience in the present moment—either as a first step or as a core part of their practice. In modern meditation theories this element of meditation has been called *open monitoring,*[22] *body awareness,*[23] or *self-awareness.*[24] Popular examples for these kinds of meditation practice are the body scan, or meditations that focus on parts of the body or physical experience like the soles of one's feet or one's breath.[25] In this way meditation can contribute to patients' learning to be present to their experience, noticing changes and spontaneous physical and emotional signals. They learn to notice and be aware of the variety of sensations and emotions that are present simultaneously. By repeated meditative practice they can learn that experiencing their emotions and physical sensations is not intolerable or dangerous, but an important part of what it means to be mentally healthy and to maintain mental health.

Isolation

In the last decades, so many people have become lonely, that it has coined the term *loneliness epidemic.*[26] On the one hand, compared to the 70 s and 80 s, modern people's loneliness has doubled.[27] On the other hand, connectedness or relational health is a strong predictor of mental health in children[28] and adults.[29] Different social developments are discussed as contributors to the loneliness epidemic, like the

[18]Gros (2021).

[19]Harris (2019).

[20]Fosha (2021).

[21]Schwartz (2021).

[22]Lutz et al. (2008).

[23]Hölzel et al. (2011).

[24]Vago and Silbersweig (2012).

[25]Swami (in print).

[26]Maté (2022); Snell (2017).

[27]Cacioppo (2015).

[28]Hambrick et al. (2018).

[29]Richardson et al. (2017).

rise of the one-person household from under 20% to up to 60% in some parts of the world.[30] Aside from societal changes, on a personal level isolation is an unhealthy coping mechanism people use to avoid experiencing shame and anxiety that ultimately contributes to more depression and anxiety.[31]

A sense of connectedness to others not only rests on opportunities for connection and social skills, but also the capacity to feel connected to others. This is another ability meditation practice can help develop and refine. In a comprehensive meta-analysis of the effects of meditation (in a non-clinical setting) Sedlmeier et al. found that the greatest changes of meditation practice partitioners observed were actually in the quality of their relationships, how connected and close they felt to others.[32] Specifically, loving-kindness meditation[33] and the element of focusing on common humanity in compassion-based meditations[34] seem effective in overcoming the burden of social isolation.

All in all, evidence strongly suggests that meditation can be an effective therapeutic tool in treating and healing from mental illness. We have discussed two possible ways meditation has positive effects on mental health in that it targets two key risk factors and mechanisms in mental illness—experiential avoidance and isolation. In addition to its benefits on mental health, meditation has shown to aid physical symptoms as well, which we will examine closer in the next section.

2.2 *Meditation as a Therapeutic Tool for Physical Illness*

It might come as a surprise, but meditation and meditative movement (like yoga, qigong, and tai chi) have even been shown to be therapeutic for physical illnesses. It has shown to have positive effects on blood pressure,[35] blood-glucose levels in patients with type 2 diabetes,[36] chronic fatigue,[37] chronic pain,[38] and sleep quality.[39] Also, there is evidence that meditation slows down aging—both in the sense of cognitive decline[40] and the aging of grey and white brain matter.[41] In a study that compared the brain ages of meditators and non-meditators, Luders et al. found that the brains of meditators were on average 7.5 years younger.[42] Summarizing three

[30] Snell (2017).

[31] Gros (2021).

[32] Sedlmeier et al. (2012, 2018).

[33] Hutcherson et al. (2008).

[34] Germer and Neff (2013), Swami (in print).

[35] Ooi et al. (2017).

[36] Xia et al. (2020).

[37] Ulrichsen et al. (2016).

[38] Veehof et al. (2016).

[39] Rusch et al. (2019).

[40] Gard et al. (2014).

[41] Villemure et al. (2015); Laneri et al. (2016).

[42] Luders et al. (2016).

meta-syntheses on the effects of meditation on health in a clinical context, Sedlmeier concludes that on average, it has a substantial beneficial effect.[43]

Much more can be discussed on this subject; however, people not only meditate to seek relief from illness. A factor analysis by Sedlmeier et al. investigating responses meditators gave when asked why they meditate revealed that on top of wanting to learn to better deal with problems and difficult feelings, amongst other reasons, they also meditate in order to calm down, understand themselves better, become more open, have spiritual experiences, connect with nature, develop personally, and be happier.[44] Next, we will discuss how meditation can contribute to fulfilling these needs and solving these problems beyond contributing to curing disease. We acknowledge that health is not only a state of being free of illness, the following benefits of meditation could still be discussed under the headline of health as defined by the WHO in the sense of "a state of complete physical, mental and social well-being and not merely the absence of disease or infirmity."[45] At the same time, meditation may be effective and relevant in different ways in clinical and non-clinical settings, and in the next section we aim to highlight mediation's benefits in non-clinical settings and mainly present studies from non-clinical contexts.

3 Beyond Illness: How Meditation Helps Achieve Human Values Apart from Curing Illness

3.1 *Calming the Nervous System*

Meditation comes in many different forms. Some kinds of meditation include movement or speech, others recommend quiet sitting or lying down. Some are more formal, others are informal. Some last for hours, others for minutes or even seconds. Some practices require focusing attention, while others emphasize open observing. Some meditators focus on an object or mantra, others on their body and/or mind or aspects of experience like their breath, thoughts, or emotions. In some forms of meditation body and mind are only observed, while in others practitioners take active influence on them and try, for example, to cultivate positive emotions.

These different forms of meditation have different goals and effects. Meditation can quickly be labeled as "just another relaxation technique," but that would do it a disservice. Meditation is more than a relaxation technique. However, some forms of meditation can help calm the nervous system. For example, a summary of EEG studies on meditation found an overall effect of slowing down the electrical activity of the brain and an increased activation of alpha and theta brain waves which

[43] Sedlmeier (2022).

[44] Sedlmeier and Theumer (2020).

[45] WHO (1948).

correspond to a restful, reflective state of mind.[46] When compared with traditional progressive muscle relaxation, a mindfulness body scan was just as effective.[47]

Relaxation, or calming down, is a subjective and objective need in all people's lives. The autonomic nervous system, which regulates various physical functions of the body, is composed of the parasympathetic and sympathetic nervous system. When the sympathetic nervous system is chronically activated, it raises a person's basic level of arousal, which contributes to the experience of stress. Techniques of calming the nervous system, specifically the sympathetic nervous system and activating the parasympathetic nervous system are required for basic energy management.

Common relaxation techniques are diaphragmatic breathing and progressive muscle relaxation. Meditation techniques that support relaxation are similar, but not equal to these approaches.[48] Mediation adds to the variety of relaxation techniques available to modern stressed people and makes it easier to choose a specific approach that is best suited for specific needs, situations, and individuals.

3.2 *Awareness, Personal Insight, and Growth*

The effects of meditation are varied and depend on the specific technique, the meditator's motivation, and the context in which practitioners meditate. In their meta-analysis on the effects of meditation on non-clinical populations, Sedlmeier et al. found that across the different varieties of meditation practices, practitioners experienced benefits in the quality of their relationships, improvements in their emotional experience (less stress, less anxiety and other negative emotions, more positive emotions), and improvements in cognitive variables like mindfulness, attention, perception, and cognition.[49] The meta-analysis also revealed that the positive effects of mediation for healthy practitioners were larger than the effects of relaxation exercises or cognitive trainings. Other studies have shown beneficial effects of meditation at the workplace[50] and for the cognitive performance of older adults (60+).[51]

An interesting contemporary Western explanation of how meditation could have these benefits can be found in the PROMISE model (Fig. 1): the process of mindfulness meditation leading to insight and equanimity model.[52] In contrast to other models, this model is based on the subjective experiences of meditators, complemented by traditional explanations and empirical evidence. Also, parts of

[46] Cahn and Polich (2006).

[47] Corbett et al. (2019).

[48] Corbett et al. (2019).

[49] Sedlmeier et al. (2012).

[50] Bartlett et al. (2019).

[51] Chan et al. (2019); Weber et al. (2020).

[52] Eberth et al. (2019).

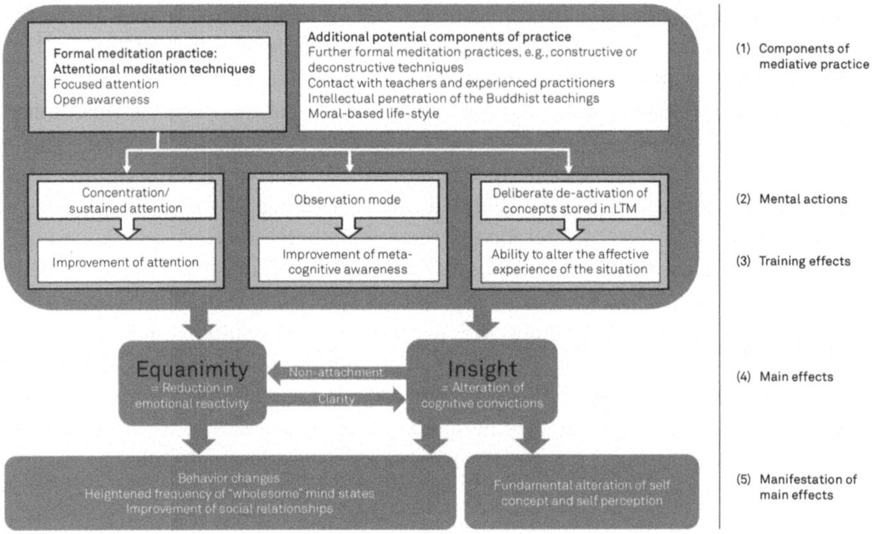

Fig. 1 The PROMISE model LTM = long-term memory Reprinted with permission from "PROMISE: A Model of Insight and Equanimity as the Key Effects of Mindfulness Meditation," by J Eberth, P Sedlmeier, & T Schäfer, 2019, Frontiers in Psychology, 10, Fig. 1 © 2019

the model have successfully been tested empirically. This model explains that the two main effects of mindfulness meditation are equanimity (reduced emotional reactivity) and insight (a change in cognitive convictions), both of which lead to positive changes in behavior and perception. This is achieved through concentration, an observation mode and deliberate deactivation of cognitive concepts stored in long-term memory (LTM).

Another model (Fig. 2) proposes a cascade of processes, starting with increased attention on the inner world, leading to heightened awareness of inner events and emotions, which leads to a different perspective on oneself and a reappraisal of situations.[53]

To sum up, evidence suggests that mediation is very helpful in pursuits of personal growth, increased awareness and insight. Many people feel stuck in old patterns and habits, defensive mechanisms and inauthentic roles that take over whenever they are triggered and overwhelmed by the body's stress response. In the context of solving personal, societal, economic, and ecological modern challenges, we need to learn to step away from emotional reactivity that is based on past hurts and learn to relate to ourselves, our companions, and nature from a perspective that is filled with awareness, insight, compassion, and self-regulation. Meditation can be a reliable guide on this path.

[53] Hölzel et al. (2011).

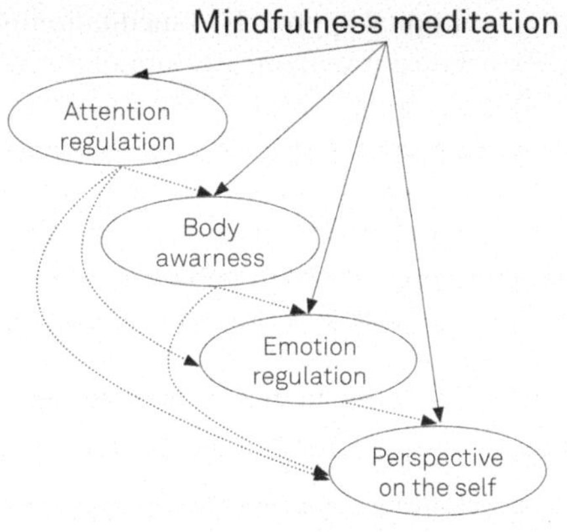

Fig. 2 Components of mindfulness meditation based on Hölzel et al. (2011) developed by Sedlmeier (2022) Reprinted with permission from "The Psychology of Meditation: Varieties, Effects, Theories, and Perspectives." By P Sedlmeier, 2022, Göttingen: Hogrefe Publ. Used with permission from The Psychology of Meditation by Peter Sedlmeier, ISBN 9780889375765 ©2022 Hogrefe Publishing, www.hogrefe.com, https://doi.org/10.1027/00576-000

3.3 Spiritual Experiences

Traditionally, spiritual experiences are the main goal of meditative practices. Also many modern practitioners aim for this goal: spiritual insight, connecting to a higher power, experiencing altered states of consciousness, enlightenment.[54] Modern practitioners are often not interested in religious dogmas, rather they seek spiritual experiences free from the baggage of organizations, institutions and their narratives. They seek the freedom to experiment with various spiritual practices, they are open to the wisdom of other cultures. They pursue peak experiences[55] and take part in the process of self-realization which C. G. Jung called "individuation."[56]

Throughout history, spiritual and religious paths have used and shaped meditation practices in order to experience altered states of consciousness. For example, meditation can be found within Christianity, Sufism, Daoism, and has a very prominent role in Buddhism and Hinduism. Traditional depictions of the kind of spiritual experiences meditation leads to often include inner clarity, compassion or love, a feeling of connectedness to others, the universe and/or a deity, and liberation from a false self/ego. Differences are mainly found in relation to shedding the false self (Does a spiritual self remain or not?) and experiences of a deity (Is he/she/they/it present or not, and if so, what is the deity like?).

[54] Sedlmeier and Theumer (2020).

[55] Maslow (1964).

[56] Jung (2001).

An example for such a traditional approach to meditation is Patanjali's eightfold path of yoga.[57] The Indian scholar systematically summarized all the spiritual paths known at the time in his Yogasutras in the third or fourth century AD. The main goal of the paths, *samadhi*, is to become aware of the spiritual self by a variety of practices, including meditation. This approach explains the effect of meditation as removing obstacles from experiencing the spiritual self by stilling the fluctuations of material cognition. With time, practitioners realize the spiritual self as autonomous from material cognition and experience the qualities of the spiritual self which is free of suffering and full of bliss.

Of the eight steps of yoga,[58] the first five prepare the ground for absorbed meditation:

1. Ethical conduct how to get along well with each other (*yama*).
2. Personal lifestyle, discipline and practice, regular beneficial observances like study of scripture (*niyama*).
3. Sitting in a steady and comfortable position (*asana*).
4. Regulating one's breath and then lengthening it (*pranayama*), allowing the mind to become calm.
5. Withdrawing the mind from external and internal engagements as much as possible (*pratyahara*).

After these five, Patanjali presents two steps that lead to increasingly focused meditation:

6. Focusing the mind on the object of meditation—an image, mantra, wisdom text, or another sacred object (*dharana*).

At this stage, actual meditation begins.

7. Effortless meditation (*dhyana*).

At this level of focus, one experiences silent inner joy, a sense of consciousness expansion, and an expanding and all-inclusive love. This is the stage where the practitioner stops focusing on formal techniques.

Finally, one reaches the goal of meditation: *samadhi*. This marks the final and eighth step. Patanjali describes it like this:

8. Trance or deep absorption—a state distinctly beyond the state where one's consciousness interacts with the world. Instead, consciousness interacts with spirit, and the experience is described as timeless (*sat*) and filled with an experience of unconstrained reality (*cit*) and boundless joy (*ananda*).

[57] Bryant (2015).

[58] Bryant (2015); Swami (in print).

Empirical researchers are also interested in the spiritual side of meditation, and a field of *spiritual neuroscience* has emerged that studies what happens in the brain during mystical experiences.[59] Studies that compared meditation practices including spiritual components with ones without found better mental and social effects such as higher cognitive and emotional intelligence.[60] Ott details that the particular gamma wave patterns in the brains of experienced meditators could be related to mystical experiences.[61]

All in all, meditation has always been and continues to be to this day a path towards the experience of transcendence. Meditation leads practitioners from a theoretical, conceptual spiritual search to actual mystical experience. This often includes a sense of deepened connection to others and coming in contact with strong, unconditional love as well as interconnectedness to a greater whole.[62] This new outlook in life is likely the basis for more empathy and prosocial behavior, something that could contribute to solving modern societal problems. In the next part of this chapter, we will discuss this effect of meditation.

4 Meditation's Contribution to Problems on the Societal and Ecological Level

4.1 Meditation, Nonviolence, Empathy

At the core of many of our economic and societal problems lies humans' potential towards selfishness. At the moment, in many parts of the world, individual economic goals seem to outweigh global wellbeing as the driving factor behind decisions on the big and small scale. This could change with more widespread empathy for our fellow companions. Furthermore, it could even foster humanity's concern for specific subgroups, all human beings as a global family, and finally, plants and animals as well.

The basis for this expansion in care towards the global societal level is a sense of fundamental connectedness, shared humanness, and empathy for others, even if they are from a different societal or ethnic group. It seems that meditation leads to exactly this experience, no matter if it specifically targets empathy, or not. A meta-analyses summarizing 21 studies by Kirby et al. found that meditations targeting compassion lead to an increase in compassion and self-compassion.[63] Another meta-analysis looked at 16 studies examining the effects of different kinds of meditation, including non-compassion specific meditations like mindfulness-based meditations, and found

[59] D'Aquili and Newberg (1993).

[60] Wolf and Abell (2003); Wachholtz and Pargament (2005, 2008); Wachholtz et al. (2017).

[61] Ott (2010).

[62] Swami (in print).

[63] Kirby et al. (2017).

an increase in compassion and empathy in the study participants.[64] Sedlmeier et al. also found an effect of meditation on empathy in their meta-analysis.[65] Additionally, changes in meditators' brain structures that are associated with empathic skills have been found in brain imaging studies, which indicates that meditation practice has an effect on the neurobiology which facilitates empathy.[66]

Judging from these studies, meditation itself seems to increase compassion and empathy. It would be interesting to see if these effects could be enhanced, if study participants also reflect on and apply the ethical recommendations traditional meditations come with. Patanjali's eightfold path, for example, starts with practices that address how the meditator interacts with others.[67] This practice includes empathy, compassion, and nonviolence. Buddha's path also includes practicing morally consistent behavior, speech and adopting a morally sound means of personal livelihood.[68] In their modern meditation framework, Vago and Silbersweig discuss three components of mindfulness meditation: focused attention, open monitoring, and ethical enhancement.[69] To them, self-transcendence is part of mindfulness meditation, which means looking beyond one's individual needs, sensing connection with others, and engaging in more prosocial behavior.

A few studies have investigated the potential added benefit of ethical components in mediation practice and the results indicate a stronger effect for empathic action (donating),[70] self-compassion,[71] and wellbeing.[72] In the face of the scarcity of these types of studies, this seems like a potential field of more elaborate research in the future.

Important to note is that traditionally the ethical components of meditation practice like empathy, compassion, and nonviolence not only address the way the meditator interacts with fellow humans, but also with plants and animals. In Patanjali's collection of do's and don'ts for meditators the foremost don't is *ahimsa:* don't be violent.[73] His commentators emphasize that this also applies to animals, insects, and trees, and that a meditator should inflict as little harm and aggression on other creatures and plants as possible.[74] Some traditional meditators, like the Jains, take this so far as to sweep the ground on which they walk, to make sure they will not step on any insects.[75] It is reasonable to assume that the ethical components of

[64] Kreplin et al. (2018).

[65] Sedlmeier et al. (2012).

[66] Engen et al. (2018); Leung et al. (2012).

[67] Bryant (2015).

[68] Sedlmeier (2022).

[69] Vago and Silbersweig (2012).

[70] Chen and Jordan (2020).

[71] Bayot et al. (2020).

[72] Matko et al. (2021).

[73] Bryant (2015).

[74] Bryant (2015).

[75] Bryant (2015).

meditation practice will also affect a meditators' environmental choices and pro-environmental behavior.

First studies are finding evidence that attests the connection between meditation and ecologically sustainable behavior. For example, Jacob et al. found a correlation between mindfulness meditation practice and sustainable household and food choices.[76] Furthermore, the practice of mindfulness meditation is associated with a greater concern for the environment and a lower environmental impact from the consumption of animal proteins.[77] Meditators have also been found to be more involved in pro-environmental behavior and more likely to support climate policies.[78] In a randomized control study, Ray et al. demonstrated that meditation increases pro-environmental behavior.[79] All in all, the understanding of the empirical connection between meditation and sustainability is still in its early stages of research.[80] While many gaps remain to be closed, these first studies show promising results and emphasize the non-negligible effect of meditation on an ecological scale.

To sum up, meditation seems to be a powerful tool to increase empathy and compassion, inner dispositions very much needed to solve modern crises on the societal and ecological level. Sometimes, however, meditation, alongside other practices for personal growth, is criticized as an egoistical pursuit. Next, we will address this objection.

4.2 *Meditation and Self-Centeredness*

The observation that meditation, self-development, self-realization, and/or self-actualization can be and is practiced in a self-centered way by some is not new. Viktor Frankl already shared this critique in the 1980s at his address to the International Family Congress in Vienna: "In principle, self-realization is desirable, however, this buzzword is often misunderstood and used as a pretext for egocentric, yes egotistical behavior."[81] He goes on to explain that this is based on the misunderstanding that self actualization could be developed without connection to another person or something larger than oneself, and that humans realize their full potential in connection to someone else or in service to their values. Pursuits of growth can be irresponsible self-centeredness when practitioners disengage from society and from solving social and political problems.[82]

If meditators become self-centered through their meditation practice, they have not experienced connection to the big mind, others, and something larger than

---

[76] Jacob et al. (2008).

[77] Thiermann et al. (2020).

[78] Loy and Reese (2019).

[79] Ray et al. (2021).

[80] Thiermann et al. (2020).

[81] quoted as in Bischof (2010).

[82] Paulus (1994); Meller (2008).

themselves, but are still anchored in small, limiting egotistic awareness. Therefore, traditional meditation practices like the ones from the yoga tradition insist that compassion needs to be cultivated by all people, at all places, at all times, and in every circumstance.[83] This is the greatest practice for a progressive life.

5 Conclusion

In order to address the environmental, social, economic, and other problems humanity faces, we require an inside-out approach. If people become satisfied within, if they develop empathy and compassion, they will be able to bring these qualities into the world. Then they will not be so dependent on external satisfaction, and the problems caused by exploitation of the limited resources of our planet will recede. Meditation can offer an invaluable contribution in that direction. It thus proves to be a transdisciplinary way of gaining insights into the indissoluble connection of man with the world. This interconnectedness quite often remain hidden from disciplinary approaches and might even be destroyed by them.

References

Bartlett, L., Martin, A., Neil, A. L., Memish, K., Otahal, P., Kilpatrick, M., & Sanderson, K. (2019). A systematic review and meta-analysis of workplace mindfulness training randomized controlled trials. *Journal of Occupational Health Psychology, 24*, 108–126. https://doi.org/10.1037/ocp0000146

Bayot, M., Vermeulen, N., Kever, A., & Mikolajczak, M. (2020). Mindfulness and empathy: Differential effects of explicit and implicit Buddhist teachings. *Mindfulness, 11*, 5–17. https://doi.org/10.1007/s12671-018-0966-4

Bischof, M. (2010). *Salutogenese Unterwegs zur Gesundheit. Neue Gesundheitskonzepte und die Entfaltung einer Integrativen Medizin*. Drachen.

Bryant, E. F. (2015). *The yoga sūtras of Patañjali: A new edition, translation, and commentary [eBook]*. North Point Press. Available: https://www.amazon.de/Yoga-Sutras-Pata%C3%B1jali-Translation-Commentary-ebook/dp/B00QRZI1XA/ref=tmm_kin_swatch_0?_encoding=UTF8&qid=&sr=

Cacioppo, S., Grippo, A. J., London, S., Goossens, L., & Cacioppo, J. T. (2015). Loneliness: clinical import and interventions. *Perspectives on Psychological Science, 10*(2), 238–249. https://doi.org/10.1177/1745691615570616

Cahn, B. R., & Polich, J. (2006). Meditation states and traits: EEG, ERP, and neuroimaging studies. *Psychological Bulletin, 132*, 180–211. https://doi.org/10.1037/0033-2909.132.2.180

Chan, J. S., Deng, K., Wu, J., & Yan, J. H. (2019). Effects of meditation and mind–body exercises on older adults' cognitive performance: A meta-analysis. *The Gerontologist, 59*, e782–e790. https://doi.org/10.1093/geront/gnz022

Chen, S., & Jordan, C. H. (2020). Incorporating ethics into brief mindfulness practice: Effects on well-being and prosocial behavior. *Mindfulness, 11*, 18–29. https://doi.org/10.1007/s12671-018-0915-2

[83] Bryant (2015).

Corbett, C., Egan, J., & Pilch, M. (2019). A randomised comparison of two 'stress control' programmes: Progressive muscle relaxation versus mindfulness body scan. *Mental Health & Prevention, 15*, 200163. https://doi.org/10.1016/j.mph.2019.200163

D'Aquili, E. G., & Newberg, A. B. (1993). Mystical states and the experience of god: A model of the neuropsychological substrate. *Zygon: Journal of Religion and Science, 28*, 177–200. https://doi.org/10.1111/j.1467-9744.1993.tb01026.x

De Abreu Costa, M., de Oliveira, G. S. D. A., Tatton-Ramos, T., Manfro, G. G., & Salum, G. A. (2019). Anxiety and stress-related disorders and mindfulness-based interventions: A systematic review and multilevel meta-analysis and meta-regression of multiple outcomes. *Mindfulness, 10*, 996–1005. https://doi.org/10.1007/s12671-018-1058-1

Eberth, J., Sedlmeier, P., & Schäfer, T. (2019). PROMISE: A model of insight and equanimity as the key effects of mindfulness meditation. *Frontiers in Psychology, 10*, 2389. https://doi.org/10.3389/fpsyg.2019.02389/full

Engen, H. G., Bernhardt, B. C., Skottnik, L., Ricard, M., & Singer, T. (2018). Structural changes in socio-affective networks: Multi-modal MRI findings in long-term meditation practitioners. *Neuropsychologia, 116*, 26–33. https://doi.org/10.1016/j.neuropsychologia.2017.08.024

Fosha, D. (2021). *Undoing aloneness & the transformation of suffering into flourishing. AEDP 2.0.* American Psychological Association.

Freitas, L., Morin, E. & Nicolescu, B. (1994). *Charter of transdisciplinarity. Adopted at the first world congress of trandisciplinarity*. Convento da Arrábida, Portugal, 2–6 November 1994, 1.

Gallegos, A. M., Crean, H. F., Pigeon, W. R., & Heffner, K. L. (2017). Meditation and yoga for posttraumatic stress disorder: A meta-analytic review of randomized controlled trials. *Clinical Psychology Review, 58*, 115–124. https://doi.org/10.1016/j.cpr.2017.10.004

Germer, C. K., & Neff, K. D. (2013). Self-compassion in clinical practice. *Journal of Clinical Psychology: In Session, 69*(8), 1–12. https://doi.org/10.1002/jclp.22021

Gilbert, P. (2009). *The compassionate mind. How to use compassion to develop happiness, self-acceptance and wellbeing*. Oakland.

Goldberg, S. B., Tucker, R. P., Greene, P. A., Davidson, R. J., Kearney, D. J., & Simpson, T. L. (2019). Mindfulness-based cognitive therapy for the treatment of current depressive symptoms: A meta-analysis. *Cognitive Behaviour Therapy, 48*, 445. https://doi.org/10.1080/16506073.2018.1556330

Gros, D. F. (2021). *Overcoming avoidance workbook*. New Harbinger.

Gustavsson, A., Svensson, M., Jacobi, F., Allgulander, C., Alonso, J., Beghi, E., et al. (2011). Cost of disorders of the brain in Europe 2010. *European Neuropsychopharmacology, 21*(10), 718–779. https://doi.org/10.1016/j.euroneuro.2011.08.008

Hambrick, E., Brawner, T., & Perry, B. D. (2018). Examining developmental adversity and connectedness in child welfare involved children. *Children Australia, 43*(2), 105–115. https://doi.org/10.1017/cha.2018.21

Harris, R. (2019). *ACT made simple - a quick-start guide to ACT basics and beyond*. New Harbinger.

Hayes, S. C. (2005). *Get out of your mind and into your life: The new acceptance and commitment therapy*. New Harbinger.

Hölzel, B. K., Lazar, S. W., Gard, T., Schuman-Olivier, Z., Vago, D. R., & Ott, U. (2011). How does mindfulness meditation work? Proposing mechanisms of action from a conceptual and neural perspective. *Perspectives on Psychological Science, 6*, 537–559. https://doi.org/10.1177/1745691611419671

Hutcherson, C. A., Seppala, E. M., & Gross, J. J. (2008). Loving-kindness meditation increases social connectedness. *Emotion, 8*(5), 720–774. https://doi.org/10.1037/a0013237

Jacob, J., Jovic, E., & Brinkerhoff, M. B. (2008). Personal and planetary well-being: Mindfulness meditation, pro-environmental behavior and personal quality of life in a survey from the social justice and ecological sustainability movement. *Social Indicators Research, 93*, 275–294. https://doi.org/10.1007/s11205-008-9308-6

Jung, C. G. (2001). *Zivilisation im Übergang*. Gesammelte Werke, Bd. 10. Olten/Freiburg i. Br.: Walter.

Kabat-Zinn, J. (2013). *Full catastrophe living – Using the wisdom of your body and mind to face stress, pain and illness (Preface by Thich Nhat Hanh)*. Bantam Books. (rev. and updated edition).

Kirby, J. N., Tellegen, C. L., & Steindl, S. R. (2017). A meta-analysis of compassion-based interventions: Current state of knowledge and future directions. *Behavior Therapy, 48*, 778–792. https://doi.org/10.1016/j.beth.2017.06.003

Kreplin, U., Farias, M., & Brazil, I. A. (2018). The limited prosocial effects of meditation: A systematic review and meta-analysis. *Scientific Reports, 8*, 2403. https://doi.org/10.1038/s41598-018-20299-z

Laneri, D., Schuster, V., Dietsche, B., Jansen, A., Ott, U., & Sommer, J. (2016). Effects of longterm mindfulness meditation of brain's white matter microstructure and its aging. *Frontiers in Aging Neuroscience, 7*, 254. https://doi.org/10.3389/fnagi.2015.00254

Leung, M.-K., Chan, C. C. H., Yin, J., Lee, C.-F., So, K.-F., & Lee, T. M. C. (2012). Increased Grey matter volume in the right angular and posterior Parahippocampal Gyri in loving-kindness meditators. *Social Cognitive and Affective Neuroscience, 8*, 34–39. https://doi.org/10.1093/scan/nss076

Loy, L. S., & Reese, G. (2019). Hype and hope? Mind-body practice predicts pro-environmental engagement through global identity. *Journal of Environmental Psychology, 66*, 101340. https://doi.org/10.1016/j.jenvp.2019.101340

Luders, E., Cherbuin, N., & Gaser, C. (2016). Estimating brain age using high-resolution pattern recognition: Younger brains in long-term meditation practitioners. *NeuroImage, 134*, 508–513. https://doi.org/10.1016/j.neuroimage.2016.04.007

Lutz, A., Slagter, H. A., Dunne, J. D., & Davidson, R. J. (2008). Attention regulation and monitoring in meditation. *Trends in Cognitive Science, 12*, 163–169. https://doi.org/10.1016/j.neuroimage.2016.04.007

Maslow, A. H. (1964). *Religions, values, peak-experiences*. Ohio State University Press.

Maté, G. (2022). *The myth of Normal: Trauma, Illness & Healing in a toxic culture*. Avery.

Matko, K., Sedlmeier, P., & Bringmann, H. C. (2021). Differential effects of ethical education, physical hatha yoga, and mantra meditation on well-being and stress in healthy participants – An experimental single-case study. *Frontiers in Psychology, 12*. https://doi.org/10.3389/fpsyg.2021.672301

Meller, S. (2008). *Salutogenese durch Selbstverwirklichung - Eine integrative und ganzheitliche Perspektive für die Gesundheitspsychologie*. Tectum.

Ooi, S. L., Giovino, M., & Pak, S. C. (2017). Transcendental meditation for lowering blood pressure: An overview of systematic reviews and meta-analyses. *Complementary Therapies in Medicine, 34*, 26–34. https://doi.org/10.1016/j.ctim.2017.07.008

Ott, U. (2010). *Meditation für Skeptiker: Ein Neurowissenschaftler erklärt den Weg zum Selbst. [Meditation for sceptics: A neuroscientist explains the way to the self]*. O. W. Barth.

Paulus, P. (1994). *Selbstverwirklichung und psychische Gesundheit*. Hogrefe.

Poissant, H., Moreno, A., Potvin, S., & Mendrek, A. (2020). A meta-analysis of mindfulness-based interventions in adults with attention-deficit hyperactivity disorder: Impact on ADHD symptoms, depression, and executive functioning. *Mindfulness, 11*, 2669–2681. https://doi.org/10.1007/s12671-020-01458-8

Ray, T. N., Franz, S. A., Jarrett, N. L., & Pickett, S. M. (2021). Nature enhanced meditation: Effects on mindfulness, connectedness to nature, and pro-environmental behavior. *Environment and Behavior, 53*(8), 864–890. https://doi.org/10.1177/0013916520952452

Richardson, T., Elliott, P., & Roberts, R. (2017). Relationship between loneliness and mental health in students. *Journal of Public Mental Health, 16*(2), 48–54. https://doi.org/10.1108/JPMH-03-2016-0013

Rusch, H. L., Rosario, M., Levison, L. M., Olivera, A., Livingston, W. S., Wu, T., & Gill, J. M. (2019). The effect of mindfulness meditation on sleep quality: A systematic review and meta-analysis of randomized controlled trials. *Annals of the New York Academy of Sciences, 1445*, 5–16. https://doi.org/10.1111/nyas.13996

Schwartz, R. (2021). *No bad parts: Healing Trauma & Restoring Wholeness with the internal family systems model.* Sounds True.

Sedlmeier, P. (2018). Meditation and altered states of consciousness. *Journal of Consciousness Studies, 25*, 73–191. Available: https://psycnet.apa.org/record/2019-06909-004

Sedlmeier, P. (2022). *The psychology of meditation: Varieties, effects, theories, and perspectives.* Hogrefe Publishing.

Sedlmeier, P., & Theumer, J. (2020). Why do people begin to meditate and why do they continue? *Mindfulness, 11*, 1527–1545. https://doi.org/10.1007/s12671-020-01367-w

Sedlmeier, P., Eberth, J., Schwarz, M., Zimmermann, D., Haarig, F., Jaeger, S., & Kunze, S. (2012). The psychological effects of meditation: A meta-analysis. *Psychological Bulletin, 138*, 1139–1171. https://doi.org/10.1037/a0028168

Singh, S. K., & Gorey, K. M. (2018). Relative effectiveness of mindfulness and cognitive behavioral interventions for anxiety disorders: Meta-analytic review. *Social Work in Mental Health, 16*, 238–251. https://doi.org/10.1080/15332985.2017.1373266

Snell, K. D. M. (2017). The rise of living alone and loneliness in history. *Social History, 42*(1), 2–28. https://doi.org/10.1080/03071022.2017.1256093

Swami, S. (in print). *Inner Stillness: 16 Meditations with Sacinandana Swami.*

Teasdale, J., Williams, M., & Segal, Z. (2014). *The mindful way workbook. An 8-week program to free yourself from depression and emotional distress.* Guilford Press.

Thiermann, U., Sheate, W. R., & Vercammen, A. (2020). Practice matters: Pro-environmental motivations and diet-related impact vary with meditation experience. *Frontiers in Psychology, 11*. https://doi.org/10.3389/fpsyg.2020.584353

Ulrichsen, K. M., Kaufmann, T., Dørum, E. S., Kolskår, K. K., Richard, G., Alnæs, D., Arneberg, T. J., Westlye, L. R., & Nordvik, J. E. (2016). Clinical utility of mindfulness training in the treatment of fatigue after stroke, traumatic brain injury and multiple sclerosis: A systematic literature review and meta-analysis. *Frontiers in Psychology*. https://doi.org/10.3389/fpsyg.2016.00912

Vago, D. R., & Silbersweig, D. A. (2012). Self-awareness, self-regulation, and self-transcendence (S-ART): A framework for understanding the neurobiological mechanisms of mindfulness. *Frontiers in Human Neuroscience, 6*, 296. https://doi.org/10.3389/fnhum.2012.00296

Van Dijk, S. (2012). *Calming the emotional storm. Using dialectical behavior therapy skills to manage your emotions & balance your life*. New Harbinger.

Veehof, M. M., Trompetter, H. R., Bohlmeijer, E. T., & Schreurs, K. M. G. (2016). Acceptance- and mindfulness-based interventions for the treatment of chronic pain: A meta- analytic review. *Cognitive Behaviour Therapy, 45*, 5–31. https://doi.org/10.1080/16506073.2015.1098724

Villemure, C., Čeko, M., Cotton, V. A., & Bushnell, M. C. (2015). Neuroprotective effects of yoga practice: Age-, experience-, and frequency-dependent plasticity. *Frontiers in Human Neuroscience, 9*, 281. https://doi.org/10.3389/fnhum.2015.00281

Wachholtz, A. B., & Pargament, K. I. (2005). Is spirituality a critical ingredient of meditation? Comparing the effects of spiritual meditation, secular meditation, and relaxation on spiritual, psychological, cardiac, and pain outcomes. *Journal of Behavioral Medicine, 28*, 369–384. https://doi.org/10.1007/s10865-005-9008-5

Wachholtz, A. B., & Pargament, K. I. (2008). Migraines and meditation: Does spirituality matter? *Journal of Behavioral Medicine, 31*(4), 351–366. https://doi.org/10.1007/s10865-008-9159-2

Wachholtz, A. B., Malone, C. D., & Pargament, K. I. (2017). Effect of different meditation types on migraine headache medication use. *Behavioral Medicine, 43*, 1–8. https://doi.org/10.1080/08964289.2015.1024601

Weber, M., Schnorr, T., Morat, M., Morat, T., & Donath, L. (2020). Effects of mind–body interventions involving meditative movements on quality of life, depressive symptoms, fear of falling and sleep quality in older adults: A systematic review with meta- analysis.

International Journal of Environmental Research and Public Health, 17, 6556. https://doi.org/10.3390/ijerph17186556

WHO. (1948). *Constitution of the World Health Organization*. WHO. http://apps.who.int/gb/bd/PDF/bd47/EN/constitution-en.pdf

WHO. (2020). *Global health estimates 2020: Disease burden by cause, age, sex, by country and by region, 2000–2019*. World Health Organization. https://www.who.int/data/gho/data/themes/mortality-and-global-health-estimates/global-health-estimates-leading-causes-of-dalys

Wittchen, H.-U., Jacobi, F., Rehm, J., Gustavsson, C., Svensson, M., Jönsson, B., et al. (2011). The size and burden of mental disorders and other disorders of the brain in Europe 2010. *European Neuropsychopharmacology, 21*, 655–679.

Wittchen, H.-U., Knappe, S., & Hoyer, J. (2020). Was ist Klinische Psychologie? Definitionen, Konzepte und Modelle. In J. Hoyer & S. Knappe (Eds.), *Klinische Psychologie & Psychotherapie* (3rd ed.). Springer.

Wolf, D. B., & Abell, N. (2003). Examining the effects of meditation techniques on psychosocial functioning. *Research on Social Work Practice, 13*, 27–42. https://doi.org/10.1177/1049731502237471

Xia, T., Yang, Y., Li, W., Tang, Z., Huang, Q., Li, Z., & Guo, Y. (2020). Meditative movements for patients with type 2 diabetes: A systematic review and meta-analysis. *Evidence-Based Complementary and Alternative Medicine, 2020*. https://doi.org/10.1155/2020/5745013

Sacinandana Swami, a monk in the Bhakti yoga tradition, embarked on a profound spiritual journey from an early age. Drawing inspiration from the ascetic practices and serenity of yogis, sages, and rishis, he discovered his life's purpose in the pursuit of spiritual enlightenment and inner transformation. Having left his affluent German upbringing behind, Sacinandana Swami made his home in ashrams throughout India and Europe. During this time, he dedicated himself to the study of the sacred texts and practices of the Vaishnava Hindu tradition, while also learning from teachers representing various other meditation and wisdom traditions. In his pastoral capacity, Sacinandana Swami has provided guidance to numerous spiritual seekers through extensive travels and lectures at international retreats. Furthermore, he has authored books that emphasize spiritual experiences, all of which is rooted in five decades of unwavering spiritual commitment. His literary contributions include articles and poems, and share profound insights. He co-founded the BrajCare charity, played a crucial role in establishing spiritual communities in Germany, and currently holds teaching positions at esteemed institutions like the Vrindavan Institute of Higher Education in India and the Bhaktivedanta College in Belgium. Sacinandana Swami is set to release a new book on meditation in 2025, accompanied by an online and in-person course. The aim of the book is to share new perspectives and practices on meditation with a broader audience.

Maika Steinborn, PhD, is a clinical and educational psychologist. In her doctoral research she studied the holistic ancient Indian mental health model of the gunas, which discusses three common kinds of coping strategies and life style choices (mentally, socially, in regard to health, motivation, and emotions). She developed an inventory to measure the mix of the three gunas in an individual (the Tri-Guna Scales), and an 8-week program to increase mental health and wellbeing that combines contemporary psychological techniques with meditation practice and values of the yoga path. After completing this research, Maika was active in personnel development as a coach and trainer for a wide range of companies based in Germany. She trained skills in the areas of mental health, emotional intelligence, stress- & conflict management, and communication. Currently she works as a psychotherapist in training and shares helpful perspectives, tools, and exercises from psychology on her YouTube channel "Dr. Maika Steinborn."

The manufacturer's authorised representative in the EU is Springer Nature Customer Service Centre GmbH, Europaplatz 3, 69115 Heidelberg, Germany. If you have any concerns regarding our products, please contact ProductSafety@springernature.com

Printed and bound by CPI Group (UK) Ltd, Croydon, CR0 4YY
01/10/2026
02225956-0012